Encyclopaedia of Nature and Environment

ENCYCLOPAEDIA OF NATURE AND ENVIRONMENT

By

M. S. SETHI
S. A. IQBAL

Discovery Publishing House
NEW DELHI - 110002 (India)

Edition - 2020

ISBN: 978-81-7141-392-8

© Authors (1998)

Published by:
DISCOVERY PUBLISHING HOUSE PVT. LTD.
4383/4B, Ansari Road, Darya Ganj
New Delhi-110 002 (India)
Phone: +91-11-23279245, 23253475, 43596065
E-mail: discoverypublishinghouse@gmail.com
sales@discoverypublishinggroup.com
web: www.discoverypublishinggroup.com

Printed at:
Infinity Imaging Systems
Delhi

Preface

This book "Encyclopaedia of Nature and Environment" is an essential foundation knowledge to answer question that arise from looking at the world around us. The book has been created to give non-specialist an understanding of the way our planet an its living organisms work.

From our planet Earth and its atmosphere, through evolution and on to the various phenomenon occurring on the planet, the text covers the broad topics in earth and life sciences. Numerous illustrations have been used to provide a clear understanding of the dynamics of the Earth and life sciences. The layman and the specialist both will find the book useful.

M. S. Sethi
S. A. Iqbal

Contents

Part I

The Green Planet

- Earth
- The Earth's Magnetic Field
- Mountains
- Continental Drift
- Plate Tectonics
- Earthquake
- Volcanoes
- Rocks
- Weathering of Rocks
- Minerals
- Mining
- Fossilization
- Fossil Fuels
- Caves
- Polar Regions
- Ice Ages
- Ground Temperature
- Radiation Balance of the Earth
- Wind
- Coriolis Force
- Jet Streams
- Thunder Storms
- Tornadoes
- Tropical Cyclones
- Tides and Waves
- Seas and Oceans
- Ocean Currents
- The Ocean Floor
- Island Formation
- Deserts
- Coastlines
- Rivers
- Lakes

EARTH

The Earth is the fifth largest planet of the solar system with a diameter of 12,757 km at the Equator, and nearly 43 km less between poles. The surface is uneven due to mountain systems and oceanic trenches. The highest peak is represented by Mt. Everest in the Himalayas (8708 m) and the deepest trench by the Mariana Trench in Pacific ocean (11,022 m). Such distinctive features of the earth make it convenient to call the shape of earth as geoid rather than a simple oblate ellipsoid. Earth has an average density of 5.42, total surface area of 510.1 million km^2 of which 360.7 million km^2 is occupied by oceans and 149.4 million km^2 by land. The earth revolves around the sun at a distance of about 149 million km, once in every 365¼ days.

The Earth is not a solid body throughout, but distinguished into distant zones from centre to surface. The centre is occupied by the inner core 2496 km in diameter and consisting of solid mass of nickle and iron. It is surrounded by the outer core, consisting of same metals but in liquid form, and 2208 km thick. The two together are often called **centrosphere**. Next to the outer core is the **mantle**, 2880 km thick. The top layer of the mantle, which is partly molten is termed as **asthenosphere**, a flexible plastic layer over which the surface portion of the earth can slide. The mantle is surrounded by crust, varying in thickness from 10-32 km.

The Earth probably began in the coming together of dust and gas particles around the newly-formed Sun about 1,000 million years after the Big Bang (the probable origin of the universe, some 20,000 million years ago). Extraterrestrial materials therefore provide a surprisingly good way of investigating the internal structure of our planet.

Meteorites are thought to be the fragmented remains of small planet-like bodies only a few tens of kilometers across. The different types of meteorites would then represent different layers in the original planetoid structure. Iron meteorites (*siderites*) are about 97 per cent metal, mainly nickel-iron, and being so dense probably represent the planetoid core. *Chondrites* are of silicate composition and may represent the lower mantle. *Achondrites* are also silica based, but with a different composition. They may be fragments of crust and upper mantle.

The Earth's crust may be upto 40 km thick under continents or only 5 km thick under the sea. The crust and the very top of the mantle form the lithosphere which drifts on a plastic asthenosphere. The upper mantle stretches down to overlie the lower mantle at 700 km depth. From the surface the temperature inside the Earth increases by 30°C/km so that the asthenosphere is close to melting point. After 100 km depth the rate of temperature increase slows dramatically. Because we know so little about what happens deep within the Earth there is a great uncertainty about the temperature at the core. Pressure also increases with depth : already in the asthenosphere the pressure is equal to 250,000 atmospheres. At the Earth's core it may go as high as 4 million atmospheres. The density near the surface of the Earth is only about 4 grams/cubic centimetre, and this does not vary by much more than 30 per cent right down through the lower mantle. The increase in density that does occur is due to closer atomic packing under pressure. There is a leap in density to 10 grams/cubic centimetre in passing to the outer core and a further leap to between 13 and 16 grams/cubic centimetre in the inner core. The heat at the core fuels vast convection currents in the material of the mantle.

Continental crust is made by many different processes and is therefore difficult to generalize. A typical cross section might well consist of deformed and metamorphosed sedimentary rocks at the surface underlain by a granite intrusion. The remaining crust would consist of metamorphosed sedimentary rock and igneous rock reaching to the mantle.

In oceanic crust ocean sediments overlie a basalt which is pillowed and also chemically altered by sea water. Below this layer the basalt is made up of dykes — small vertical intrusions that are packed so close together they intrude each other. *Gabbros* are coarse-grained igneous rocks with a very similar composition to basalt. They are sometimes layered.

The inner core of the earth is the hottest region with a temperature of around 4000°C, and from this central core the heat constantly moves out. The heat currents circulates throughout the mantle keeping it in continuous movement and the earth's crust in unstable condition. The cracks in the earth's crust often throw up the molten lava, resulting in volcanic activity. Such cracks in the oceanic basin result in the development of oceanic ridges due to the rapid cooling of lava under oceanic water.

Over 2000 different minerals have been identified from the earths crust, made up of more than 100 elements oxygen (47%) and Silicon (28%) are the most abundant in the earth's crust. The other common elements are aluminium, iron, calcium, sodium, potassium and magnesium. Quartz, composed of oxygen and silicon is the most abundant mineral of the crust.

Volcanic Information

Volcanic outpourings also give us an insight into the Earth's hidden depths, but not as much as might be expected because, although lavas are generated well below the surface, the actual depth is shallow when compared with the Earth's radius of 6,370 kilometres (3,950 miles). The basaltic lavas that issue through the ocean ridges are created from the Earth's mantle by differentiation (a number of processes by which different rocks can be produced from a single parent magma). This suggests that the mantle itself is far denser than these basalts, because material that is less dense tends to rise to the surface.

Shocks to the System

Another major source of information about the Earth's structure is the study of the seismic, or shock, waves produced by earthquakes. One of the main measurements determined by the paths of these waves is the extent of the Earth's core, since shear waves do not pass through liquid material and are absorbed by a zone which starts at a depth of 2,900 kilometres (1,800 miles). From here to the centre of the Earth is the planet's core, although the core is not liquid throughout.

The boundary that separates the core from the overlying mantle is called the Gutenberg Discontinuity, and is the region where, over a few kilometres, the mantle rocks change to matter of much higher density. From the behaviour of seismic waves and from meteorite and other evidence it is believed that the inner core is solid iron and nickel at a temperature over 4,000°C. The outer core is less dense because it comprises liquid metal.

The mantle has also been investigated seismically. The upper limits of the mantle are where the seismic waves are slowed down at the Mohorovicic Discontinuity (*Moho*). This marks the change from mantle to less dense crystal rock. Within the mantle, density of the rock material increases rapidly with depth.

Scanned Images

In recent years the mantle has, in addition, been investigated by seismic tomography, a method similar to CAT (Computer-aided tomography) used in medical scanning.

The shock waves of more than 10,000 earthquakes are monitored every year. The actual velocity at which the waves travel in the mantle is the most significant factor. "Slow" and "fast" areas in the mantle are determined, and from an intersecting network of seismic waves a three-dimensional model of the mantle can be plotted. Hot parts of the mantle transmit waves more slowly than areas that are cooler, which allows the mapping of the heat-transporting convection cells in the mantle, which cause movement in the continental plates.

THE EARTH'S MAGNETIC FIELD

The Earth's magnetic field has been getting weaker for the last 2,000 years and may temporarily disappear altogether within the next 2,000 years. But this is only a part of a cycle that has already been enacted many times in the Earth's history, as the magnetic field slowly reverses itself, and north becomes south. Despite the constant variation, people have been using compasses to navigate for millennia. Animals like pigeons, bees and salmon — which contain grains of magnetic material in their bodies — may have been using magnetism to navigate for much, longer.

The Earth's magnetic field is similar to that of a simple bar magnet. Because of its two opposing poles — north and south — it is known as *dipole* field. But the Earth obviously does not have a giant magnet stuck at its centre. A dipole field can in fact also be created by a coil of wire carrying an electric current. The north pole is at the end of the coil where the current appears to be turning anticlockwise. Therefore scientists speculate that the Earth's field is due to charged particles — like electrons in an electric current — moving about in the molten centre of the planet. The Earth's core is made of iron and nickel, and in fact the inner core is solid. But the temperature here — some 5,000°C — produce convection currents in the molten metal surrounding the central core. These currents, it is argued, create magnetic fields that will continue for as long as the heat and the Earth's rotation are sustained to cause them.

The Origins of Magnetism

The Earth's magnetic field was probably initiated by the Sun's magnetic field when the Earth was still a swirling ball of dust and gas, billions of years ago. As electrically conducting material in the dust cloud swept through the solar field, the Sun's magnetism exerted a force on its electrons. The electrons began to move, which created an electric current which in turn created a magnetic field that was the origin of the field of today.

The magnetic field of the Earth extends far out into space as the magnetosphere. Streams of charged particles from the Sun — called the solar wind — bend the magnetosphere out of the normal ball-shape of a dipole magnet and into the shape of a teardrop. The magnetic field of the Earth deflects the dangerous solar wind, which is mainly protons, electrons and alpha-radiation, and keeps the particles from the planet's surface. Without it, life on the Earth would be impossible. Some solar wind particles do enter the atmosphere at the polar cusps.

The Permanent Record

For a number of reasons the Earth's magnetic field varies in strength. There are daily variations in the overall magnetic field which may be caused by disturbances in the atmosphere. Much more locally, changes in the rocks near the surface produce variations called magnetic anomalies, which are measurable with sensitive electronic apparatus.

There are three main types of magnetic materials. *Ferromagnetic* substances like nickel, or iron, will become magnetized in a magnetic field and will still be magnetized after they are taken from the field. *Paramagnetic* materials like copper or even oxygen will be magnetized by a field but will lose their magnetism when taken out of it. When *diamagnetic* materials are placed in a magnetic field they become magnetized in the opposite direction to the field.

The rocks affected most by magnetism are those that contain iron-bearing minerals such as magnetite and titanomagnetite. The latter is the most common magnetic mineral. Dark volcanic rocks such as basalt are over 100 times more likely to acquire magnetism than sedimentary rocks such as sandstone. When basalts are molten, the tiny crystals of magnetite forming in the liquid lava may take on a magnetic polarity aligned to that of the Earth's magnetic field at the precise time.

Geologists can map rock magnetism to show where basalts and related rocks occur in intrusions such as sills and dykes. Iron ore can also be found in this way, and when a mining company wishes to prospect a new region, one of the initial stages is often an aeromagnetic survey. This rapid and inexpensive method can be used over land or sea, and was used extensively in the preliminary exploration of the North Sea oilfields.

The Earth's liquid outer core is 2,200 km (1,400 miles) thick, and flows between the mantle and the solid inner core. It is a circulating mass of nickel-iron alloy, but is not intrinsically magnetic, because no material could retain its magnetism at the temperatures at the Earth's core. Instead the Earth's magnetic field is generated by the circulation of the electric charges in the molten mass. The matter moves because of convection currents that are driven by the difference in heat between the inner and outer core (about 2,500°C). The coriolis forces that arise from the Earth's rotation — and, for example, cause cyclonic storms on the surface — could twist these into "rollers" of fluid. The rollars — nobody is sure exactly how many would be aligned along the Earth's axis of rotation — act like the coil of wire in a dynamo, to generate a dipole magnetic field.

The magnetic poles do not coincide with the geographic poles, and in fact they do wander as the Earth's magnetic field varies. Yet if the position of the poles is plotted back in time by studying the magnetized rock on different continents, it seems that the poles have wandered many thousands of kilometers more than seems plausible. The magnetic pole positions seems to become increasingly farther away from the geographic poles with the increasing age of the rocks studied. But the apparent polar wandering is a product of continental drift and not the actual large-scale movement of the magnetic poles.

As lavas that are formed at mid-ocean ridges cool, the magnetically susceptible minerals, like magnetite, that form within them get magnetized in the direction of existing magnetic field. As the polarity of the field reverses with time, "strips" of normally magnetized and reverse magnetized rock get laid down successively by the spreading sea bed. Of course the pattern of magnetic strips is not very neat, and in reality shows many irregularities and breaks. Rocks that cool more slowly, deeper down, record the direction of the average magnetic field throughout the time that the magma was solidifying.

MOUNTAINS

All the great mountain ranges were formed at the edges of the Earth's huge, mobile plates of crust and upper mantle — the lithospheric plates. There, rocks either bend or break under the compressive and extensional forces that result when the plates collide. Folds, which come in all shapes and sizes, mainly form deep within the continental crust. Rocks tend to fault at the shallower levels.

A section through the mountain belt might start with flat-lying sedimentary rocks that gradually, over a distance of a few kilometres, become more and more folded to create alternating arch-shaped *antiforms* and trough-shaped *synforms*, as in the Jura Mountains of France and Switzerland, or the Valley and Ridge province of the Appalachian Mountains. In other instances, for example the eastern Rocky Mountains in Canada, rocks may be stacked on top of each other by a series of low-angle thrust faults.

In the central part of mountain ranges like the Alps or Himalayas, where the deformation is more intense, the rocks are not only folded, but the folds have been flattened and sheared so that the upper part is moved forwards, in some instances by many kilometres, in structures called nappes. In such mountains, which are formed by the collisions of two continental plates, a few very deep major thrust faults slice packets of rock and pile them on top of each other. These faults cut deep into the basement of old, strong, crystalline rocks.

The Collapsing Crust

But mountains are not static features. As soon as they start to rise, the slow process of erosion starts to wear them down, and some times when the horizontal forces that thicken the crust stop — for example, if subduction of one plate under another ceases — then the thickened crust starts to spread. Evidence of this kind of crystal extension is plentiful in the Basin and Range province to the west of the Rocky Mountains in the United States, where the alternating basins and upward-titled ranges are bounded by normal faults.

As the crust spreads apart, blocks drop along the faults to form basins. One Death Valley is now below sea level, but at one time it may have been at an altitude of several kilometres. Some geologists speculate that the Andes, and the Tibetan plateau, which lies to the north of the Himalayan range, may also one day collapse in a similar way because these mountains have deep roots of lighter buoyant material which can spread out if it is not squeezed by horizontal forces. Rifts, such as those found in the North sea, the East African Rift or the Rhine Valley, are another type of extensional zone in which a long, narrow block has sunk down between two parallel normal faults.

Erupting Giants

But the biggest mountains are formed by volcanic activity, not structural deformation. Two volcanoes in the Hawaiian Islands — Mauna Kea and Mauna Loa — have bases over 140 kilometres (88 miles) across. Although single volcanic mountains occur, they can also be grouped in clusters or chains like the Cascades of the American North-West, which contains Mount St. Helens.

The Himalayan mountains are nearly twice as high as the Alps, even though both chains of mountains were formed by the self-same process — the collision of two continents. The answer is partly because the European lithospheric plate is only about half as thick and consequently half as strong as the Indian plate. One theory of how mountains are prevented from sinking into the Earth holds that the weight of the mountains bends the lithosphere and so spreads the load across a broad area. An older theory assumes that the higher mountain ranges sit on top of areas where the crust is much thicker.

This means some mountains, like the Andes or the Tibetan plateau, are buoyed up by deep low-density rock, so that the mountains float on the crust in much the same way that an iceberg floats on the sea.

The Alps exhibit most of the common forms of mountain folding. Their geological history extends back over 250 million years, but before the Cenozoic era, which began some 65 million years ago, there was very little evidence of the extensive crumpling and dislocation that can be observed today. The Alps arose from the collision of the Eurasian and African continental plates. Deformation of the western Alps began about 45 million years ago. The eastern Alps, formed in the Italian prong of the African plate, was already deformed by folding. Firstly, thrusting began near the suture between the plates, but successive faults lifted up previous thrust slices and brought deep crustal rocks to the surface.

CONTINENTAL DRIFT

It is now widely agreed that continents are not fixed, but have changed their position.

especially over last 2000 million years. The oceans, similarly increased or decreased in size. The idea was first mooted in the form of theory of continental drift, by a German meteorologist and geophysicist, Alfred Wegener, in 1912. According to Wegener, there was earlier one super continent, shaped like a biscuit; he named it **Pangaea**, surrounded by primeval pacific ocean. Some 200 million years ago, it broke up into pieces which drifted away. He later proposed, that Pangaea first broke up into two large pieces, one northern which he named **Laurasia** and one southern the **Gondwana.** Whereas Laurasia which, gave rise to northern land masses showed lesser breakage and drifting the **Gondwana** showed breakup into at least five large pieces, which drifted apart and formed South America, Africa, India, Australia and Antarctica. The drifting away of South America from Africa opened up Atlantic ocean. Indian ocean appeared by drifting away of India, Antarctica and Australia.

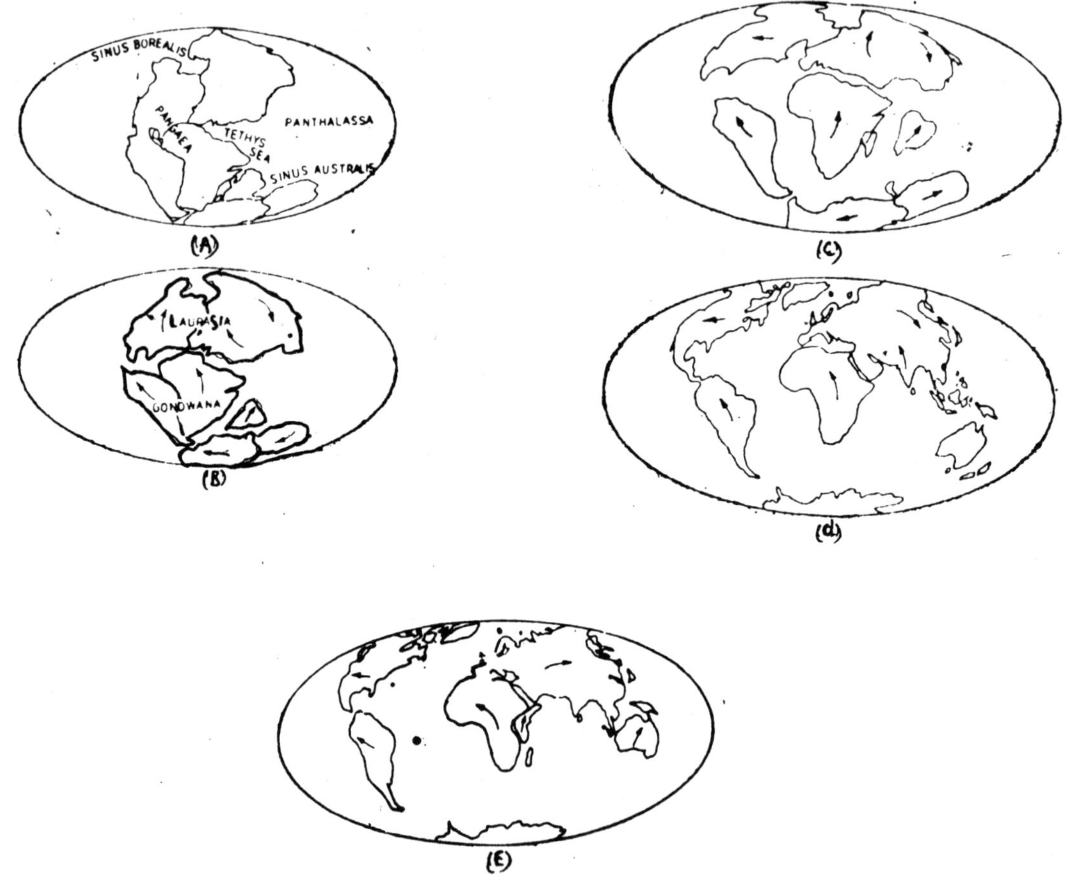

Continental drift. Arrows show general direction of movements: the tectonics plates that bear the continents do the actual drifting. For simplicity the plate boundaries, which are largely under the oceans, are not shown here. (A) The super-continent of Pangaea, 200 million years ago. (B) Pangaea breaking up into a northern supercontinent, Laurasia, and a southern supercontinent, Gondwana, 100 million years ago. (C) As the dinosaurs disappear, the South Atlantic opens up, 65 million years ago. (D) The Earth today. (E) The future : 50 million years from now, the Atlantic widens and the Pacific shrinks.

The major support to the theory of continental drift comes from the east coast line of South America fitting perfectly into western coastline of Africa, the similar glossopteris fossils of India, Australia and Antarctica and the fact that oldest mountain ranges in Western Europe, continue on the other side of the Atlantic Ocean into Canada.

Although, Wegener failed to explain the mechanism of continental drift, the researchers carried with the help of advanced technology from Early 1960s, accumulated information that has adequately explained the mechanism of continental drift, formation of oceanic ridges, shrinking or expanding of oceans and throwing up of mountain ranges. These processes are now better explained by the theory of plate tectonics, based on the assumption that the earth's crust consists of a number of movable slabs or plates of different sizes.

PLATE TECTONICS

The processes of continental drift, sea floor spreading and mountain formation are now better explained by the concept of plate tectonics. It is now believed that the earth's surface consists of a number of movable slabs or 'plates'. There are about 12 major plates and several smaller ones. The largest of these plates, the pacific plate, supports the whole pacific ocean, while one of the smallest plates supports Turkey.

The idea of plate tectonics, which has now emerged as a new branch of geology, gained momentum from the studies of Hess, of Princeton University in America on mid-Atlantic ridge. He proposed that this long underwater mountain range was caused by the deep crack in the crust of the earth, splitting the crust. The molten material from the asthenosphere pushes through the crack, pushing the old crust sideways. As the American plate extends from mid-Atlantic ridge to the west coast of South America and the African plate from Mid-Atlantic ridge to the other side of Africa, the result is the drifting away of the two continents and the widening of Atlantic ocean. The cooling of material along the crack resulting in the underwater mountain range. In areas when this range arose out of water, islands, such as the Azores were formed.

Since the volume of the earth is constant, the movement of the plates have relative action. As the two continents are moving away the two may be approaching near, as this has happened between India and Asia. When an oceanic crust meets a continental crust, the former being denser slides along the edge of the plate and under the continental crust, collides with the mantle and melts to become its part. When the two continental crusts collide, high mountains are thrown up, like the Himalayas which came up when Indian plate collided with Asia some 40 million years ago.

When the material from the mantle pours out of a crack in the crust, this is compensated by pushing back of oceanic crust at some other place to form the oceanic trenches. These activities maintain the net volume of the earth.

The slow movement of crustal plates is made possible by underlying asthenosphere which acts like a conveyer belt, being less solid and more plastic than the crust above.

EARTHQUAKE

An earthquake is shaking of the ground by seismic waves. It ranges from a faint tremor to a wild motion capable of tearing buildings apart and causing gaping fissures in the ground. The energy causing the earthquake is suddenly released from the focus and travels through surface layers of the earth in widening circles. These seismic waves travel outward in all directions, gradually losing energy. Whereas the earthquake **focus** lies beneath the surface, sometimes as deep as 650 km, the point directly above the focus on the surface is called **epicenter.**

The magnitude of an earthquake is measured by instruments called **seismographs**. The Richter scale measures the actual amount of energy released by an earthquake on a scale of 0 to 9. The local damage caused by an earthquake is measured on Mercalli scale recognising 12 levels of intensity. Amongst the most severe earthquakes recorded are Honshu in 1933, Assam 1950 and Alaska, 1964 all measuring 8.4 on the **Richter Scale** closely approaching the highest ever recorded 8.6 which released energy nearly 3 million times that of the first Atomic bomb on Hiroshima (5.7 on Richter scale). In terms of human loss severe earthquakes occurred in: Tangshan, China in 1976 (655,000 deaths) Yokohama and Tokyo, Japan in 1922 (150,0000 deaths) and Lisbon, Portugal in 1755 (50,000 deaths).

Earthquakes commonly result from faulting in the earth's crust, occurring because the upper rock layers are unable to withstand pressures applied by tectonic forces. If the slippage, or fault plane is nearly vertical, one side is raised relative to another. This is called **normal fault**. If dislocation is primarily horizontal it is termed as **transcurrent fault** or **strike-slip fault**.

When an earthquake occurs in the floor of an ocean, seismic seawaves are generated and spread in widening circles that may reach the coast after several hours. The seismic seawaves, called tsunami in Japan have inflicted great damage to coastal cities in Japan and Hawaii. The tidal surge is often upto 9m, arriving at intervals. The catastrophic coastal flooding that occured in Japan in 1703, killing 100,000 persons was probably caused by a tsunami.

The effort of seismologists is now to have advance prediction of earthquakes using sensitive instruments which record seismic waves and detect the changes in wave severity which give an indication of approaching earthquake. Attempts are also made to find ways to reduce intensity of an earthquake by injecting lubricating fluids along the fault line to make fault blocks slide gently instead of jerking violently.

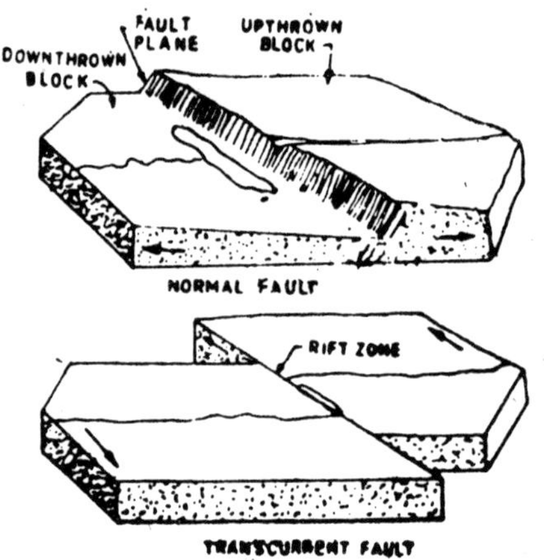

Normal and Transcurrent Faults.

VOLCANOES

The upward movement of magma through lithosphere termed volcanism or vulcanism is a major geological activity by which mountains are formed. The other activity leading to the mountain formation is the tectonic activity. Two types of volcanic activity are encountered: intrusive and extrusive. **Intrusive volcanism** results when magma cannot escape the surface of the crust and solidifies as intrusive igneous rocks. Such large deposits deep beneath the surface constitute **Batholith**, but when near the surface it generally forces the rock layers above to arch upward to form a structural dome called **laccolith**. The magma may sometimes accumulate in the form of sheets between layers of sedimentary rock when called **sills**, or send finger like projections through cracks in rocks, termed **dikes**. **Extrusive volcanism** occurs when magma escape out of the crust in the form of a lava which flows out like a river, or ejecta consisting of solid fragments and gasses. The resultant structure is called volcano.

A volcano can take a variety of forms. When thick lava flows out of the vent of the volcano, on subsequent solidification, it forms a shield or dome like structure called **shield volcano.** If the eruptions occur as a blast due to blockage in the vent, the rocks and ashes will be flung high and on deposition build a steep sided structure called **cinder-cone volcano.** When eruptions come in the form of alternating lava flows and solid ejecta the resultant cone like volcano will have alternate layers of lava and solid ejecta, and is termed **composite volcano.** The very fine volcanic dust generally rises high into the troposphere and stratosphere and may remain suspended for years.

The volcanoes are mostly located along the edges of crustal plates, being highly concentrated in the circumpacific ring extending from Andes in South America, through Aleutians, Japan, East Indies and extending to New Zealand. The region is approximately named as pacific ring of fire.

A volcanic eruption commonly results in a large depression at the top of a volcano, due to blowing away of the top , or its collapse. This depression is called a **caldera.**

History has recorded many violent volcanic eruptions. The one in Mount Vesuvius in A.D. 79 destroyed the city of pompeii in Italy. The volcanic eruption of Korakatoa, west of Java, caused 40,000 human death. Mount Etna in island of Sicily erupted repeatedly, almost 400 times since 500 B.C., in 1669 A.D. alone killing some 20,000 people. Nearly 40,000 people in city of Saint Pierre lost their lives in volcanic eruption of Mount Pelee on island of Martinique in West Indies.

ROCKS

A rock is a naturally occurring aggregate of minerals in solid state. It can be made of only one mineral (such as gold) or a combination of two or more. Granite, for example is composed of quartz, feldspar and mica. The minerals generally retain their identity within a rock, and can be separated out. The rocks are generally classified into three types : **igneous rocks**, **sedimentary rocks** and **metamorphic rocks**.

Igneous rocks are also called magmatic rocks, are formed from the solidification of molten rock called **magma.** A magma that moves up from asthenosphere, but cannot escape the earth's surface, cools within the crust to form **intrusive igneous rock.** These are also termed **plutonic rocks.** A large accumulation of intrusive igneous rocks is called batholith. Granite is the commonest example of intrusive igneous rocks. When magma escapes the crust in the form of lava, and solidifies on surface of the earth, **extrusive igneous rock** or **volcanic rock** is formed. The magma that splashes out of crater and cools quickly forms rock with glassy texture as seen in basalt. On the other hand, the magma ejecting out of the volcano with force is frothy and filled with air when cool. This forms the **pumice**. The ashes that are ejected out of the volcano, do not form a solid rock, but a loosely compacted mass called **tuff.** Three types of igneous rocks are often differentiated on the basis of mineral composition : felsic igneous rocks which are rich in silica aluminium and potassium but poor in iron and magnesium (*e.g.*, granite rock), mafic igneous rocks which are rich in magnesium and iron (*e.g.*, basalt rock) and ultramafic igneous rocks having higher proportion of magnesium and iron, but lack aluminium.

Sedimentary rocks are derived from the sediments composed of sand, silt and clay originating from disintegration of any type of preexisting rocks, usually igneous rocks. They may also be formed from compaction of organic remains of the once-living organisms. Sedimentary deposits of rock origin are termed clastic sedimentary rocks and may be formed from compaction of silt and clay (*e.g.*, fine grained rock shale) or cemented sand grains (*e.g.*, sandstone). Organic sediments form organic sedimentary rocks. Common examples of organic sedimentary rocks are calcareous limestone made up shells and coral and chalk made from microorganisms of the past. Coal is another organic rock, not strictly sedimentary, derived from carbonized vegetation of distant past. The chemicals within the rock layer, suspended or dissolved in water may, likewise precipitate out commonly through evaporation of water to form chemical sedimentary rocks, as for example table salt and gypsum.

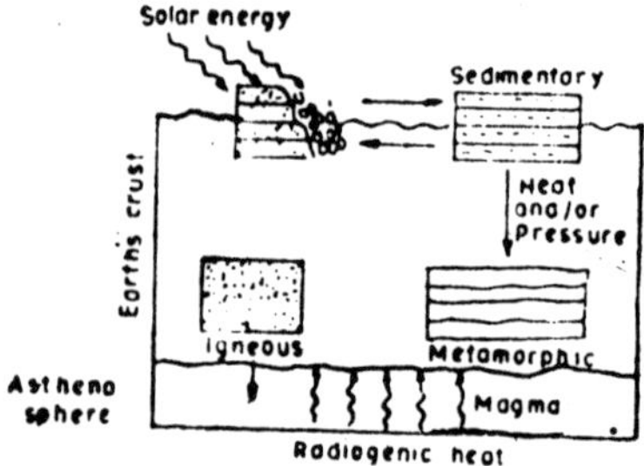

Metamorphic rocks are derived from either igneous or sedimentary rocks and are formed under the influence of high temperature and/or tremendous pressure. The original texture and mineral are altered and new rocks with different physical and chemicals characteristics are formed. Common examples of metamorphic rocks are diamond formed from pure carbon, marble formed from limestone, state formed from shale and qneiss formed from granite.

A distinct rock cycle has been in operation, through millions of years propelled by radiogenic heat generated from within the earth and solar energy. The magma maintained in liquid form by radiogenic heat on reaching the crust forms igneous rock, which in turn may breakup under the influence of solar energy and sediments ultimately compact into sedimentary rocks. Both igneous and sedimentary rocks, under the influence of heat and/or pressure form metamorphic rocks. Metamorphic rocks may, when exposed to solar energy on surface may likewise form sedimentary rocks. All the three types of rocks under the extreme influence of radiogenic heat melt and get converted into magma, again.

Sedimentary Rocks

The vast white chalk cliffs of Britain and Europe are made from countless billions of microscopic plates that were formed from plankton. Yet each individual plate is only a few thousandths of a millimetre long. This is only one of the many ways in which sedimentary rocks have been made. Some were formed, grain by grain, of pre-formed rock that was weathered, eroded and deposited elsewhere either by wind, rivers or glaciers. Others settled as vegetable matter or dead organisms on river or sea beds. And yet others arise through chemical reaction.

Sedimentary rocks comprise separable particles, often quartz or rock fragments, cemented together by minerals. They are the products of weathering and erosion of pre-formed rocks of all types. Weathering is the decomposition of rock by chemical reaction (as in rain action) or by mechanical means (as in ice expansion in rock joints). Local climates determine weathering rates and types: erosion is the movement of weathered particles.

Most *fluvial* (river) erosion will deposit at the coast, but inland deposits, particularly at gradient discontinues and in the river's senile stages, are also significant. Fine desert-weathered particles (vulnerable because of a lack of binding vegetation), frost-pulverized glacial outwash and soil from mismanaged farm land are often blown away and deposited as loess. Huge glacial deposits, or *till,* are common in glaciated zones of the last 2,000,000 years; iceberg calving off glaciers have also deposited thousands of square kilometres of sediment far out to sea.

Sediments with large rounded particles — some boulder-sized — are *conglomerates*; angular particled sediments are *breccias*. Particle shapes often betray origins, rounded ones being water-formed and angular ones the result of frost action or mountainside scree deposition. Smaller-grained rocks like sandstone are usually formed in deserts, river or seas.

Generally, the finer the grain size the weaker the current that carried the sediment, and, often, the deeper the water in which it was deposited, Clays and mudstones are often found in deep oceans, where river-borne sediment is rarely encountered. Much material deposited so far from land is volcanic dust and other material carried long distances by the wind. It is now possible to study sedimentation processes at great depths, thanks to advances in submersible technology.

Sediments often exhibit feature that indicate how they formed. For example, a layer of sandstone that has ripple-marks and desication cracks was evidently once located in shallow water and later appeared above the water level for some time. Sands deposited as a stratum by the wind often show large-scale dune-bedding where the layers are deposited in sweeping curves and ripples.

Three main processes form sedimentary rocks. All three are shown in a coastal context in the stylized diagram opposite. Organic sedimentary rocks are formed from once-living matter. For example, corals extract lime from seawater to make their skeletons. A reef's living coral is found only underwater near the reef top, and beneath an atoll may be hundreds of meters of dead corals. Bits of coral, broken off and eroded away by the waves, and shells of dead crustaceans left on the sea floor will form reef limestone, a typical organic rock.

Some sediments are formed by chemical rather than physical means. Economically important deposits such as rocksalts, gypsum and potash are known as evaporites, and are deposited by precipitation from saline water, often in warm, arid climates. This is common in the tropics, where the seawater in a partially enclosed basin evaporates, depositing its salts to form these chemical rocks. Another chemical process that gives rise to sedimentary rocks is that in which calcium carbonate is precipitated around minute nuclei to form spherical particles called oolites, which are the constituents of oolitic limestone. Oolites tend to form where the wave action is strong, by a "snowballing" build-up of calcium carbonate. Individual oolites tends to grow up to about a millimetre in diameter.

The rocks arising from the more familiar mechanical process — which are produced as a result of erosion from older rocks — are called clastic types. Whether flowing from mountains to adjacent lower land or when reaching its ultimate destination at the sea, a river tends to first deposit stones and large particles, dropping finer material later. As a result of this process a gentle slope of increasingly fine sediments is created downstream or out of the sea. Inland, this feature is called an alluvial fan; if conditions are right, river mouth sediments form deltas. The fine clays farther out will form mudstone, which has no internal structure, or, if buried by new sediments, will be squeezed into shale, which has its flat crystals well aligned, and so breaks easily into thin slices.

Loose sediment is turned into rock by three basic processes. Sandstone results from cementation, where water percolates between sand grains and deposits thin layers of binding iron oxide, calcium carbonate or silica. Clay is turned to mudstone by compaction, when water between the grains is squeezed out by the weight of more sediment accumulating above. Marbles often show the marks of colossal mountain-building forces, which cause the rock minerals to recrystallize in a solid mass with no space between them.

Fossils coral reefs containing numerous shellfish and other fossils cemented by lime mud occur in many parts of the world. These limestones are of great economic significance, used in building and the making of cement. Chalk is made of innumerable microscopic shells, and coal is fossilized vegetable matter.

WEATHERING OF ROCKS

The surface of earth, constituting the land masses is rocky. The bare rocks can support a very small fraction of life forms, mainly lichens and mosses. The first step in the process of pedogenesis is the breakup these rocks to form particles of various sizes — gravel, sand and clay. The process of breakup of rocks is termed weathering. Three types of weathering are encountered : **Physical weathering**, **chemical weathering** and **biological weathering**.

Physical weathering results from mechanical processes, without altering the chemical composition of rocks. The unequal heating of outer and inner layers of the rock and their consequent expansion alternated by their contraction during night usually develops stress, relieved by cracks, and disruption of rocks. The phenomenon also termed thermal weathering is prominent in temperate climate with wider temperature fluctuations. Frost action is another common process in cold climates. The water enters through rock crevices, and upon freezing expands, causing the rocks to breakup along fracture lines. This ice formation may be in the form of blocks or ice crystals, especially in very small crevices. These crystals slowly build up to ultimately cause rock disintegration. The growth of salt crystals is a common process in dry climates, when water containing dissolved minerals in crevices evaporates leaving behind salt crystals. With continued evaporation, salt crystals build up to cause rock disintegration along weaker points. Rocks are also broken by abrasive action of running water and winds.

Chemical weathering, also termed as decomposition of rocks, involves chemical changes in rocks. Oxidation, the combination of oxygen with metallic element is common in iron-carrying rocks. Carbonation involves dissolving of carbon dioxide in water to form carbonic acid, which corrodes or dissolve certain minerals such as lime stone, often forming large underground caverns. The water may take minerals directly into solution, or the water enters into the chemical make up of the mineral-water of hydration — or bring about the chemical change of the mineral (hydrolysis). These processes develop certain weak spots in the rocks and may help in their ultimate breakup.

Strictly speaking there is no biological weathering as such — it is either a physical or chemical weathering through the agency of living organisms. Mechanical weathering is caused by roots of plants entering crevices of rocks, enlarging with growth to ultimately force break up of rocks. Strubs and small trees, with long woody roots are prominent cause of such a weathering. Chemical weathering is brought about by CO_2 released during respiration of roots and rhizoids, corroding the rocks through carbonation. This corrosive action can be easily detected when a layer of plants, commonly mosses, carpeting a rock is removed. The rough surface of rocks, will show the scars of corrosion. The bacterium *Thiobacillus thiooxidans,* oxidizes sulphur to sulphuric acid, which can corrode the rocks.

The processes of weathering are primarily concentrated near the surface of the rock with the result the relatively intact rock, called bed rock lies at the bottom, followed by a layer of partially disintegrated rock pieces, loosely arranged to form regolith. The solum, or true soil with smaller fractions of rock material in form of gravel, sand, silt and clay are distinctly arranged under the influence of soil forming processes. Solum covers the regolith, unless transported away by various erosional forces.

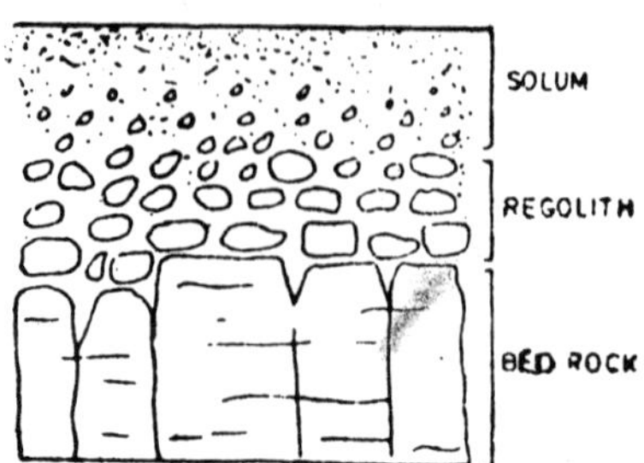

Weathering of Rocks.

The weathered parent material would depict typical zonation only if undisturbed, developing residual soils. It may, however, get transported by gravity (colluvial deposit), water (alluvial deposit), wind (eolian deposit) or glaciers (glacial deposit) and give distinctive characteristic to the soil.

MINERALS

More than 90 elements naturally on the Earth, but just eight of them — oxygen, silicon, aluminium, iron, calcium, sodium, potassium and magnesium — account for 98 per cent of the weight of the Earth's crust. And two of them , oxygen and silicon, make up close to 75 per cent of the crust. The silicates, which contain these elements, are the most common group out of all the 2,000 minerals that are formed when elements combine. Minerals vary hugely depending on their atomic structures, from slippery talc to diamond to the fan-shaped crystals of molybdenite.

Silicates and most other minerals form by crystallizing out of some sort of liquid. Salt forming from pools of sea water on a hot day, quartz slowly crystallizing in a cavity a few hundred meters into the Earth's crust, and crystals that form when slowly cooling magma freezes many kilometres below the surface all grow in similar fashion. But the form the crystal takes depends on the environment in which it grows. The milky white lumps of quartz occurring in veins that cut across the natural grain of many rocks at the surface are the same as the beautiful hexagonal prisms of quartz that are usually only seen in museums and books. The difference is that the prisms grew where the chemistry of the solution, cooling rate, temperature, and space were all just right.

The Crystal Maze

Well-formed crystals occur in a bewildering variety of shapes, from the neat cubes of pyrites, or fool's gold, to the curved fans of silver-purple molybdenite. But all these shapes can be placed into categories which reflects the different regular arrangements of the constituent atoms.

Slippery talc is the softest mineral whereas diamond is the hardest, all because of their atomic structures. Another crystal characteristic related to atomic structure is *cleavage*, or the ability of a mineral to split more easily along certain planes. Natural glasses, like the volcanic glass obsidian, break along curved irregular surfaces because they are amorphous solids that have no underlying regular structure. They are said to have a *conchoidal* fracture because the broken surfaces look a bit like a curved sea shell. But in some minerals the planes of weakness are so pronounced that the crystal breaks much more easily in one direction than another. Crystals of calcite, the main constituent of limestone, always break along three characteristic surfaces to form parallel-sided rhombohedra.

Because minerals physical properties depend on their structures, many of them, like cleavage or hardness, are different in different directions. The characteristic, called *anisotropy*, means that when energy, such as light, or sound, passes through most crystals it travels at surprisingly different speeds when moving in different directions. When white light falls on a mineral some of the wavelengths are absorbed and some are reflected. A mineral's colour corresponds to the wave-lengths that are reflected. Sometimes this depends on the mineral's chemical composition, which is why copper compounds are usually blue or green. In other instances, for example diamond, the colour is related to the crystal structure. But in quartz, colour is related to the presence of chemical impurities, which give the lovely purple amethyst, pink rose quartz and brown smoky quartz, although pure quartz is colourless.

So crystal colour can relate to a number of different factors. But if the mineral is ground up, the intrinsic colour is revealed. Haematite and magnetite, two minerals composed of iron and oxygen, can be difficult to tell apart, but if they are rubbed on a piece of unglazed procelain (a process called a streak test)., the haematic gives a reddish streak wheras the magnetite leaves a black mark.

A mineral derives its physical properties not only from its chemical composition, but also from the way its chemical subunits are arranged. For example, the hardest naturally occurring mineral known to man — diamond — and one of the softest — graphite — are both made of the same element — carbon. In graphite [A], the carbon atoms are arranged in parallel layers. Within the layers, strong chemical bonds join the carbon atoms into interlocking hexagonal rings: but the layers are joined to one another by weak bonds, so they tend to "slide" across each other, a property that makes graphite a good industrial lubricant. In diamond [B], all the atoms are linked together by strong chemical bonds into a single, resilient molecule. Each atom in the molecule is joined to four others, all of which are an equal distance away from it. This is known as tetrahedral linkage, and it creates a very dense and closely knit crystal.

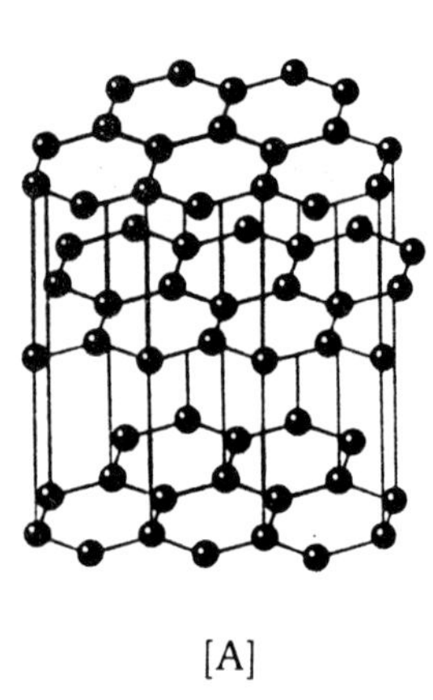

[A]

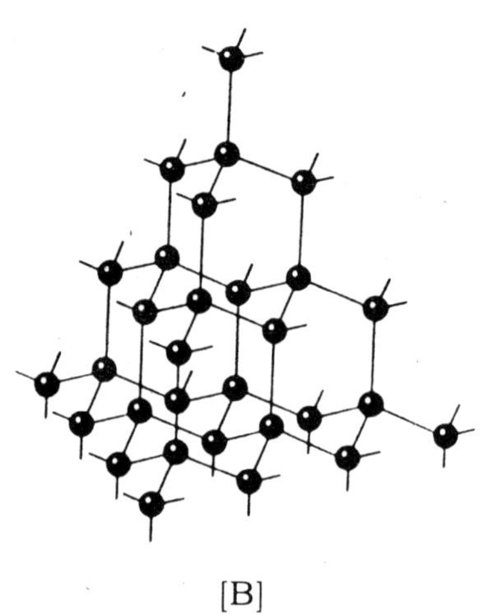

[B]

Diamond from Coal

Most attempts to create diamonds have concentrated on reproducing the high temperatures and pressure found beneath the Earth. An Early scheme involved melting carbon at the centre of a ball of iron, then abruptly cooling the iron so it would contract and squeeze the melted carbon. Explosives have been used to generate high pressure and temperature, but because the pressure falls immediately any diamonds formed quickly revert to graphite. "Grit" diamonds for industrial use can be made by heating specially prepared carbon compounds up to 2,700°C (4,900°F) at a pressure of 105,000 kg/cm^2 (1,500,000 lb/in^2).

MINING

Removal of metal ores, coal, or other minerals from surface or underground deposits. Quarrying is the removal of crushed stone or dimension stone without the intent to extract a metal.

Three general types of mining operations exist :

1. underground mining,
2. surface mining (open pit, open cut, auger, and strip mining, with or without prior removal of earth or rock overburden), and
3. placer mining among alluvial sands and gravels.

Placer mining is a wet form of mining surface deposits of dense minerals weathered from rock and carried into a stream bed, where dredges may be used to excavate material that can be sorted on sluices. In hydraulic mining, another wet mining method, a powerful stream of water is used to remove overburden from an ore body. The water was washed out into streams or other bodies of water.

The sediments from all forms of mining constitute one of the serious environmental problems of mining.

FOSSILIZATION

Only a tiny fraction of all the billions of plants and animals have ever lived on the Earth are preserved as fossils. These ancient remains hint at the diversity of past life and give us a unique insight into the process of evolution. Fossils range in size from huge dinosaur skeletons to the casts of bacteria visible only under a powerful microscope. The composition of a fossil depends on how it was formed: it may be an entire organism encased in rock, amber or ice, or the faintest impression of a dragonfly wing in soft mud, later compressed into mudstone.

The process of fossilization begins when the remains of an organism, or traces of its passing, such as footprints, are buried in mud or sand. Then, given the right circumstances, the buried animal or plant fragment may undergo a number of changes as the mud and sand become compressed into rock. These changes make the fragment physically and chemically similar to the surrounding rock, allowing it to be preserved indefinitely as a fossil.

Although fossilization cannot be predicted accurately in space or time, it is favoured by certain conditions. Various rock types contain fossils, but none more so than the sedimentary rocks, consisting of strata of sand, mud or clay, formed on or near the Earth's surface. Of these, limestones are possibly the most productive, and commonly contain a wide variety of well preserved bivalve mollusces, other hard-shelled creatures and corals. Fossils are also occasionally found in shales and clays, although they are often crushed by the great pressures that build up within these rocks unless they are protected within rounded nodules or concretions.

Fossils in rocks formed from shales and clays are of particular value because the sediments are fine-grained and able to preserve delicate, soft-bodied organisms in great detail. The burgess shale, high in the Rocky Mountains of British Columbia, is a classic site of this type.

Probably the most significant single factor promoting fossilization is rapid burial. Speed reduces decay, scavenging by other organisms and physical destruction, and it is for this reason that most fossils are of sea-dwelling creatures, for sediment is most rapidly deposited in marine environments. The remains of terrestrial organisms are rapidly eroded and are less likely to be buried by sediment. But this does not mean that land plants and animals are totally absent from the fossil record. Given the right conditions — for example, a flooded river that traps many creatures beneath great masses of mud excellent fossils can be formed.

Candidates for Fossilization

A creature is more likely to become fossilized if it has some hard parts, such as a shell or bone structure, than if it is made up only of soft tissues. However, not just large, bony organisms are preserved. Soft-bodied creatures can leave a characteristic black carbon film on a stratum, and tiny microfossils (visible only through powerful microscope) are very common in many strata. These microscopic fossils have recently become the focus of considerable attention because they enable geologists to accurately determine the age and relative positions of rock layers that are useful — or even critical — indicators in the search for oil and gas.

Occasionally, an entire creature, or part of its original tissue, may be preserved unaltered: an insect may crawl into the fragrant resin from a pine tree which later hardens into amber, or an animal may become trapped in a naturally occurring pool of tar. Such rare fossils are preserved because they are locked away in an oxygen-free environment where decomposition cannot take place.

The type of fossil formed in a rock is dictated by the chemistry of both the organism and the rock itself. The harder parts of animal and plants contain a number of materials that do not decay, such as phosphates in bone and calcium carbonate in shells. During fossil formation, such minerals are often replaced by others that are better able to survive the rigours of the subterranean environment. For example, iron pyrites, hematite or quartz commonly replace minerals in a shell or a bone, molecule by molecule, so a calcium-based coral may be preserved as a hematite fossil. The organic part of the shelled creature is always the first to decay. This leaves a gap in the shell which is often filled with sediment. The shell is more soluble than the rock, and slowly dissolves, leaving a sedimentary rock cast of the inside of the shell. Under the right circumstances the shell material is replaced by some other substance — often a type of silica — to give a cast of the shell's exterior. The "replacement" may even be the shell material, recrystallized. Sometimes, no sediment fills the empty shell. The solutions that permeate the strata may totally dissolve a buried shell, leaving a "mould" behind in the rock. A replica of the original shell may later be cast, as another mineral or sediment is deposited in this mould.

The tracks of living creatures are often fossilized, as well as their bones or shells. Typically, a footprint — or the burrow of prehistoric worm, for example — is left behind the soft mud, which partially hardens to form a cast. If the mud is flooded by a river, or the sea , sediment is laid over the mud especially quickly helping to percieve the shape of the footprint. Over the course of time, the mud and the overlaid sediment become compressed and turn to rock. The original mud-based rock forms a mould of the footprint. The sediment-based rock forms a cast.

Many of the animals that become fossils clearly died — and were preserved — suddenly and unexpectedly. They were eating, being eaten, or, like this ichthyosaur, in the act of giving birth. Fossil evidence like this provides valuable clues about the dinosaur physiology. Scientists were puzzled about how the dinosaurs could return to the water when they needed to lay their hard-shelled eggs on land. However, evidence such as this shows that at least some dinosaurs produced live young.

FOSSIL FUEL

The modern world depends on plants and animals that died millions of years ago. Our industries, powered by coal, oil and gas, are burning the fossil remains of prehistoric life. Over 40 per cent of the energy consumed in Western Europe comes from coal, and what took millions of years to create will be consumed in centuries. Predictions about how long fossil fuel reserves will last are coloured by political expediency, as are government responses to the charge that fuel combustion builds up sulphur-rich gases and carbon dioxide which threaten the environment.

Oil, gas and coal are all formed from the decay of once-living organisms under heat and pressure. Over 80 per cent of the oil and gas currently exploited formed in Mesozoic or Tertiary strata between 180 and 30 million years ago, from marine micro-organisms deposited as sediment on the sea bed. The basic components of oil and gas are created when the organic remains are not completely oxidized, leaving a residual mass of carbohydrates, hydrocarbons and similar compounds. As layers of sediment bury this residue, temperatures and pressures increase and the liquid hydrocarbons are segregated into pore spaces in the rock. Coal deposits come from many epochs, but the best and most abundant are from forests in the warm swampy river deltas of the Carboniferous period (360-280 million years ago). The process by which forest peat is turned to coal is complex. Initially the carbohydrates and waxy materials of the plant are attacked by swamp bacteria and fungi to give volatile gases like methane and carbon dioxide. Gradually the mixture gets richer in sulphides and hydrocarbons. Layers of strata build up, gradually squeezing out water from those lower down.

Cooking up the Perfect Fuel

The temperature of the Earth's crust increases with depth by about 1°C for every 30 metres. The increasing temperature causes chemical reactions that turn the peaty matter into coal. A temperature of 200°C — meaning burial at a depth of over 5 kilometres — is needed to make top-quality coal. Variations in temperature give a range of coals of different "rank" or carbon percentage, from the peaty brown coals or *lignites* at 60 per cent carbon, through the *bituminous* coals at 92 per cent to the *anthracites* at over 92 per cent.

Optimum Oil Production

The quantity of oil and gas formed depends on temperature and on the speed of subsidence of the strata. Ideally, up to 3 kilometres (2 miles) of overlaying strata are needed. If subsidence is too rapid the temperature may rise too high and oils and gas are lost. Neither is abundant below 6 kilometres (4 miles). In fact they are rarely found in the rocks where they were formed, either they migrate up the gas and lighter oils first with heavier bitumens staying closer to their source until they are trapped by overlaying impermeable rocks like shale and clay. Considerable quantities of oils and gas, however may be lost by migration to the surface, where they oxidize and evaporate.

Oil and gas form as layers of plankton are buried under thick piles of sediment. Increasing heat and pressure at depth first cause fats and oils from the bodies of the marine organisms to "link up" into a thick compound called kerogen. As temperature increases with burial, long chains made of hydrogen and carbon atoms break away from the kerogen, giving a viscous heavy oil. With even more heat, valuable light oils and natural gas are formed. The oils accumulate in "reservoir" rocks — permeable rocks such as sandstone which hold the oil like a sponge. To form an oilfield, the oil must be trapped between layers of an impermeable rock, like shale. Faults where such rock has sheared to form a seal, can trap oil and gas, as can convex domes. The making of most coal was begun in the middle of the Carboniferous period. The Earth's equatorial regions were hot and wet, and lush tropical forests grew in extensive swamps. In these types of environment, thick layers of peat were laid down typically they were sandwiched between layers of sediment, such as shale deposited when the waters temporarily retreated. The seas receded during the Permian period and many of the tropical coastal plains turned to desert. Other sedimentary rocks, such as sandstones, were laid down over the shale and peat. With increasing temperature and pressure, the buried peat began its metamorphosis into coal

Petro-chemistry to Petro-dollars

Petroleum fluids are a complex mix of organic compounds with a range of specific gravities and boiling points. Natural gas is essentially methane and can be piped directly from the source or may be liquefied for transport. The oils are classified as light as distillates, medium distillates and heavy residuals. Light distillates are the motor fuels like gasoline and benzine. Medium distillates make up paraffin, diesel, jet-engine and power station fuels. Heavy residuals are used to fuel power-station and ships.

Fuel Resources

The predicted future life of the world's coal reserves was once put as low as a few decades, but the real figure is actually nearer 300 years. The Soviet Union and the United States have the greatest resources. Western Europe, India, China, Brazil, South Africa and Australia also have large stocks of top-rate coal from the Carboniferous or Permian periods. Brown tertiary lignites can be found in Central Europe and the Ukraine. Antarctica also has large unused reserves of coal, which could only be exploited at disastrous environment cost. The best estimates are that, without a major switch to alternative fuels, the world's oil reserves will last a little over 40 more years.

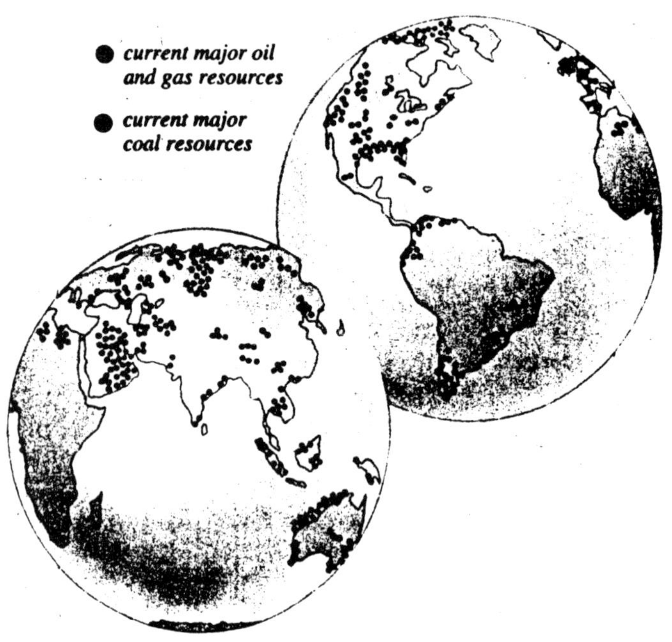

CAVES

A room more than 550 meters long and 30 metres high would be a remarkable feature in any building. It is even more remarkable 220 meters below ground. But this is the size of the Big Room — the main chamber of the Carlsbad Caverns in New Mexico. The Caverns were discovered in 1901 when a cowhand on his way home investigated a curious black cloud that appeared to be emanating from a hole in the ground. The cloud was in fact a huge swarm of bats. The hole in turn was the entrance to the vast caves that now form part of a National Park.

Carlsbad's infinitely varied stalactites and stalagmites are almost all composed of calcium carbonate in the form of calcite. Their graceful shapes attest to the crucial role of water in decorating the caves. But water is not essential to building up such fantastic forms in the caves, it is also usually one of the basic factors in the process by which the caves themselves are created.

Water hollows caves out of the calcium carbonate that makes up a limestone strata by dissolving it away. But water, by itself, is by no means able to do it all: four litres (nearly a gallon) of pure water can dissolve only 0.028 gram (0.001 ounce) of limestone. What enables water eventually to dissolve proportionally between 10 and 60 times more limestone than this is the fact that as the surface water seeps downwards through the soil it passes over decaying organic matter. The plant and animal remains give off carbon dioxide, which dissolves in the water and forms carbonic acid. The acid completes the first step in the process of cave formation.

But if the dissolution of limestone by pure water and the formation of carbonic acid were in turn all that happened, the reactions would grind to a halt once completed, when equilibrium was achieved. There is, however, a third chemical reaction, which uses the products of these first two reactions to continue the process of erosion.

Dissolved in water, calcium carbonate breaks up into charged particles (*ions*) of calcium and carbonate, and carbon dioxide combines with water to form bicarbonate ion and a hydrogen ion. The third reaction occurs when the carbonate combines with the hydrogen to form more bicarbonate: it is this reaction that allows the groundwater to dissolve up to 2.5 millimeters (a tenth of an inch) of limestone in a year.

The opposite happens when water rich in limestone drips off the ceiling of an empty cavern. The carbon dioxide gas comes out of solution, the reaction go into reverse, and calcium carbonate precipitates out.

Stalactites and stalagmites form in caves that have been drained of water following a fall in the level of the water table. Drops of water, charged with calcium bicarbonate, hang from the ceiling of the cave and lose carbon dioxide, so depositing a tiny amount of calcium carbonate. Further drops deposits more layers of calcium carbonate in the same place, and a stalactite slowly develops. Growth rates vary greatly, but stalactites have been known to increase by 7.6 cm in 100 years. They are brittle and rarely reach great lengths. Drops of water may fall to the floor and lose carbon dioxide, leaving deposits of calcium carbonate that grow upwards into stalagmites. These may meet to form continuous columns.

Acidified surface waters percolates through tiny fissures in the limestone, widening these channels before they reach the water table — the surface of the water—saturated part of the ground. The waters then flow horizontally towards a natural outlet — in this case a river — dissolving away limestone in their path.

Karst Landscapes

As acidic rainwater works on a limestone area, certain typical surface features become apparent. Regions that display these distinctive formations are know as Karst landscapes, named after the Karst region in the Dalmatian coast of Yugoslavia. Underground, complicated systems of shafts and caves can extend through hundreds of meters of limestone. On the surface, the terrain is peppered with sink holes which often coalesce into large depressions known as *poljen.*

The ground resembles a pavement, with blocks of bare limestone separated by enlarged joints. The collapse of caves may leave behind irregular pits surrounded by pillars of rock, or even bridge-like structures. Through deforestation and soil erosion, vegetation is often sparse, and restricted to dips and hollows.

Agents of Erosion

Additional factors may have a significant effect on the creating of caves. Impurities in the limestone may cause inherent weakness in the newly forming stratum: to be strong enough to form a roof over large cavern chambers, the limestone has to be fairly pure. If the rock contains too much sand and clay the stratum will be flaky or clumpy caves occur only in limestone that is more than 50 per cent pure. Even then, the limestone must also have the quality of impermeability so that the flow of water is concentrated along isolated fissures and cracks.

Cave Varieties

Other agents help to create caves. The surface of volcanic lava-flows commonly solidifies rapidly to form a rocky conduit through which molten lava flows. After all the molten lava has drained away, a curving network of tunnels is often left behind. Elsewhere, hydrogen sulphide brines rising from oil and gas fields may generate sulphuric acid, which eats away limestone, and, on the coast, the action of wind and water may erode caves.

POLAR REGIONS

It may take several thousand years for a snowflake falling on the Greenland ice cap to reappear as ice on the coast of Greenland. Both the Arctic and Antarctic regions are in a constant state of change — a dynamic balance between growth and shrinking. The ice coverage of the Arctic increases by a third in its winter, whereas the ice coverage of the Antarctic grows seven-fold. Despite their apparent similarities, the two regions are fundamentally different — the Arctic is largely ocean, while the Antarctic is a high and mountainous continent covered in ice.

At the north pole the summer temperature is near freezing and the winter can get as cold as —67°C. But this is a real scorcher by the standards of the south pole. There, an observer could expect winter temperatures of —89°C or even lower. The Antarctic is much colder than the Arctic because it is so much higher. At the north pole you would find yourself at sea level on slowly shifting ice under a cloudy sky. Near the south pole you could be standing on about 4,500 metres of ice, and the sky would probably be clear. In addition , the Arctic ocean helps to warm up the northern polar region. But not only is the Antarctic colder than the Arctic, it is also much bigger. Using the *Antarctic Convergence* — a zone, separating the cold waters of the Antarctic from the warmer, less biologically productive mid-latitude waters — as the boundary of the south polar region, the Antarctic has an area greater than 50 million square kilometres (20 million square miles). The boundary of the Arctic is taken to be the line where mean temperatures rise to 10°C in July. Its area then is 25 million square kilometres of which two-thirds are sea and one-third land.

The Glancing Sun

But these comparisons do not explain why the polar regions are so cold in the first place. The most basic reason is that, at the Earth's extremities, the Sun never rises very far above the horizon. It remains low in the sky so that less of its radiant energy reaches the polar regions. Secondly, during their six-month-long sunless winters, the polar regions lose much more heat than they gain during the equally long six-month summer sessions.

They year-round ice caps are the third factor that keeps the Arctic and the Antarctic so cold. They reflect away 95% of the solar energy that reaches them, so what little warmth the polar region receive is sent straight back into the atmosphere. In fact the polar regions would be much colder then they are now if it were not for the oceans and the atmosphere, which moderate the polar climate and warm things up by transporting heat, eventually, from the equatorial regions.

The Antarctic ice cap and the Greenland ice sheet have been very stable in recent history. Their growth through the addition of new snow and ice and shrinkage through *sublimation* (the passage of ice directly into water-vapor), evaporation, melting and the formation of the icebergs are more or less balanced. Whether they remain stable, grow or shrink depends on how the Earth's global climate changes.

Ice Giants

Each year about two inches of ice accumulates on the Antarctic ice cap. It falls as either snow or as hoar frost. Glaciers, the "rivers of ice" which move faster than ice sheets as a whole, usually originate in areas of higher than average precipitation. The very low precipitation of the high Arctic means that the glaciation may be less around the shores of the polar sea than in mountains regions farther south. Much of the Peary Land, in Greenland, for instance, is ice-free even though it is the most northerly land on Earth.

Antarctica is really two main land-masses combined, but the visible shape of the continent's surface is the shape of its ice covering. The ice-shelves of the seaward margin may stretch many kilometres out to sea, and be up to 1,000 m thick. They are anchored to features like the headlands of drowned bays, and rarely extend far over deep water. Thinner perennial bay ice may extend from the shelves. The ice sheet and ice shelves are in turn surrounded by pack ice, which varies seasonally but covers a maximum of 25.5 million sq km in September.

Glacier bergs break away from so-called tidewater glaciers, or glaciers that end in the sea. Up to nine-tenths of a glacier berg may be under water. Tabular iceberg break off from large ice shelves, and have around five-sixths of their total volume submerged.

Polynyas are open areas of water among sea ice. Because there us no insulating layer of ice, they act as heat vents from the ocean to the air. The convection currents set up by this heat loss influence the pattern of local ocean circulation. As well as open ocean polynyas, there are coastal polynyas. As well as causing, convection currents these polynyas affect ocean circulation because their surface water has a high concentration of salt — left behind when the salt water freezes. This denser water helps the formation of the bottom water currents.

The **Greenland ice cap** constitutes 90 per cent of the ice in the northern hemisphere (but only 9 per cent of Earth's total), and provides 90 per cent of the hemisphere's icebergs. The ice-belt around the north pole is distorted by its ocean currents. Because of its own insulation, the ice in the north polar sea can get little thicker than 3-4 m except where it is layered as a result of movement. Over much of the 6 million sq km of permanent pack ice, ice coverage is over 70 per cent of the sea's surface. Icebergs also float along with the pack ice, and with the ice floes that surround it.

ICE AGES

Even in the severest ice ages the global temperature drops by no more than 4-12°C (7-21°F). But this may be enough to plunge 40 per cent of the Earth's surface into a 100,000-year winter. There have been as many as 20 ice ages during the past 2.5 million years. At the peak of the last of these glacial periods, around 20,000 years ago, one third of the world's land resembled the ice-covered landscapes of Antarctica. The breaks between ice ages typically last for only about 10,000 years, so the warm period of the last 10,000 years may be ending, and the next ice could be on its way.

Ice ages are known as *glacials,* and the periods between them as *interglacials.* The periodic fluctuation between glacials and interglacials in the past 2.5 million years is thought to be caused by long-term variations in the Earth's orbit and in the behaviour of its axis. In the 1920s, the Yugoslavian scientist Miluti Milankovitch computed how orbital changes alter the seasonal amounts of solar energy received by each latitude. He proposed that the strength of the Sun at middle latitudes in the Northern Hemisphere is the Key to ice-sheet growth and decay because the quantity of energy received in the form of heat from the Sun, determines whether a permanent snow cover can persist or not.

Once summers have become cool enough favour a persistent snow cover, ice accumulates, and glaciers and ice sheets expand rapidly because the ice itself reflects solar energy and lowers regional and global temperatures even further.

Alternative Theories

Although the Milankovitch theory of ice ages is widely accepted and is, in fact, supported by mathematical analysis, other scientists have put forward their own alternatives theories. Some, for example, believe that ice ages are caused by huge volcanic eruptions releasing massive amounts of volcanic dust into the atmosphere, which subsequently blocks out sunlight and lowers the surface temperature of the Earth. Others believe that the Earth's natural radiation output fluctuates. This is thought to be brought about by the Earth's core heating up or cooling down. If this is true, then it is possible that the core cools sufficiently to give rise to ice ages. Other theories include clouds of cosmic dust blocking out the Sun's rays, or even a drop in the Sun's solar radiation output.

Ice Power

The land buried beneath an ice sheet is transformed dramatically by the power and weight of the ice. Each summer, raging torrents issue forth from the ice sheets carrying tremendous amounts of sands and gravels. In winter, the ice's progress is unstoppable. Strong winds flowing from the tall ice sheets carry away with them some very fine silts and clays, which are then laid down in distant lands as thick layers of fertile soil.

The astronomical theory of ice ages, the "Milankovitch Model", is based on the three changes in the Earth's movements through space which affect the distribution of the Sun's heat on the Earth's surface. The Earth's orbit around the Sun is constantly changing. It varies from an almost perfect circle (known as an orbit of low eccentricity) to a distinct ellipse (high eccentricity) and back again. This whole cycle takes between 90,000 and 100,000 years. Currently the orbit is becoming increasingly circular. There is also a 400,000 - year cycle over which maximum eccentricity increases and decreases. When in a circular orbit, the Earth receives an even spread of heat from the Sun. However, in the more elliptical shape, at certain times the Earth is farther from the Sun, and therefor cooler. The second change involves the tilt of the Earth's axis in relation to the plane of its orbit. At present the axis is titled at about 23.5°C off the perpendicular. This means that the pole tilted towards the Sun receives more hours of sunlight than the pole pointing away from the Sun — this is why we have seasons. However, at the moment the tilt is decreasing, which means the seasonal changes will become less dramatic. The Earth takes about 40,000 years to go from the minimum tilt of 21.8° to the maximum tilt, 24.4° and back again. The third variation is called the "circle of precession". This is the imaginary circle made by the "wobbling" of the Earth's axis. It takes about 22,000 years for the Earth's axis to describe a complete circle. The combination of these three variants — and particularly a high eccentricity — are thought to create the pre-conditions for severe ice ages.

About 20,000 years ago permanent ice sheets covered 42 million sq km compared with the present 15 million sq km (6 million sq miles). They locked up vast amounts of water, causing world sea level to fall by 120 metres (400 ft), revealing many new landmasses.

Evidence of the Ice Ages

Advancing ice sheets distort or destroy the signs of any previous advance so it is difficult to determine exactly how many glacials have occured. Recent research has concentrated on investigating deep ocean floors, which were unlikely to have been distributed by the ice. Cores drilled from ocean sediments contain fossils of tiny planktonic organisms called *foraminifera*. The preference of each species for warm or cold waters allows a continuous record of past climate to be inferred. Such evidence reveals 13 long-lasting glacials to have occured in the past 1 million years alone. This has now been confirmed by a range of other indicators of past environmental conditions such as cave stalactites, coral reefs, pollen and insect preserved in sediments in peat bogs and fossilized fauna in loess deposits.

Retreat of the Ice

When the ice sheets eventually retreat, plants and animals begin to colonize the newly exposed lands, returning from their refuges in milder, ice-free climates. During the last glacials, for example, islands off the coast of British Columbia remained ice-free (because of dry, cold air which did not bring snowfall) although the mainland was covered by ice upto 2,000 meters

(6,500 feet) thick. Not only did the islands provide sanctuary for flora and fauna but they formed a migrations corridor during the glacial for the early entry of humans' into North America from Asia via the Bering Strait. Subsequently, the land-bridge between the continents was submerged as the ice sheets melted and caused sea levels to rise again. Today, in some countries such as Sweden, new lands are emerging from the sea as the land recovers from the effects of being depressed under the vast weight of ice.

GROUND TEMPERATURE

The energy for the increase in ground temperature comes primarily from solar radiation, although a small fraction of heat originates from the centre of the earth(geometrical energy)) and is conducted to the surface. Decomposing organic matter also contributes a very small fraction of heat. These two, however, have negligible influence over ground temperature in comparison to solar radiation. Ground temperature is of great importance for plant and animal life. It influences seed germination, root and shoot growth and activity of soil micro-organisms, especially those involved in decomposition of organic matter.

The temperature of the ground increases due to absorption of solar radiation, and falls due to re-radiations or ground radiation. As the balance between the two undergoes diurnal fluctuations, the Midday presents a net temperature gain of ground surface, which consequently is hotter than the lower depths of ground, as also the atmosphere above. At night, with no solar radiation there is cooling of the surface. The heat from the surface is transmitted by conduction, which being a slower process, the increase in depth is accompanied by a temperature lag, thus resulting in a particular depth attaining its maximum temperature, a few hours after ground surface. On an average 15 cm depth shows a lag of 4 hours, 1 m depth a lag of 80 hours. Lower layers often exhibit seasonal lag, especially in temperate climate, a 3 m depth showing a lag of about 5 months. This lag accounts for the water of covered wells and hand pumps being warmer in winter than in summer.

Ground temperature is influenced by a variety of factors. Dark coloured soils show rapid temperature rise than a light coloured soil. The vegetational cover tends to lower the daily maximum temperature level, loose coarse textured soils show sharper increase, whereas the increased water content, owing to high specific heat of water, decreases the rate of warming as well as the cooling of ground.

RADIATION BALANCE OF THE EARTH

Heat for the atmosphere comes from the **radiant energy** given of by the sun. The sun's energy is emitted as electromagnetic rays of different wavelengths from short wave X-rays to longwave radiowaves. However, only a part of the sun's radiant energy reaches the earth. Some energy is reflected into space by gas molecules, by clouds, and by the earth itself. Some of the energy is absorbed by the ozone layer. Of the radiation that reaches the atmosphere from the sun, only 30 to 80 per cent reaches the earth's surface. The amount that reaches the earth depends largely on cloud cover and altitude. Areas near the equator receive the most energy because the air envelope is thinnest. The sun's ray reach the earth at nearly a 90 degree angle at the equator. At the poles, the sun's rays strike the earth at about a zero degree angle. Energy received by the earth decreases from the equator to the poles.

Earth's surface features influence the amount of energy absorbed by the earth. Polar ice reflects the sun's rays and little energy is retained. Oceans also reflects much of the sun's energy when the sun is close to the horizon. Sand and snow are other good reflectors. Vegetation, dark rocks, and asphalt tend to absorb the sun's rays.

Most of the sun's radiant energy that reaches the earth comes from the visible light spectrum. Some sunlight is used by green plants in their food-making process. Some sunlight is reflected by the earth. But light rays that are absorbed by the earth are changed to heat waves. These heat waves are the major source of heat for the atmosphere. When the sun sets, the earth and atmosphere lose heat to outer space. Land areas lose heat more quickly than oceans do. Deserts and mountain-tops also lose heat rapidly. Earth would have great differences in day and night temperatures if it were not for the atmosphere. Both carbon dioxide and water vapour absorb the heat given off by the earth. This heat then is returned to earth. The effect is like a greenhouse. In the greenhouse sunlight comes in through the window. Heat, however, is prevented from escaping because of the screening effect of the glass. Heat loss to outer space is least wherever large amounts of water vapour are in the air.

The earth would get constantly hotter if an energy flow were not directed from it to the universe, whereby the energy absorbed from the sun is emitted again into outerspace. There is thus an exchange of energy between the outer space and earth, as a result of which the solar inflow and the thermal outflow tend to balance one another.

Before the solar radiation reaches the earth's surface, it is diverted by air molecules, with the result that part of the incoming solar radiation is reflected back immediately into outer space. The earth's surface also reflects back part of the radiation; and in a cloudy sky there is also the reflection from the cloud layer. About 20 per cent of the energy produced by solar radiation is absorbed by the molecules of the gases (water vapour, carbon dioxide, ozone, etc.) present in the atmosphere, and this results in a rise in the temperature. Considering the average for the whole earth, about 34 per cent of the incoming solar radiation is reflected back into the space by air molecules, clouds and the surface of the earth — which are collectively referred to as *albedo*. Absorption by the ozone, clouds, water, carbon dioxide, other gases and dust accounts for 20 per cent, so that only 46 per cent of solar radiation is absorbed by the land and the oceans of the earth.

The earth loses 14 per cent of the heat directly to the outer space in the form of long wave radiation. The remainder 32 per cent passes into the atmosphere in three forms: long wave radiation which heats up the atmosphere by greenhouse effect, convection and the latent heat of the water vapour evaporated from land and ocean surface. This increment of 32 per cent plus the 20 per cent of the short wave radiation absorbed by the atmosphere from the solar radiation is ultimately lost back to the outer space as long wave radiation. The total loss of 66 per cent thus balances the short wave radiation that had entered the earth's atmosphere. This delicate balance tends to maintain the steady temperature of our planet year after year. The periods of positive heat balance result in increased annual temperature, whereas a negative heat balance results in decreased annual temperature.

Over the last few decades the rapid industrialisation seems to have modified the environment of earth by introducing pollutants at an alarming rate, burning of fossil fuels related activities that threaten to modify the heat balance of earth, and possible being in a cycle of global warming with dangerous consequences.

WIND

Wind broadly speaking is air in motion. In the strictly scientific sense, the airflow must be directed, and in addition it must have a specific speed of flow over a considerable distance. Usually, only air movements in a horizontal direction are described as wind, but vertical air movements are by no means without significance.

Wind arises as a result of differences in air pressure in the atmosphere, from which the air acquires velocity. The particles of air then follow the decline in air pressure from high-pressure areas to low. The direction of the decline is pressure follows the pattern of the air pressure distribution.

Apart from the acceleration due to the drop in air pressure, the air particles also acquire acceleration (in the case of wide-ranging movements over long distances) from the divergent effect of **coriolis force** generated by the earth's rotation. The Coriolis acceleration plays an important part in the movements in the atmosphere. The Coriolis force directly affects the directional movements of the air particles and increases as the speed of the particles rises.

In the layer of air near the earth's surface, the direction and speed of the wind will also be determined largely by the friction of the air particles on the earth's surface and deflection caused by the features of the terrain. So the speed of the wind is less on land than over open water. The direction of the wind can also be deflected by mountain chains and valleys.

The long distance winds generally, follow a fixed pattern as observed in north-eastern trade winds or easterlies of the tropics, westerlies of the temperature region and polar easterlies of the polar region. Long distance winds showing seasonal change are represents by the Monsoons. Short distance winds are generally short lived and show considerable variation in speed and direction, due to the development of local phenomenon. Wind speed is commonly represented on the Beaufort scale ranging from 0 to 12 as given below:

Beaufort Number

0. *Calm Air* : Wind speed less than 1.5 km/hour. The air is still and smoke rises vertically. Water surfaces are smooth and mirror like.

1. *Light Air* : Wind speed 1.6-5 km/hour. Smoke drifts downwind but vanes remain still. Water surface shows ripples or small wavelets.

2. *Light Breeze* : Wind speed 5-11 km/hour. Wind is felt on face, leaves and vanes move gently. Distinct wavelets in water are short and do not break.

3. *Gentle Breeze* : Wind speed 12-19 km/hour. Leaves and twigs move constantly and a small flag is extended. Large water bodies like lakes and seas show large wavelets, beginning to break, glassy foam forming occasional white horses.

4. *Moderate Breeze* : Wind speed 20-29 km/hour. Wind raises dust and small paper and moves small twigs and branches. Small but somewhat longer waves with frequent white horses are observed in the sea.

5. *Fresh Breeze* : Wind speed 30-38 km/hour, Small trees begin to sway, crested wavelets form on inland waters. Sea shows pronounced waves, distinctly elongated with many white horses.

6. *Strong Breeze* : Wind speed 39-49 km/hour. Telegraph wires start whistling, umbrellas are difficult to control and large branches start moving. Large breaking waves with extensive white foam crests and spray are encountered in the sea.

7. *Strong Wind* : Wind speed 50-61 km/hour. Walking becomes somewhat difficult and whole trees move. Sea heaps up and streaks of white foam are seen moving downwind.

8. *Fresh Gale* : Wind speed 62-74 km/hour. Walking becomes very difficult and tree twigs break off. Sea shows moderately high waves with crests of considerable length, streaks of white foam are well marked and spray is blown from crests.

9. *Strong Gale* : Wind speed 75-86 km/hour. Slight structural damage occurs, loose tiles and bricks come off. Sea shows high waves, rolling sea and dense streaks of white foam spray, reducing visibility.

10. *Whole Gale* : Wind speed 87-100 km/hour, cause much structural damage and trees are uprooted. Sea encounters heavy rolling with very high waves, over hanging crests and dense streaks of foam.

11. *Storm* : Wind speed 101-120 km/hour. Causes wide spread damage on land, where it is very rarely experienced. Sea encounters extraordinary waves and deep troughs, streaky foam and the visibility is impeded by a strong spray.

12. *Hurricane* : Wind speed higher than 120 km/hour. The sea surface is entirely white with air full of foam and spray.

The storms and hurricanes, the two strongest forms of wind disturbance often result from a collision between advancing cold and warm air masses. The upward rush of heated air spirals up and eventually develops into a cloud. The resultant precipitation and its down draft within the cloud often creates high electric charge, ending up in lightening flashes as the warm air rushes up past the cold air. Such thunderstorms soon end up into heavy rain for a short period, for lack of continuous supply of warm air. The violent wind storms that spin off from thunderstorms as a tornado occasionally in the form of a funnel, circling wildly in counter clockwise motion, can cause huge destruction when they reach the ground level. The storms which develop in the coastal areas of the tropics and commonly called tropical cyclones often result in heavy rainfall and immense destruction.

When an air mass descends from a higher to lower elevation it gains temperature due to compression. Such hot winds blowing across an area are known by different names such as Afghanets which blow from the Pamir mountains of Afghanistan to Soviet Uzbekistan, Santa Ana winds which move down the Great Basin region of Nevada into Santa Ana in California, Chinook winds of the Rockies, Foen winds of Europe and Samoon winds of Iran.

CORIOLIS FORCE

Winds move from high-pressure to low-pressure areas, governed by pressure gradient. The rotation of earth, however, produces a force, called Coriolis force or Coriolis effect, which tends to turn the flow of air. Named after the French Scientist who first observed it, the force causes any moving object to deflect to its right in Northern Hemisphere and to its left in Southern Hemisphere.

Circulation of air caused by convection alone is in a north-south direction. But the earth turns towards the east, and this causes winds to appear to curve away from a straight course. For example, a particle of air that moves directly north appears to be deflected toward the west in the Southern Hemisphere. A particle of air that moves directly south appears to be deflected towards the west in the Northern Hemisphere. When the air particle reach the equator, the point toward which it was moving has already moved some distance eastward. Thus, the direction of travel appears to be from northeast to southwest in the Northern Hemisphere; from southeast to northwest in the Southern Hemisphere. The Coriolis force is maximum at poles and nearly absent at equator, with little horizontal air movement. Coriolis effect is more pronounced away from the ground surface. Close to the surface, the moving air encounters friction with land, partly counter acting the Coriolis effect. The Coriolis effect also causes the ocean currents to veer to their right Northern Hemisphere and to their left in the Southern Hemisphere.

Coriolis force is also responsible for the development of easterlies and westerlies. The easterlies at low latitudes, also called the *trade winds* blow from about the 30° latitude to the doldrums in both Northern and Southern Hemispheres. Trade winds blow northeast to southwest in the Northern Hemisphere. They blow from Southwest to northwest in the Southern Hemisphere. Since winds are named for the direction from which they blow, the trade winds are called *easterlies.*

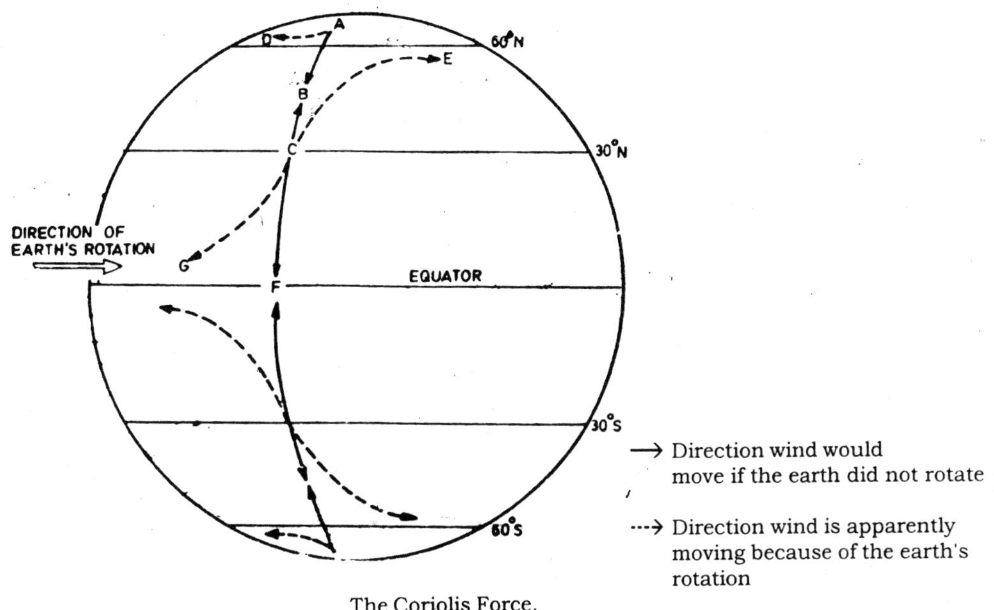

The Coriolis Force.

These easterlies provided a popular sailing route for early trading ships. Today, airplane pilots take advantage of the easterlies when travelling toward the west in the tropics. These trade winds are more appropriately known as northeast trade winds in Northern Hemisphere and southeast trade winds in Southern Hemisphere.

The *westerlies* zone lies between 30° and 60° latitude. At 30° latitude, air curretns descend to the earth's surface. Part of this air turns southward as the trade winds. Part of it flows northward as the westerlies. Winds blowing towards the poles are deflected towards the east because of the Coriolis force. Westerlies are the air currents responsible for the movement of weather across the United States and Canada.

Beyond the 60° latitude the surface winds move towards warmer latitudes, Coriolis force causing their deflection towards west. These surface polar winds are known as *polar easterlies.*

JET STREAMS

Jet Streams are narrow bands of fast moving air following broadly curving tracks and are often called "swiftly flowing rivers of air." At least three such high-altitudes bands circle the earth in each hemisphere near the tropics, near the subtropics and along the polar fronts. These are located at altitudes of 10 to 12 km and may attain speeds upto 480 km/hour. The occurrence of jet streams came to be known around 1920 when a balloon expected to land in New York after 2 weeks took only 4 days. During World War II American bombers attempting to avoid detection on mission to Japan, took higher altitudes but were surprised to see their speed reduced from normal 720 km/hour to 250 km/hour. Jet streams have little or no vertical movements and blow in horizontal direction. They develop near tropopause where warm westerlies collide with the cold polar easterlies.

Jet streams take the form of flattened tube, about 1.5 km thick and upto 150 km wide, with wind speeds ranging from 50 to 480 km/hour, whereas the adjacent wind may be travelling at mere 15 km/hour. The greatest speed is attained off the eastern edge of Asia and near Greenland. The tube is, however, not continuous along the globe but broken into sections of varying length. A jet stream formed during the summer is called *subpolar jet* and one formed during winter as *subtropical jet.*

Jet stream is an important factor in the operation of jet aircraft. The aircraft flying to Europe from North America make use of this jet stream to save fuel, and those making a return trip avoid it. Jet stream is avoided when its turbulence level, designed as clear air turbulence (CAT) is severe.

THUNDERSTORMS

Thunderstorms is an intense local storm formed by strong upward convectional wind currents, creating a tall dense cumulonimbus cloud from which short heavy rainfall and lightening results. Violent surface winds may occur at the onset of storm. Occasionally hail also accompanies a thunderstorm.

Thunderstorms commonly occur after a few days of warm spell. The warm air near the ground surface, creating low pressure region rises up in cyclonic flow (counter clockwise in northern hemisphere and clockwise in southern hemisphere). The warm air, alternatively, may rise as succession of bubble like air masses, instead of a continuous flow. As the warm air rises, outside air is drawn in to take its position. Upon condensation at higher altitudes, the tall cumulonimbus cloud is formed, which may result as high as 16 km, when its top flattens out to form anvil top. As the temperatures within the cloud decreases, precipitation occurs in the form of rain, snow and hail. The precipitation creates a downdraft within the cloud and the subsiding air drags behind it. As the warm air races past the cold air and clouds become highly charged, lightening flashes and thunder result. Downdraft of the cloud cuts off the supply of warm air, with the result the thunderstorm loses its vitality and soon disappears.

Occasionally a single thunderstorm consisting of several adjacent units, becoming active in succession may yield many heavy bursts of rain. A thunderstorm precipitation though heavy is generally of short duration and patchy in distribution so that an area of a few kilometres across may receive heavy rainfall whereas adjoining areas may have none during the same duration.

Thunderstorms are a dominant form of precipitation throughout the year in equatorial zone. In subtropical and temperate zone they are largely formed in summer. These are rare in arctic and polar regions.

TORNADOES

A tornado is a small cyclonic storm, usually short lived, but spiralling at tremendous speed and often immensely destructive. It is a commonly a spin off from a thunderstorm and appears like a dense funnel cloud, hanging from a dense cumulonimbus cloud and 100 to 450m in diameter at its lower end. It circles wildly and reaches for the earth. On touchdown the funnel causes immense destruction. Tornado destruction results from great wind speed as also from sudden reduction of air pressure in the vortex of the cyclonic spiral. Wind speeds of a tornado are estimated as high as 400 km/hour. As the tornado moves across the area it writhes and twists. The end of the funnel brings destruction where it sweeps the ground. It may then rise in the air and leave the adjoining ground unharmed.

Several tornadoes may be formed from one thunderstorm at intervals as it sweeps across an area, but not all of these reach the ground. The path of a tornado is usually very narrow, not more than 0.4 km wide but may attain lengths upto 500 km. Tornadoes originate where turbulence is greatest. They are the commonest in spring and summer. They occur very commonly in central and south-eastern states of United States of America. The worst tornado ripped across Missouri, lower Illinois and Indiana for three hours on March 18, 1925 killing 689 people and causing great damage to the property. Tornadoes are rare in mountainous and forested regions. National weather service in U.S.A. maintains a tornado forecasting and warning system. People residing in the path of tornadoes can thus be warned in time to take shelter in underground built storm shelters.

TROPICAL CYCLONES

Perhaps no other environmental hazard has been the consistent cause of such a great loss of life and property as tropical cyclones, particularly in the coastal areas of Bay of Bengal. Cyclonic storms are of common occurrence in tropical climate and known by different names such as *hurricanes* that form in the Atlantic Ocean southeast of Carribean, *typhoons* that form in oceans of east Asia, *cyclones* that form in Indian ocean mainly Bay of Bengal and *bagnios* encountered in phillipines. Coastal areas of Bay of Bengal, particularly the Bangladesh have faced the burnt of cyclonic storms over the centuries. The area broken by the distributaries of Ganges and Brahamputra has numerous low lying deltaic islands that are readily submerged. The major disasters occured, in 1822 (40 thousand deaths), 1876 (1 lakh deaths), 1897(1 lakh and 75 thousand deaths) and 1970 (3 lakh deaths).

Tropical cyclones are not encountered near the equator where the air is relatively calm with weak coriolis effect. They are very active between 8° to 15° in both hemispheres. The warming of air in this belt results in low pressure center which deepens. As the heated air rises it develops a spiraling vortex upward. The spiraling circular storm that eventually develops, moves westward as tropical easterlies at a speed ranging from 30 to 60 km/hour and spiraling winds within the storm may attain 200km/hour, and the storm itself may attain a diameter of upto 500 km. The resultant cooling of the rising air eventually results in thunderstorms and heavy precipitation. On lashing out a land surface, huge destruction of life and property is caused. A characteristic feature of a tropical cyclone is its central eye also called 'eye of the cyclone'. The eye is the cloud-free vortex resulting from rapid spiraling of storm. In the eye air is calm, and flocks of birds may be found flying in circles, not daring to move into violent winds around, waiting patiently for the storm to die. The eye may extend upto 15 km in diameter. The eye often also gives a temporary respite to storm-tossed ship or a battered city. Eventually, however, the air descends from high altitude through the eye, gets a diabatically warmed and a fresh activity of the cyclone is initiated with winds spiraling in opposite direction. The area receives battering again.

The great human loss caused by cyclonic storm is primarily because of storm surge, a sudden rise of water level as the cyclone moves over the coast line, carrying the huge wave on the land. Storm surge can be as high as 12 m.

In an effort to lessen the fury of hurricanes (American name for tropical storm), project *Stormfury* was initiated as a cooperative effort of the U.S. Navy and National Oceanic and Atmospheric Administration (NOAA)) in 1962. Plans were laid out for cloud seeding of Hurricanes, under the direction of National Hurricane Research Laboratory (NHRL). Cloud seeding of the Hurricane that appeared in Atlantic ocean in 1969 was done using silver iodide particles using five aircraft. Within five hours of seeding winds decreased by 30 per cent. During second day wind speeds again increased, but a second seeding decreased winds by 15 per cent. Though the results were encouraging, the project was terminated in 1972. The benefits of such studies reaching the area most ravaged by cyclones in tropical Asia, however, seems a far cry.

A tropical cyclonic storm in the Bay of Bengal hit the coastal regions of Bangladesh; feared to be the biggest recorded natural disaster of the world, it claimed around 5 lakh human lives. The cyclone hit the coast line of Bangladesh on April 29, 1991 with wind speeds ranging from 210 to 235 km/hour. Most islands were submerged under 3 to 4 metre of water. At places water rose to 6 m. The worst hit was Swandip island, where alone about 1 lakh deaths occurred. The cyclone battered the 650 km coastline of Bay of Bengal. Besides the human toll (official figure 1,25,730), nine lakh of cattle were lost, 6,42,553 houses destroyed and 5,64,371 dwellings partially damaged. Tidal surge was upto 6 metres.

TIDES AND WAVES

Tides are not unique to the sea. The force of gravity exerted by the Sun and the Moon can raise the Earth's crust by over half a metre (18 inches), and cause widespread cracking in the planet's surface. These colossal forces have variable effects. On some coasts, spring tides can cause the sea to rise and fall by a vertical distance of over 16 metres (50 feet), while some shores experience not tides at all. Both tides and waves contain vast amounts of potential energy. If harnessed, the available power could more than satisfy the total demand for electricity in many countries.

Tides rise and fall — flow and ebb — as a result of the gravitational pull of the Moon and the Sun. The extent of the tidal fluctuations varies on both a monthly cycle, corresponding to the orbits of the Moon around the Earth, and on an annual cycle, because of the way in which the Earth orbits the Sun. The height of the whole body of water in the oceans fluctuates, but it is generally only at the margins — the places where the sea meets the land — that such variation is apparent in the form of the tides.

Global Variations

Tides are affected by the shapes of ocean floors and land masses. In Vietnam and the Caribbean there is a one tidal high and one tidal low every lunar day (slightly longer than a normal day): each is known as *diurnal* tide. In places such as the North Sea, however, there are two high tides and two low tides every lunar day: this is called a *semidiurnal* tide, and the two high—and two low water marks are normally much the same per day. The angle between the Moon and the equator varies between 28°N and 28°S, tilting the tidal bulge at an angle to the Earth's rotation and causing different heights in the tides Diurnal tides generally occur where the smaller of the two high tides is insignificant.

Some areas of the world experience the characteristics of both diurnal and semi-diurnal tides: the pacific and Indian Oceans coasts mostly have these mixed tides. At the other extreme, there little tidal fluctuation in the nearly enclosed Mediterranean and Baltic Seas. A feature of estuaries into a high body of water with a wall-like front.

For effective tidal power generation, a constant tidal range in excess of 3 metres is necessary. To this limitation must be added the fact that suitable sites for harnessing tidal energy tend to be environmentally sensitive areas. Barrages inevitably affect tidal movement, and pollution may become an increasing menace.

Waves and the Weather

Understanding of the mechanism by which the tides are generated is not yet complete, but enough is known to allow relatively accurate tidal predictions. One complicating factor, though is that atmospheric and climatic conditions can have strong positive or negative effects. Fierce winds or extremes of atmospheric pressure may alter a tidal height by upto 3 metres. The wind may actually blow water towards or away from the shore, whereas high atmospheric pressure weighs the water down and prevents it from rising, and low pressure allows water to rise up more than it otherwise would.

A difference of 1 millibar (up or down) in atmospheric pressure can make a difference of about 1 centimetre in the height of the sea. The maximum tidal surge caused by atmospheric fluctuation is of the order of 50 centimetres.

The Swelling Seas

Waves are mainly caused by the wind blowing over a *fetch*, or open stretch of water. They may persist — in the form of a swell — long after the forces that created them have subsided. All waves, including those that reach the shore, are the product of interactions of the swell from distant weather systems, oceanic currents and prevailing winds.

The swell that reaches the Californian coast could have come from the stormy region south of New Zealand, and waves that began off Cape Horn may crash against the coasts of western Europe, 10,000 km (6,000 miles) away. Thus ocean waves are major transporters of energy around the world.

As the Earth and Moon rotate about their common centre of gravity, they are in effect continually falling towards — and past — each other through space. The gravitational pull of the Moon makes one tidal bulge. But there is a puzzling second bulge making it look like there is also an "anti-gravitational" force. The other bulge exists because the Moon's pull on the Earth is stronger than its pull on the farthest body of water. So the Earth accelerates faster towards the Moon, leaving the water behind as a second bulge that appears to be the result of a pulling-force. In fact no such force exists. The tidal water-bulges rotate at roughly the same speed as the Moon, once in 29 days. But the Earth is also independently rotating on its own axis, within this parcel of water, once every 24 hours. So once a day a single piece of coast will pass through two high and two low points on the parcel: the high and low tides. However, in the day it takes the Earth to revolve once, the Moon has moved on and dragged the water parcel after it, through an angle of some 12°. The Earth therefore has to travel a little further to catch up. This is why the tides occur some 50 minutes later every day.

The Sun's gravitational force on the Earth is less than half that of the Moon, but it still has a major influence on the tides. The spring tides (spring meaning "to rise up", and not the season) occur when the Sun, Moon and Earth are in a straight line, so the forces exerted by the Sun and the Moon add up. The forces are less effective when acting at 90° to each other. Then we get the lower, neap tides. The tides do not coincide exactly with the Moon's phases, however. There is a time-lag of as much as a day and a half —called the age of the tide— between the Moon's position and the tidal range. The angles between the Sun, the Moon and the Earth also vary. At the equinoxes (21 September and 21 March) the Sun is over the equator. Then the Sun-Moon-Earth line is very straight and the spring tides are highest. At the solistices (21 June and 21 December) the Sun is at its greatest angle, and the spring tides are smallest.

SEAS AND OCEANS

If life in the Oceans stopped, the carbon dioxide content of the Earth's atmosphere would triple. This is because of the vast numbers of marine plants that use carbon to form their bodies, thus reducing the carbon dioxide in the surface water and hence in the air. In fact, the sea is the major source of atmospheric carbon dioxide — it has dissolved within it 45 times as much carbon as there is in the atmosphere. The sea covers over 70 per cent of the surface of the Earth, and most of its incredible volume is still little known and unexplored.

The ocean is divided into two distinct zones, or reservoirs. The upper reservoir is warmed by the Sun and stirred by the wind. Water from the sea's surface evaporates, condenses and falls as rain. Rain water that falls on land then percolates through the land, dissolving and carrying into the ocean some of the chemical constituents of sea water. Volcanic activity, on land and beneath the ocean, supplies others. In the upper reservoir microscopic phytoplankton convert the Sun's energy into organic matter which supports a food chain of animals and bacteria throughout the oceans. The systems that control the distribution of carbon between the sea and the air involve the components of sea salt that are used by life in the sea: oxygen, calcium ions, phosphates and three carbon-based mixtures: carbon dioxide, carbonate ions and bicarbonate ions. Marine animals use oxygen for respiration and carbon and phosphorus for various biochemical functions. Some surround themselves with shells made of calcite, or calcium carbonate. Somehow, through a series of interconnected cycles, the ocean chemically balances the demands of marine organisms with the input of material from erosion or volcanic activity and the loss of material by burial as sediments on the ocean floor.

The Thermal Barrier

The upper reservoir, which accounts for only about 2 per cent of the oceans' volume, is separated from the lower reservoir by the main *thermocline*, a zone in which water temperature decreases rapidly with depth. The temperature starts to decrease at depths ranging from 100 metres to 300 metres, and continues to drop for the next few hundred metres down to just a few degrees centigrade. Since cold water has a higher density than warm water, the thermocline acts as a very effective barrier to prevent mixing between the two layers. So efficient is the thermocline that a water molecule starting off in the lower reservoir takes about 1,000 years to enter the upper layer.

The Elemental Store

The lower layer comprises most of the oceans' volume and here the temperature of the water is generally constant and lies within the range of 1 to 5° C. The waters of the lower layer originate in the polar regions where the surface water is denser than water at the lower latitudes, because it is colder. These waters of the lower layer also have a constant concentration of dissolved material or *salinity*. Most of the naturally occurring elements are included in the 5,000 million tonnes or more of solids that are dissolved in the sea — If only in minute amounts — and sea water is the only source of one element, iodine, which is absolutely essential for human life.

Although the ocean's composition is often thought to have been constant through time, there is now strong evidence that significant changes have occurred in the concentrations of oceans salts. Some changes appear to be related to major cycles of glaciation, which changed the volume of the ocean, while others may relate to the change in the locations of the continents caused by plate tectonic couplings and splits.

The Bab-el-Mandeb sill divides the Red Sea from the more typical open-ocean waters of the Gulf of Aden. This explains the very uniform temperature profile, which lacks a thermocline. Open-ocean basins are filled with cold water like that found on the bottom of the Gulf, with a thermocline. The thermocline disappears wherever this cold bottom water reaches to the surface, as it does in some parts of the Atlantic. The sill also contributes to the Red Sea's very high, uniform salanity. The high evaporation, low rainfall and presence of salts released from the sea bed are also factors. Salinity measures the concentration of material in solution — it need not be salt. Salinity is expressed as the number of parts per thousand, by weight, of the constituents dissolved in the water. In the deep layer of the ocean basins, salinity is a fairly constant 34.5 — 35 parts/thousand.

Shades of the Sea

The Earth is often called the blue planet because the oceans give it a blue colour as seen from space. Clean, clear ocean water appears blue because the shorter wavelengths of blue light are absorbed less and travel farther through the salt water then the long red wavelengths. Thus the blue colour is due to a greater proportion of blue light being scattered and returned back to the surface without being absorbed. But even in the clearest seas, only 1 per cent of light — of all wavelengths —penetrates a few hundred metres. Phytoplankton in their teeming millions make ocean water appears green and decrease the depth to which light penetrates. The yellow-brown colour of many shore waters can be caused by various pollutants dumped by man, or by sand and mud that has been stirred up by the action of the waves and tides.

OCEAN CURRENTS

There are many awesome currents on the Earth's surface. Venezula's Angel Falls has a 1 kilometre (0.6 mile) vertical drop, and the Amazon River pours forth 200,000 cubic metres (4 million gallons) of water every second. But even this is dwarfed by the deep-ocean cataracts that sweep across the sea floor. Perhaps the biggest of these submarines waterfalls lies below the Denmark trait that separates Greenland and Iceland. Here, 5 million cubic metres (1 thousand million gallons) of water fall 3.5 kilometres (2 miles) into the North Atlantic Ocean every second.

Deep-ocean currents are fed by water from the polar regions, which sinks and flows beneath the warmer, lighter water at lower latitudes. Because these currents are driven by the density differences caused by the temperature and salinity of the waters they are called thermo-haline currents. The cold bottom water that eventually cascades over the Denmark Strait starts its journey in the North Atlantic. As it moves northwards near the surface it becomes colder.

Eventually it enters the Norwegian Sea, where the formation of sea ice leaves a greater concentration of salts in any water that has not frozen. The increased salinity means that the water can continue to cool to below 0°C without freezing. This makes the water dense and so it sinks and begins return trip across the ocean floor. As the dense water flows towards the equator it hugs the sea floor and plummets over cliffs at the bottom of the ocean, dropping down just like a waterfall and displacing the resident, warmer water upwards.

Planetary Influences

Oceanic circulations, like atmospheric circulation, is controlled by the rotation of the Earth and the effects of the Sun's radiation. Density differences drive the deep ocean currents, but surface currents — like the Gulf Stream, which carries warm water from the southern tip of Florida northwards along the east coast of North America as far as the Grand Banks of Newfoundland — get driven by the major wind circulation systems.

The trade winds of the Northern and Southern Hemisphere both drive westward flowing equatorial currents. These currents would circle the Earth, but the continents block the flow of water and force the currents into circulation cells called *gyres* which are centered near the subtropical atmospheric high-pressure cells. Water tends to move in towards the centres of the major ocean gyres. It piles up and produces a pressure gradient to oppose the Coriolis force that result from the Earth's spin. The forces balance out — giving so-called *geostrophic* currents — and water-flows continues around the gyre. Warm currents flow polewards on the western sides of the oceans in relatively narrow bands of fast-moving water, while the cold currents in the sea are broader and more slow moving.

Water-rings

The surface currents, like all ocean currents , are not smooth steady flows. At their margins the currents give rise to massice eddies, hundreds of kilometres in diameter, which break away from the current. In the Gulf Stream the eddies from when the current starts to wind or meander as it flows north. If the meander becomes a loop it pinches itself off from the main flow and form a ring. These big rings are important because they transport water from different regions across frontal boundaries, like the Gulf Stream. The oceans are continuously in motion: the wind and submarines disturbances, whip up waves and eddies and swirls in addition to the powerful currents at the top and bottom of the sea. All this movement which still awaits thorough investigation transfers vast amounts of heat, and helps to moderate and control the Earth's climate.

The rings of water that form around the Gulf Stream can have warm-water cores or cold-water cores. The cold-core rings of the Sargasso Sea may be 300 km across and stretch to the sea floor 5,000 m down. The warm-core rings of the slope waters are shallower. The Gulf Stream is the fastest surface current in the north Atlantic gyre, travelling around 200 km in a single day. It carries warm water from the North-east coast of North America, but beneath it flows a large cold current heading south from the Arctic.

The Corkscrew Current

A wind will tend to drag ocean surface water after it. However, as soon as the water starts to move, the Coriolis force throws it off at an angle. Each successive thin sheet of water beneath the surface layer begins to move because it is linked by friction to the layer above. But each layer is thrown even farther off course by the Coriolis force. So the surface current moves at an angle of 45° to the wind at about 2-3 per cent of the wind speed. Lower layers move more slowly, at an increasingly eccentric angle.

The resulting vertical profile of the motion is the Ekman spiral. At a depth called the Ekman depth, the water flows in the opposite direction to that of the surface current and at 0.043 times the speed of the surface current.

THE OCEAN FLOOR

The greatest mountain chains on Earth are beneath the sea, some stretching almost all the way from pole to pole. So are the mightiest volcanoes and deepest valleys. From the undersea trenches to the continental shelves, the ocean floor has features as striking as any landscape on Earth. The basaltic ocean floor is the youngest part of the Earth's crust, at most only 200 million years old; the oldest rocks on land are over 4,000 million years old. The sea bed also holds some of the richest mineral and metal deposits — although reclaiming them is a hazardous operation.

Extensive volcanic mountain range cross the vast abyssal plains on the ocean floor between the continental shelves. From many, lava still issues, creating fantastic shapes as it cools. A few mountains broach the sea's surface, as in Iceland and the Azores; beneath the water the flanks of the mountain chains may plunge more than 5 kilometres and the ridges may be clearly 1,000 kilometres wide.

In some oceans, especially the pacific, clusters of volcanoes have been formed where the moving floor of the ocean has crossed a hot plume of lava rising from the Earth's mantle. In Hawaii, for example, the greatest volcano on Earth, Mauna Loa, rises some 9,100 metres from its undersea base to 4,250 metres above the sea.

Much of the sediment that filters down to the ocean plains is wind-blown dust (often volcanic), rocks and sand carried out to sea on icebergs, space debris and microscopic skeletal remains of plankton. This all generally settles into a soft, fine-grained clay. Depending on the water depth, pressure and temperature, such clays vary from being calcium-rich above about 4,500 metres, to silica-rich at greater depths. The cut-off line is the *Calcite Compensation Depth* (CCD), below which calcium carbonate dissolves in the carbon dioxide-rich sea water. The CCD varies from 4,300 metres to 5,200 metres from ocean to ocean. Iron-rich red clays, from volcanic dust borne on the wind, cover enormous areas and occur in the deepest waters.

Below the Abyss

Even deeper than the abyssal plains are the ocean trenches. These can be more than 10 kilometres deep. One of the best known is the Peru-Chile trench which, in common with other trenches, has been formed by the relentless subduction of the ocean floor below the adjacent continent.

Such trenches may be over 1,000 kilometres wide and thousands of kilometres long. Great thickness of sediment accumulate in the trenches, derived from the attrition of rocks on nearby land masses and from the silt of river estuaries.

On and off the Shelf

Between the ocean planes and the coastline are two further important regions: the continental slope and the shelf. Climbing up from the deep is the continental slope. This rises from plains at a depth of about 3.5 kilometres (2 miles) to a depth of only about 150 metres below the sea level.

Thorough investigation of the shelves has shown that the solid rocks of each shelf are similar to those of the continent and very different from the basalts of the ocean floors.

Two types of the continental shelf have been recognized. The "Atlantic" or passive type tends to be wide — extending over 100 kilometres from the coast — stable and covered with sand and mud; the "pacific" or active type is much narrower and prone to earthquake activity. Both description derive from the names of the oceans in which they tend to be most common. The Atlantic type occurs around the oceans where sea-floor spreading is taking place; the Pacific type has subduction zones nearby.

Mineral Riches

Hydrothermal vents in the sea bed create local areas of hot water. The mineral-rich (iron, zinc and copper sulphides, among others) columns of water that shoot from the vents — at temperatures up to 350°C (660°F) — are called "black smokers". Potato-sized manganese nodules (below), especially common in the Pacific, average 18 per cent manganese, alongside iron, zinc, copper, cobalt and nickel. They have an onion-skin structure, precipitated from the sea water at only 1-4 mm (0.04-0.16 in) thickness/million years around a nucleus such as a sand grain, or fish tooth.

Chains of volcanoes or volcanic islands are often found at active margins. The sediment scraped back when the oceanic plate subducts beneath the continental plate forms the —often highly structured — accretionary wedge. Submarine canyons are more common on passive margins, but can be found here, too. They were probably caused by surface erosion during periods of low sea level, and subsequently widened by turbidity currents, or "submarine avalanches" of sediment. Many parts of the ocean floor are highly fractured, especially around the mid-ocean ridge. There are many volcanoes. Those that never emerge are seamounts. Guyots are volcanoes with their tops cut off flat, probably by wave erosion when the sea level was lower. Their great weight has sunk them into the ocean crust.

ISLAND FORMATION

Not a single one of the 1,300 coral atolls that make up the Maldive Islands is higher than 9 metres (30 feet) above the sea level. Many are much lower. If the sea were to rise just a few metres they would all vanish. Rising seas or sinking continents can create islands — such as Britain, for example, where the dry land connection with continental Europe was severed by rising water. The long thin islands off the coast of Yugoslavia are the tops of a flooded mountain range, and the Thousand Islands in the St. Lawrence River of Canada are the remains of a drowned river valley.

Chains of islands like those in Japan and Hawaii, and isolated landmasses like Australia — the "island continent" — are all the result of fundamental plate tectonic processes. The production of molten magma at hot spots and subduction zones — where one plate is squeezed beneath another — builds up volcanic islands and chains, which of course can also be destroyed at the subduction zones. The movement of plates can also slice fragments from a large continent and carry them away, thus creating an island. This is what is happening in Western North America, where the San Andreas Fault is removing a piece of the continent.

Unsinkable Islands

If an island is too buoyant to be subducted, it will eventually be accreted — or slapped on — to a continent. The western Pacific Ocean is a site where accretion appears to be taking place today. As the Pacific plate moves west and the Australian-India plate moves north, islands like

Fiji, Java and the Philippines will probably eventually be joined to continents, so forming what is called a terrain. A similar process is responsible for the western margin of North America. It is composed of hundreds of different exotic terrains, some of which have travelled distances of over 5,000 kilometres before finally being accreted.

On a larger scale, plate movement resulted in the final splitting of Australia from Antarctica during the Eocene epoch, between 55 and 38 million years ago.

A Necklace of Volcanoes

Volcanic islands chains that are formed when one tectonic plate is subducted beneath another are called islands arcs, because they form graceful curves a few thousand kilometres in length. Island arcs occur about 100 -200 kilometres (60 -120 miles) landward of the trench where the subducting plates meet. Along its length the arc usually consists of separate groups of about five or ten volcanoes, each of which forms a more-or-less straight line around a few hundred kilometres in length. Bends occur between the straight segments. Islands are volcanoes usually occur about 50-70 kilometres apart. Which is also about the thickness of the lithospheric on which they form. The volcanoes erupt above a site where the subducting plate is about 100-200 kilometres below the Earth's surface. Here, the water and carbon dioxide in the sediments dragged down with the subducted plate act as a flux that lowers the melting temperature of the rocks, so creating fluid magma. The water and carbon dioxide also cause the volcanoes above the site to erupt explosively. The volcanoes of such an arc-like the Aleutians off southern Alaska are all of the same age.

The Hawaiian Island — Emperor Sea Mount chain extends for over 6,000 km (3,700 miles), with more than 107 volcanoes. The volcanoes become older from the south-east to the north-west and the islands to the north-west are lower and more eroded. Rocks on the island of Hawaii are less than 1 million years old, rocks from Oahu are 2 to 3 million years old, and the lava flows of Kauai are about 5 million years old. The islands in an island are, such as the Aleutians, are by contrast all of the same age, having been created by the same process — subduction — at the same time.

The island age as they do because they are formed on a sort of plate tectonic conveyer belt. A plume of hot material from deep in the Earth's mantle rises beneath the present location of the island of Hawaii.

Hawaii consists of two large volcanoes, Mauna Kea and Mauna Loa, one of the most active on Earth. The plume, or hot spot, is the source of the magma that feeds the Hawaiian volcanoes. It is stationary. The Pacific plate, on the other hand, is moving to the north-west, driven by ocean floor spreading, so that after the islands form they are carried away by the plate motion.

Conditions for Life

Rain that falls on an island may form a lens-shaped reservoir of fresh groundwater beneath the surface. The freshwater lens floats in the denser sea water. To maintain a stable freshwater zone beneath it, an island must be about 400 metres across and receive sufficient rainfall.

DESERTS

The largest and driest desert in the world is the Antarctic. Most of the Greenland, parts of Canada and Siberia, and many high-altitude mountain ranges are also technically desert. Deserts may be defined by low precipitation, high evaporation, or by the landforms or vegetation they contain. But most people think of deserts as rocky or sandy wastes, stretching under a burning Sun. There are deserts in all the continents except Europe, and an estimated 20 per cent of the

Earth's surface is covered by arid land. And human activities are helping to expand these barren environments every day.

Among warm deserts there are huge variations to be found in landscape, vegetation and other natural factors. Deserts can be completely bare rock (called *hamada*) with, perhaps, a sprinkling of pebbles and stones (*desert pavement*); or they can be covered with sand-dunes. Their surfaces can consist of ash, lava or salt; and they either have very little vegetation of any kind, or plants that are classified as "desert" types. These either have very deep tap-roots, or are xerophytes)plants such as cacti, specially adapted to life in a dry environment). Lichens, algae and fungi may also grow under stones.

All of the warm deserts are dry, lying within the arid and hyper-arid regions of the world. Average annual precipitation is under 25 centimeters, and evaporation rates are high. Dry lands are not necessarily desert lands, however: they may be productive, like the vineyards of Southern Spain.

Arid and hyper-arid lands cover between 15 and 30 per cent of the world's land area (depending on how they are defined). Hyperarid regions may not receive any rainfall at all in a year, and are thus indisputably deserts. Semi-arid and arid regions may be classified as "deserts" according to the amount and type of their vegetation.

The Shifting Sands

Sandy deserts, in which the surface is composed of wind-blown dunes, are often devoid of vegetation because the sand particles are constantly being moved on by the wind, and plants have nothing on which to anchor. Although the Sonoran Deserts in Arizona and northern Mexico is teeming with plant life and supports a magnificent desert ecosystem, deserts normally have a lower *biomass* (the weight of living plants and animals) than any other landscape. Typically, a sand-dune desert has a biomass under 250 kilograms/hectare, compared to a savanna's biomass of 10,000 kilograms/hectare, or a tropical forest's biomass of over 250,000 kilograms/hectare.

The Pressure Cooker

Climatic factors are the natural creators of arid and hyper-arid lands. Prevailing dry conditions are often the results of atmospheric stability associated with the presence of great high-pressure cells around 30° latitudes. Only occasionally do rain-bearing depressions spread into these zones. Deserts also form in the lee of mountains, or where land is subject to very dry continental winds, or even at the coast, when sea breezes do not release their moisture until they get farther inland. When rain does come, it is frequently heavy.

The huge sand seas that cover about 25 per cent of the world's desert areas were largely formed by the water erosion of sandstone, shells or whatever else the sand particles are composed of. Sand need not originate where it is found. Frequently it is created on the borders of desert areas and is carried by wind or water into low-lying regions. In many cases it may have been created during earlier, wetter climate cycles. The pebbles, cobbles and bedrock of desert pavement are often smooth and shiny from the polishing effect of particles in the wind.

Desertification

Human activities over recent centuries have caused and are causing extensive tracts of arid lands to be degraded. Changes normally occur slowly, although sand-dunes can rapidly encroach on farmland if they are not anchored by fast-growing, drought-resistant trees, or stabilized by waist-high windbreaks across their crests. Shelter belts of trees and shrubs can also cut down wind-erosion, which, along with water-erosion, is exacerbated by overgrazing, deforestation and bush-burning. If the topsoil is blown away, degradation of the land is permanent. Even irrigation can cause problems. Water evaporates quickly in arid lands, leaving behind any salt that it may carry. This salt is not leached away into the deep soil, so its concentration rises rapidly in the upper layers, until at last the land may have to be abandoned.

COASTLINES

Over 100 million years ago the Mississippi delta started at Cairo, Illinois, but since its birth the mouth has migrated south to the Gulf of Mexico — over 1,600 kilometres (1,000 miles) away — at a rate of more than a centimetre a year. Much of the beach that runs up the eastern coast of North America, from the Gulf of Mexico to New Jersey, is every bit as mobile. It consists of 295 barrier islands — sedimentary deposits flanking the coast — that have migrated, split up and re-formed under the action of storms for the last 6,000 years.

On some barrier islands as much as 8 metres of land a year can be eroded. But as long as there is a surplus of sediment, the islands will remain. And there is always a surplus of sediment dumped into the Atlantic by rivers like the St Lawrence and the Hudson. Beaches, however are not just made of sand. Whatever sedimentary material is available from rivers, eroded cliffs, glacial deposits or coral and shell fragments can wind up concentrated on a beach. In Hawaii, where the only material is broken lavas, many of the beaches are black, and in western Scotland where the old eroded roots of the Highlands expose garnet-bearing metamorphic rocks, some of the beaches are pink.

Winds and Waves

Cliffs are attacked by waves and suspended sediment in the area between high and low tides. A notch is usually cut in the base of the cliff, and as it deepens the cliff-face above it collapses. Eventually a wave-cut platform is created between the high and low tide marks, and is visible at low tide. Cliff debris may be carried along the coast or deposited offshore, but the removal of debris is rarely continuous, and destructive wave action usually alternates with periods when the waves deposit material on the shore.

Although storms do the major work of modifying beaches, each wave that runs up the beach also plays a role. Breaking waves throw sand and suspension and fling it further up the beach before the wave retreats. In addition, because waves approach the coast at an angle, the wash and backwash is asymmetrical and although most of the movement is up and down, some of it is along the beach — a process called *longshore drift*. The angle of approach of the waves determines the longshore drift of sand and this angle is in turn determined by the location of the sea storm from which the waves emanate.

The longshore current produced as a result of the angular approach of the waves and the return of the backwash water can transport large quantities of sand. This is also the current that moves swimmers rapidly down shore. But if waves approach with their crests parallel to the beach the longshore currents are weak. Instead, under the pressure of the incoming waves, the backwash piles up and is periodically released as *rip currents*, which flow for tens on hundreds of metres out to sea as dangerous and often lethal undertows.

Delta Deposits

Rivers are the main transporters of sand and before the sand finds its way to a beach it often form a delta. When a river meets the ocean the characteristics of the delta that forms are controlled by the speed of the river, the size of its channel, depth of the water and of course the nature of the sediment. As the moving river water enters the sea the main process that occurs is the mixing of still water with fast water at the edges of the current. When fresh water enters the sea, it often floats out over the top of the denser salt water. However, if the amount of sediment in the river is high enough, the river water and its suspended sediment is sufficiently dense that no salt wedge forms and the muddy water flows directly into the sea.

A wave's height and its power depend on the strength of the wind blowing it and on the distance of open sea — the fetch — over which the wind has blown. Although a wave moves forwards, the particles of water that carry the wave just rotate in circles.

When waves enter shallow water, their crests pile closer together and topple over causing the water to race forwards. Coming up a gently sloping shore, the front of a wave gets steeper and its crest starts to spill over the top. This is a spilling breaker, and creates an uprush, which tends to deposit material on the shore. On a steep shore the wave front also steepens violently. An unstable hollow appears in front of the wave and the crests curls over and plunges down into it. A plunging breaker will create a backwash that scours sand and other materials away from the shore. As they enter shallow water, waves are slowed by different amounts according to the contours of the bottom. This drags the line of the waves around to follow the contours. The waves have been bent, or refracted. Refraction causes waves to concentrate their force on headlands. Although headlands are outcrops of especially resistant rock, this concentration tends to wear them away and, over time, "flatten out" the coastline. Cracks in a rock are easily eroded, forming caves. Caves on opposite sides of a headland may eventually meet to form an arch, and when the roof of the arch collapses a stack, or isolated finger of rock, is left behind. Sometimes a cave will connect with a hole to the surface, caused by a collapse of the cave roof. If the incoming waves regularly erupt through this, it is called a blowhole. The entire roof of the cave may also collapse to form an inlet. Human activities like dredging and dumping can also affect the shapes of waves and coastlines. Mudflats form at estuary mouths. Longshore drift deposits sediment as spits or ridges, which may be curved by wave action. When the drift is strong and any flow into the bay is weak, the spit may extend right across the bay or even cut off the bay, as a bay-mouth bar. A tombolo is a ridge of sediment linking the mainland to a rocky island. The shape of a delta depends on whether the sea water is quiet or active. It may be modified or curtailed, as shown, by tides or currents. The Mississippi delta is unaffected by strong tides or currents and consequently consists of many lobes. On the other hand, rivers like the Congo and the Amazon have no surface deltas because strong ocean currents and waves remove the sediment before such a delta can form. Sediments may also generate barrier islands, which will eventually grow to isolate lagoons.

Fjords

Fjords are deeply glaciated valleys partly filled by an arm of the sea. These sinuous and elongated basins reach depths of over 300 metres (1,000 feet) and give evidence of deep glacial erosion. Because the effect of the glacier is greater further inland, fjords are often more shallow nearer the sea. The Sognefjord in Norway is over 900 metres (3,000 feet) deep, but near its seaward end it shallows to only 1400 metres (450 feet). Fjords are so deep because the glaciers that made them were descending to a lower sea level than today. Glacial erosion determines the U-shape of the fjord but the former ice load also helped to cause the high steep-sided cliffs in another way. During glaciation the weight of the ice causes the land to sink. When the glacier is gone, the land rises at a rate too fast to be rounded off by erosion.

RIVERS

Both the River Nile and the Amazon are more than 6,400 kilometres long, but whereas the Nile has few tributaries, the Amazon has around 15,000, four of which are themselves more than 1,600 kilometres long. In its course, the Amazon descends no fewer than 5,400 metres, but the Nile actually descends for less than a quarter of that height, despite its comparable length. The great and evident differences between these waterways and others around the world belie their underlying similarities.

The principle feature of all rivers is unidirectional, down hill water movement. The rate of flow of the water determines the physical, and hence the biological, characteristics of each particular region of the river as it wends its way to a greater waterways or to the sea.

Rivers most often begin at a mountain spring or lake, from which one or more headstreams flow. Such streams typically are swift-running, have a steep gradient, follow a relatively straight course, and are subject to stretches underground or to waterfalls. Waterfalls result when a river's course passes over strata of markedly differing hardness: where the water passes over hard and then soft material, the latter is eroded more quickly, causing a sharp change in gradient at that point. These fast-flowing headstreams tens to have a stony bed on which few plants can survive; they tend also to have a typical but limited fauna adapted to withstand such arduous conditions.

The Merging Water

Rivers gather strength as their headstreams merge, combining their flows to form wider, deeper, swifter courses. The large amounts of sediment being carried by these waters cannot settle on to the bed on the fast flow, and the water that tumbles over the rocky bed is rich in oxygen and maintains a comparatively regular temperature throughout the year. Again, few plants can survive these conditions, and only trout and migrating salmon make their nests at this level.

As the gradient of the river valley declines, the river broadens and the speed of flow diminishes. At this level, the relatively horizontal course results in a meandering path and the selective deposition of suspended solids. The river bed becomes less stony and more muddy, particularly on the insides of meanders where the river flow is least and accumulation takes place. Because of the slow-moving water, and because the muddy deposition usually has a high nutrient content, such areas are ideal for plants.

Finally, the river continues at a much more stately pace to the coast, where it flows into the sea. Here it may form an estuary, or a delta. Estuaries are protected arms of the sea into which the river flow. The most common are drowned river valleys. Deltas mostly arise where the offshore water are shallow and sheltered, so that the transported sediments may be deposited far out to sea. But there are exceptions, such as the Mississippi delta.

Most rivers systems are fed by rain falling on high ground. The amount of water that reaches a river depends on the nature of the underlying rock. Steep mountains of impervious rock deliver about 75 per cent of the total rainfall to a surface river system, whereas gently undulating porous rock may deliver as little as 15 per cent. The pattern of precipitation may also influence the size and shape of the river. Many rivers are seasonal, and remain dry for part of the year. Even on permanent rivers, seasonal variations far upstreams can affect the down stream course. This can be seen in the criss-cross pattern of old and new channels in a braided river. Braiding occurs when periodic increases in water levels, perhaps due to annual snow-melt, cause the river to overflow its existing banks on to the flood plain. While the plain is inundated the current cuts new channels into which the river retreats when the water level falls.

River flow very slowly across flat plains, and small variations in the speed of flow can alter the river's course. Water flowing at the outside of the bend moves a little faster, and cuts away the bank as it picks up sediment. On the inside of the bend, the water moves more slowly and tends to deposit sediment. Over time, these variations form a migrating meander, an exaggerated looping curve. Subsequently, the river cuts an alternative course across the neck of the meander. Sediment builds up at the ends of the loop, which eventually becomes separated, forming axis bow lake.

Tides of Change

Few, if any, major rivers remain that have not suffered the ravages of humankind. Their waters have been diverted or dammed for power generation, their lower channels have been dredged for navigation, their waters polluted by industrial chemical cocktails. Activities on land have also changed the nature of many rivers: improved drainage and the increasing use of fertilizers have meant river flows are less controlled — floods occur more frequently and the water are generally more acidic. Although the impact of these influences is finally being recognized, it is important that the rivers system from source to end is seen as a whole that is indivisible, and remedial actions devised accordingly.

LAKES

Almost a fifth of the world's fresh water is contained in just one lake: Lake Baikal in eastern Siberia, which is in parts 1.7 kilometres (just over a mile) deep. Not only is it the deepest, it is also the oldest lake on Earth. The vast rift valley that the lake fills began to form 80 million years ago, during the age of the dinosaurs, and the valley has been full of water for about 25 million years. Most lakes are geologically temporary phenomenon, lasting a maximum of hundreds of thousands of years and sometimes, in the instance of a small beaver pond, for only a few days.

A lakes is a body of water collected in a depression in the earth's crust. They occur everywhere, the highest being Lake Titicaca between Peru and Bolivia at an altitude of 3,810m, and lowest as Dead sea which is 395 m below sea level. The largest lake is the Caspian Sea, a salt water lake in the Soviet Union having the surface area of 395,400 square kilometres, and a volume of 78,800 cubic kilometres. Lake Baikal of Siberia, a freshwater lakes, is the deepest of all the lakes, having a depth of 1620 m. Only one other lake, Lake Tanganyika in Africa (depth 1,470 m) has a depth more than 1,000 m. There are not more than 20 lakes around the world with depths more than 400 m. The water in the lake generally standing, with negligible current. Smaller bodies of standing water are called **pond and pool** with no clearly defined limits. Similarly, there is no unanimity whether the Baltic is a sea or a lake, or for that matter Maracaibo in Venezula, a lake or a lagoon. A coastal water subject to influx of salt water from the ocean is appropriately, a sea. A **lagoon** though isolated from the sea due to an intervening barrier island, receives frequent flushes of tidal water.

Lakes may be formed in a variety of ways: tectonic lakes (formed by tectonic activity forming deep clefts in earth's crust — Lake Baikal, Caspian Sea, Lake Victoria between Tanzania and Uganda, etc.), glacial lakes (formed in hollows created by glaciers — Lake Superior of America, Hemkund, Tarsar and Ahlpathar lakes in Himalayas), volcanic lakes (formed in hollows left by a volcano — Nemi lake near Rome), fluvial lakes (formed by the activity of a river or stream — called Oxbow lakes because of their shape) and artificial lakes (man-made lake by collecting water of a river or stream by construction of a barrier or dam — Sukhna lake in Chandigarh, India).

Apart from lakes that form in meteor craters, lakes are created by various geophysical processes that occur above and below the Earth's surface. Glaciation during the last ice age accounts for most lakes. A cirque lake forms as water collects in the hollow left at a glacier head. At the snout of a glacier, a lake may form as meltwater is trapped by the terminal moraine, a bank of rock and debris left behind as the glaciers retreats. Lakes may also form behind lateral moraines at the glaciers sides. Blocks of ice buried in glacial debris eventually melt, leaving hollows that become kettle lakes.

Lakes may also be created by volcanic activity, as when larva flows into a river and solidifies into a dam. Subsequent cooling of the volcanic cone may permit a crater lake to form in the caldera. An oxbow lake forms when accumulated sediment cuts off an exaggerated loop of a river. In regions where the underlying rocks is limestone, a surface lake may form in the sink-hole created by the collapse of an underground cave system. Large-scale earth movement may form a lake creating surface depression in which water can accumulate. Faulting of the rock strata tends to produce deeper lakes than folded rock strata.

Depending on the quality of water two main types of lakes are recognised: fresh water lakes and salt water lakes (saline) lakes. A saltwater lake is generally formed by seepage of underground sea water or due to higher rate of evaporation in arid climates. The latter generally have a very high salinity, the Dead Sea having a salinity of 275 parts per thousand as compared to 34.5 of an average sea. Some of these in the arid region oftendry out in dry season, as is the case with Lake Eyre of Australia.

Water is densest at 3.94°C and thus sinks to lake bottom, where it cannot cool further without becoming less dense and trying to rise again. Thus lakes freeze from the surface downwards, allowing living organisms to survive in warmer water below the ice. Shallow lakes may freeze all the waydown. In temperate regions, lakes over 10 m deep show seasonal cycles of thermal layering. In early spring, water at all depths (apart from the bottom) is at a similar temperature. In summer surface heating creates three distinct thermal zones. In winter, surface ice leaves water below unfrozen.

The six great lakes — *Lake Superior, Lake Michigan, Lake Huron, Lake Erie, Lake Ontario* and *Lake Baikal* — contain about 40% of the world's liquid fresh water. Lake Baikal in Siberia contains as much water as all five of the American Great lakes combined despite having only about one eight of their surface area. The reason for this is that Lake Baikal formed in a very deep rift valley, and as a result is more than four times deeper than Lake Superior.

Saltwater Lakes

Most salt lakes, such as the Dead Sea of the Great Salt Lake in Utah, start as fresh water that becomes saltier with time. But in the Caspian Sea, the lake with the largest surface area in the world at 360,700 sq km (139,000 sq miles), the water is getting fresher. About 290 million years ago the Caspian Sea was cut off from the Mediterranean, since when the Volga River has been diluting it. Lakes usually become salty when a geological upheaval cuts off the lake's outflow and slows the streams feeding it. When lake water evaporates, only fresh water is lost — any streams remain. The water of Lake Natron, in the African Rift Valley, are salty not just through evaporation, but because the lake is fed by volcanic springs which produce hot solutions rich in sodium carbonate, which in places is thick enough to walk on.

Lakes grow old and die because they eventually fill up with sediments and wastes. The lifetime of a lake is controlled by its origin, size and shape, together with the climate and most importantly, the area drained by the lake (its watershed). The natural ageing process that affects lakes is called *eutrophication*. Although the process is gradual and continuous, it can be divided into stages according to the amount of nourishment the lake offers any potential life. At first, when the lake is "young", there is *oligotrophy* — nourishments is sparse; next, *mesotrophy*, when there is moderate nourishment; and finally, *eutrophy* or abundant nourishment.

Barren Waters

Young lakes in temperate climates have, at all depths, clear oxygenated water that is poor in the nutrients essential for plants to grow, such as nitrogen and phosphorus. This lack of fertility limits the growth of *phytoplankton,* the tiny plant organisms that form the basis of the food chain. As a result, relatively few plants and animals grow and live here. Rapid replacement of the water in a lake slows eutrophication, because the discharged water carries the phytoplankton away with it. In some lakes in northern Canada the inflow and discharge rates are so great that the entire volume of water is replaced in less than a week. At the other extreme, Lake Tahoe in California takes 700 years for complete water exchange.

Lakes, natural as well as artificial are highly susceptible to deterioration in quality by human activities. These include large sediments deposited into lakes and pollution of its water through sewage, often leading to rapid eutrophication of the lakes. The pesticides which leach out of the agricultural surroundings often pollute the lake water and cause immense damage to its natural ecosystem.

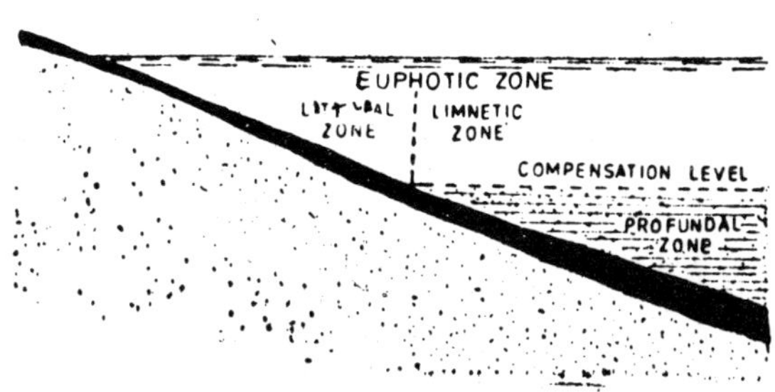

Biological Zones of a Lake.

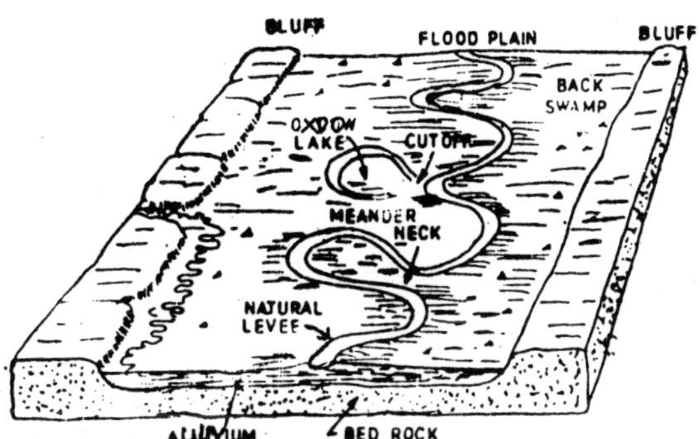

Floodplain Landforms of an Alluvial River.

A lake is generally divided into a number of zones: **euphotic zone** dominated by photosynthetic organisms, **profundal zone** where light penetration is insufficient for producers and only concumers can live. The boundary between the two is set at a depth where 1 per cent of the solar radiation can penetrate, thought to be the lowest limit for photosynthesis. This is called **compensation level**. Horizontally the portions of the lake shallower than the compensation level are called **littoral zone** and those deeper than this to include both photic and profundal zone as **limnetic zone.**

Based on temperature regimes a lake may be divided into upper **epilimnion** showing more or less warm uniform temperature, a middle zone **thermocline** showing a sharp decline in temperature with depth and a cold lower layer called **hypolimnion**.

Bedrock Variations

Lake Tahoe's water changes only slowly, but it is still pure and nutrient-free because it lies on crystalline rocks. Algae and phytoplankton require 21 different elements in order to flourish. In general, small amounts of these elements are carried into the lake by rain and snow, but the main source is the lake's watershed. Rivers and streams that flow into the lake leach these elements form the rocks and soil. Where lakes occur on hard crystalline rocks, the supply of nutrients is low and the lake stays "younger" for longer. In lakes formed on a sedimentary terrain or on glacial deposits, there are more nutrients and ageing takes place more quickly.

As a lake becomes old, its fertility increases at an accelerating rate. In old lakes, plants grow luxuriantly near the shore: mats of algae cover the surface in green slime. At depth, the water of a eutrophic lake are cloudy with decaying organic matter - and the decomposition of the organic matter uses up the oxygen dissolved in the lake water. At the surface, interaction with the air replaces the oxygen. However, at the lower levels of the lake especially in deep lakes where different well-defined thermal layers prevent the mixing of the surface and bottom waters during the summer, the water becomes permanently oxygen-depleted. In the absence of oxygen, *anaerobic* bacteria attack any organic matter, releasing hydrogen sulphide gas. Eventually the lake turns into a marsh, which in turn will finally disappear.

Lakes remain numerous in regions that were once glaciated, such as northern Canada and Scandinavia, but far more numerous are the ones that formed, existed for a period —and have since filled in and died.

Part II

The Power of the Atmosphere

- The Atmosphere
- Atmospheric phenomenon
- Air Circulation
- Land and Sea-breezes
- Solar Radiation
- Atmospheric Cooling
- Atmospheric Warning
- Temperature Inversion
- Ozone Layer
- The Ozone Hole
- Weather and Climate
- Cloud
- Monsoon
- Precipitation (Rain formation)
- Lightning and Thunder
- Greenhouse Effect
- Hurricanes and Tornadoes
- Antartica
- How Seasons Change

THE ATMOSPHERE

How the Atmosphere was Formed

Life on Earth is supported by an extraordinary thin shell of atmospheric oxygen — at an altitude of only 20 kilometres even the world's highest-flying bird, the Ruppell's vulture, would not only be unable to breathe in the low-pressure air, but its blood would boil in its veins. However, the Earth is unique in the solar system in having an atmosphere capable of supporting any life at all. In fact, the evolution of life was intimately linked with the evolution of the atmosphere, most critically 2,000 million years ago, when plants began producing free oxygen.

The cloud of dust and gases that formed the Earth 4,600 million years ago created a dense molten core enveloped in cosmic gases. This proto-atmosphere, composed mainly of hydrogen, carbon dioxide and carbon monoxide, did not last long before it as stripped away by a tremendous outburst of charged particles from the Sun. As the outer crust of the planet solidified, a new atmosphere began to form the gases outpouring from gigantic volcanoes and hot springs. This created atmosphere of carbon dioxide, nitrogen oxides, hydrogen, sulphur dioxide and water vapour. As the Earth cooled, water vapour condensed into highly acidic rainfall which collected to form lakes and oceans.

Birth of the Ozone Layer

For the first half of the Earth's existence, only trace amounts of free oxygen were present. But then green plants evolved in the oceans and they began to add oxygen to the atmosphere as a waste gas. The addition of large amounts of oxygen was critical for the further evolution of life because of the role that ozone (tri-molecular oxygen) plays in protecting plants and animals from lethal ultra-violet radiation. Because the early ozone layer was very thin and close to the surface, living organisms had to rely on alternative protection, and could only develop under about 10 metres of water.

The Evolving Atmosphere

As oxygen increased to l per cent of the atmosphere, the required depth of protective water became only 30 centimetres and complex ,multi-cellular marine life-form could develop. The atmosphere itself continues to evolve but human activities—with their highly polluting effects— have now overtaken nature in determining the changes. Because the stratospheric ozone layer absorbs ultraviolet radiation from the Sun, it creates a warm lid to the troposphere beneath. Consequently, virtually all moisture is trapped below about 8-16 kilometres and this, in turn, sets the altitude limit for clouds and precipitation. However, special clouds do occur. Around sunrise and sunset mother-of-pearl nacreous clouds may sometimes be seen at around 20-30 kilometres. They are composed of spherical droplets either too small to freeze at the exceptionally low temperatures or prevented from freezing by the presence of sulphuric acid released from volcanic eruptions.

The atmosphere is an insulating blanket of air around the earth. Without it the temperature at the equator would rise to 82°C during the day and drop as low as —140°C at night. It burns up meteors that would bombard the surface of the earth from space. The atmosphere does many things to make life possible on earth. It is a reservoir of oxygen needed by animals and of carbon dioxide essential for plants. It serves as a distribution system for moisture. Without the atmosphere there would be no sound and no flight. There would be no conventional long-distance radio communication, because this is dependent on the electrons in the upper atmosphere. Without air there would be no lightning, no clouds, no wind, no rain, no snow, and no fire. The surface of the earth would be as bleak and sterile as the moon.

The atmosphere shields the earth from lethal concentrations of ultraviolet radiation. It selectively filters the sun's rays so that only two small segments of the electromagnetic spectrum penetrate to the earth's surface in appreciable amounts. One segment is the **optical window**, consisting essentially of the visible spectrum of light, from near-ultraviolet to near-infrared. The other is the **radio window**, consisting of radio waves from about 1 centimetre to 40 metres in length.

It is impossible to define the limits of the atmosphere because the atmosphere becomes progressively thin with increasing distance from the earth. There is no boundary between the atmosphere and the void of outer space. However, 75 per cent of the earth's atmosphere lies within 16 km of the surface, and 99 per cent of the atmosphere lies below an altitude of 30 km.

The presence of a gaseous atmosphere around the earth may be due to a balance existing between the forces of gravity, which hold gases to the earth's surface in the same way as other masses and the characteristic motion of molecules, which left to themselves would let the gases evaporate rapidly into outer space by diffusion. The molecular motion is greater the smaller the molecular and atomic masses and the higher the temperature. If the gases have low molecular masses, at constant temperatures they will attain higher velocities and so have a greater chance of escaping into outer space against the pull of gravity.

Except as regards its water vapour content, which is subject to sharp variations, the atmosphere consists of a gaseous mixture that changes very little in its composition. Only in the layer of atmosphere within 10 to 50 metres of the earth's surface do minor variations appear, as a result mainly of the influence of organic life. The principal constituents of the atmosphere are the gases nitrogen, oxygen, and argon.

Clean, dry air at sea level contains, by volume, 78 per cent nitrogen gas, 21 per cent oxygen, 1 per cent argon, and .003 per cent carbon dioxide, plus small amounts of other gases. The latter (in approximate order of decreasing concentration)) include neon, helium, methane, krypton, sulphur dioxide, hydrogen, nitrous oxide, xenon, ozone, nitrogen dioxide, iodine, ammonia, and carbon monoxide (Table 1).

The nitrogen gas is biologically inert for most organisms, except under special conditions. It normally takes no part in the role of animal life except as a diluent for the oxygen and carbon dioxide, but it is ultilized by some bacteria and some plants. Oxygen is essential to the production of biological energy in the burning of cellular fuel. Carbon dioxide, despite its low concentration (less than 1/600 that of oxygen), plays a vital role in the cellular processes of animals and plants. Its most important, use is by plants in photosynthesis, by which the green pigment chlorophyll uses carbon dioxide, water, and sunlight to synthesize sugar and starches. In animals, carbon dioxide is a waste product, but it also regulates the rate of breathing. Carbon dioxide also performs an important function in controlling the acidity—alkalinity balance of the blood by its presence in the form of carbonic acid, identical to the product in carbonated water.

Table 1 - Composition of Clean, Dry Air at or Near Sea Level

Component	*Per cent by volume*
Nitrogen	78.084
Oxygen	20.9476
Argon	0.934
Carbon dioxide	0.0314
Neon	0.001818
Helium	0.000523
Methane	0.0002
Krypton	0.000114
Sulphur dioxide	0 to 0.0001
Hydrogen	0.00005
Nitrous oxide	0.00005
Xenon	0.0000087

(Contd.)

Table 1 (Contd.)

Component	*Per cent by volume*
Ozone	Summer 0 to 0.000007 Winter 0 to 0.000002
Nitrogen dioxide	0.000002
Iodine	0 to 0.000001
Ammonia	0 to trace
Carbon monoxide	0.1 ppm — 0.2 ppm

However, all of the major, and some of the minor, atmospheric components are biologically important. The participation of each in the living processes is in some cases critically sensitive to slight changes in concentration.

Major Regions of the Atmosphere

Troposphere

The zone nearest the earth, the troposphere, derives its name from the Greek *tropein* (to turn, to rotate, to change). One property that changes in the troposphere is the temperature. It drops to—56°C by the time an altitude of about 16-18 km is reached. The rate of temperature drop is about 6.4°C per 1000 metres. This is called the **normal environmental lapse rate.** The region through which this lapse rate occurs defines the troposphere. At the equator the troposphere is about 18 km thick but at the poles it is only about 8 km thick. Essentially all the atmosphere's water vapour is in the troposphere, which therefore encloses essentially all the storms and precipitation of the earth.

The upper boundary of troposphere, termed tropopause is market by temperature showing no further fall with increase in altitude, but remaining constant at about —56°C. There is little or no vertical movement of winds. Tropopause, however, has horizontal winds, called jet streams moving at tremendous speed, often reaching 480 km per hour.

Stratosphere

This zone begins at the top of the tropopause, extending to an altitude of 50 km. Here the temperature shows a gradual increase with increase in altitude. The stratosphere includes much of the **ozone layer**. The warming trend in the stratosphere is caused by a cycle of chemical changes in the ozone layer which converts all the incoming high energy ultraviolet rays to heat. If these rays reached the earth's surface, life as we know it would be impossible. The **stratopause** marks the narrow zone at the top of the stratosphere where the temperature begins to fall again with increasing altitude as the mesosphere is entered.

Table 2. Major Regions of the Atmosphere

Region	*Altitude range (km)*	*Temperature range(°C)*	*Significant chemical species*
Troposphere	0—8 to 18	15 to —56	N_2, O_2, C) O
Stratosphere	8 to 18—50	—56 to —2	O_3
Mesosphere	50—85	—2 to —92	O_2+NO+
Thermosphere	85—500	—92 to 1200	O_2^+, O^+, NO^+

Mesosphere

The mesosphere, above stratosphere, extends roughly to 80 km and its temperature drops from—2°C to between—80 and —94°C at the mesopause, a thin zone where the temperature stabilizes and soon starts to rise again as the **thermosphere** is ascended. The upper boundary of mesosphere records the lowest temperature of our atmosphere.

Thermosphere

This zone is above 80 km altitude. The temperature becomes very high, but the air has such a low density that it can hold very little heat. Any denser object in the thermosphere will be extremely hot in sunlight but very cold at night.

The zones just discussed are defined by changes of temperature with altitude. Changes in chemical composition or activity also occur with altitude. Up to about 80 km the proportions of components given in Table 2 remain nearly constant, and the long zone up to that altitude is sometimes called the **homosphere**. Within it is a region called the **chemosphere** extending from the upper troposphere where a huge number of chemical changes occur powered by solar radiation. Ozone production is one such activity. Above the homosphere lies the **heterosphere**. So named because various gases are not mixed up as in homosphere, but occur in distinct layers and are insufficient in quantity, not dense enough to form air. Thermosphere includes heterosphere and perhaps a small upper boundary of homosphere. For its first 40 km, the **ionosphere**, the very thin concentration of atoms and molecules, exists largely as electrically charged particles called ions. This globe enveloping band of ions reflects outgoing radio waves back to earth, making radio transmission possible well beyond the horizon. Television waves are not reflected, and a TV transmitter's ranged does not extend beyond the horizon. Meteorites entering the earth's atmosphere generally burn out in the ionosphere.

The region above ionosphere includes **exosphere** or outer atmosphere, which fades out into outer space. Located in the exosphere are two belts of high-energy radiation particles, one at 4,000 km and another 16,000 km. These belts called **van Allen radiation belts** protect us by absorbing corpuscular energy of sun.

The Atmospheric Thermometer

From an average surface value of 15°C , the temperature of the air falls to around—60°C at the top of the troposphere and warms to 10°C at the top of the stratosphere. It then plunges to—120°C in the mesosphere, after which it warms through the thermosphere, where a layer of oxygen around 200 kilometres high absorbs some solar ultraviolet radiation and is heated in the process.

ATMOSPHERIC PHENOMENA

An Article explorer once reported seeing nine Suns in the sky. He had fallen victim to one of the many optical illusions caused by light rays being bent or reflected as they pass through the atmosphere. The towering, spectacular Fata Morgana is another such illusion, which fools sailors into seeing mountain ranges floating over the surface of the sea. The atmosphere provides an ever-changing backcloth of colour, with dynamic vistas of blue sky, white clouds, grey storms and red and yellow sunsets. But one of its eeriest phenomena is the albino rainbow sometimes seen in fog.

Molecules of air scatter the shorter wavelengths of light (which are at the blue end of the spectrum of colours) more than the longer wavelengths (at the red end)—which is the reason that the sky appears blue in colour. The cleaner the air, such as after a rain shower, the darker the shade of blue the sky appears. When the Sun is near the horizon its light passes through large dust particles, which scatter longer wavelengths, giving a red sky at night and in the morning. Fog and cloud droplets, with diameters larger than the wavelengths of light, scatter all colours equally and make the sky look white.

A green flash from just above the sun's disc is sometimes seen at the moment the disc disappears below the horizon at sunset. It occurs because different wavelengths of light are bent (*refracted*) in the atmosphere by differing amounts. Because green light is refracted more than red light by the atmosphere, green is the last to disappear (blue light having been scattered by air molecules and lost from view).

Rainbows

When sunlight enters a raindrop, refraction and reflection take place, splitting white light into the spectrum of colours from red to blue, and making a rainbow. Rainbows come in pairs. Red is on the outside bend of the stronger, primary rainbow, but the secondary rainbow has its red bands on the inside. Between the two rainbows, little light is scattered and this creates a dark area known as Alexander's dark band. As many as four faint pink, green or violet bands are sometimes visible on the inside of the primary rainbow. They are called supernumerary arcs and are caused by interactions of light waves arriving at the observer's eye. When light waves are out of phase (so that the crest of one wave coincides with the trough of an adjacent one) they cancel each other out, but when in phase (with crests and troughs matching) colour intensity is enhanced.

Primary rainbows are caused by light that, after it is split by refraction into the spectrum of colours, is reflected only once from the back of a water droplet. The light for a secondary rainbow is reflected twice within the droplet, giving it a greater angle and reversing the spread of colours.

Living Haloes

Interactions of light waves can produce a "glory". People standing on a mountain, with the sun on their backs, may cast shadows on the fog in the valley which appear to be surrounded by coloured haloes. Similarly, the shadow on the cloud-tops of an aeroplane flying above the clouds may be surrounded by rings of coloured light. The glory is caused by light entering the edges of tiny droplets and being returned in the same direction from which it arrived. These light waves interfere with each other, sometimes cancelling out and sometimes adding to each other. A glory can seem even more impressive when the shadow of the observer is enlarged (a Brockenspectre).

Coronas are caused by the bending of light around small cloud droplets. These bent or diffracted waves of light interfere with the undiffracted waves to give alternating rings of light and dark. As the different wavelengths corresponding to the various colours of light bend slightly differently a red ring may appear in the dark region between two blue rings. The size of the rings of light depends on the size of the droplets. The smaller the droplets, the larger the rings. The more uniform the droplets, the more regular the rings. Typical coronas have a $4^\circ - 10^\circ$ radius, corresponding to cloud droplets of 4—10 μm.

Multiplying Suns

Reflection and refraction of light by ice crystals can create bright haloes in the form of rings, arcs, spots and pillars. Sun dogs (mock Suns) may appear as bright spots 22° to the left and right of the Sun. Sun pillars occur when ice crystals act as mirrors, creating a bright column of light extending above the Sun. Such a pillar of bright light may be visible even when the Sun has set.

Mirages

The mirage is an image created when rays of light from a distant object are bent (refracted) by a difference in air density which is caused by varying temperature. The observer imagines the light has travelled in a straight line, and sees the object displaced. When the air temperature is greatest near the ground, the image of a distant object is displaced downwards, creating an "inferior" mirage. An image of the blue sky on a hot day can seem to cover a road with a shimmering pool of water. When temperature increase with height, as is common over a cool lake on a summer afternoon, the image is displaced upwards, creating a "superior" mirage. Sometimes the light from the top and bottom of a distant object are refracted by different amounts, giving magnified or compressed images.

AIR CIRCULATION

Air in the troposphere is constantly in motion. Horizontal movements of air are called **winds**. Vertical movements of air are called **currents**. Winds and currents together carry heat from the equator to the poles. This heat comes directly from the earth but indirectly from the sun. Winds and currents determine the major circulation pattern of the air as they move air from high pressure areas to low pressure areas (Fig.).

High pressure is associated with cold air. Low pressure generally is associated with warm air. The density of air is determined by its temperature and the amount of water vapour per unit volume. A given volume of water vapour has less mass than the gases it displaces. Thus, moist air is less dense than dry air.

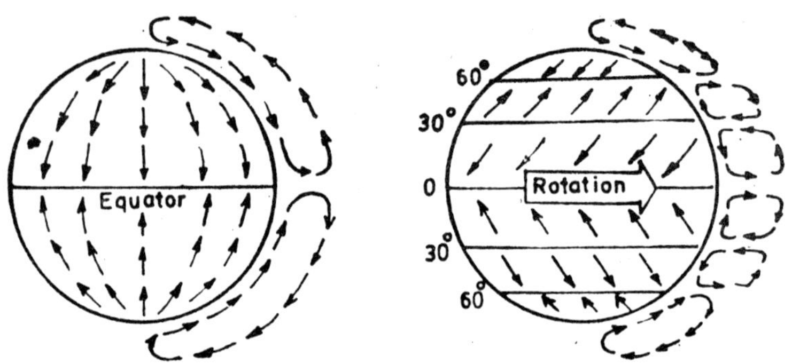

Major north-south convection currents in the atmosphere are broken into eddies by the earth's rotation, forming the wind patterns of the globe.

The portion of the earth between latitudes 23.5° north and south of the equator receives the most radiant energy. Therefore, air at the equator is warmer and less dense than air at other latitudes. Cool surface air flows toward the equator and forces currents of warm air to rise. In the upper atmosphere, warm air flows toward the poles. This air gradually cools and at the poles it is cold, dry air. Cold air sinks to the earth's surface in the polar region. From the poles, the air flows toward the equator. This circulation of air caries heat from the equatorial region to the rest of the earth. Two circulating cells of air, one in each hemisphere are called **Hadley cells**, named after George Hadley who postulated their existence in 1735.

The circulating air pattern in each hemisphere, however, gets broken into three units due to subtropical Highs and sub-polar Lows. The air mass rising at equatorial region moves poleward, but as it reaches subtropical zone, it splits into two, warmer mass continuing poleward movement, whereas less warmer mass descends down at around 30° latitude building up a high-pressure subtropical belt, popularly called **subtropical Highs** or "**Horse Latitudes**." At low levels a large portion of this air mass moves towards equator, a smaller portion moves towards poles. At around 60° latitudes this comparatively warmer air meets the cold, dense surface air coming from poles, and is forced to rise creating low-pressure zone called **sub-polar Lows**. This results in a situation where is surface air between equator and 30° latitudes moves equator-ward, but between 30° to 60° it takes poleward direction. Again between 60° and the poles it again moves equator-ward. The upper masses, however, maintain their poleward direction.

The eastward deflection of equator bound surface air due to Coriolis effect of earth rotation develops polar easterlies and tropical easterlies. Similar westward deflection of pole-ward surface air develops westerlies in each hemisphere.

LAND AND SEA BREEZES

People who live near a lake or on a seacoast are familiar with local winds blowing towards the land during day and towards the water body during the night. During the day when insolation gets warmer much faster as compared to water. The heated land air rises creating low pressure area, thus inviting cooler air from water body to rush onshore. This is called **sea breeze**. During night, with no solar insolation, land loses heat equally faster, becoming cooler than the water body. It is now the turn of cooler land air to move towards the lake or sea, which is relatively warmer at night. This offshore movement of the air from land to the water body constitutes a **land breeze**.

Similar breezes are also encountered between a mountain slope and a valley in the form of upcurrents during the day and downcurrents during the night. This is also influenced by diurnal sun's radiation. The air immediately flanking a mountain slope gets warmed more rapidly by the sun's rays than that in the open atmosphere adjoining the slope. In this way it gets lighter and rises up the slope as an upcurrent. The upcurrents carry moisture-laden warm air from the valley up the hill side and so contribute to cloud formation.

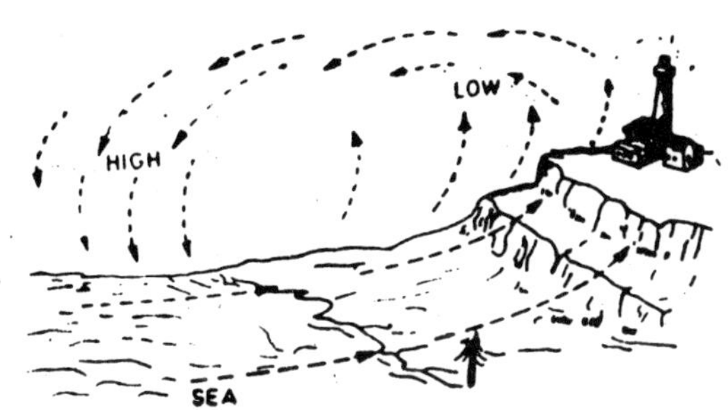

DAY — Sea breezes blow On— shore

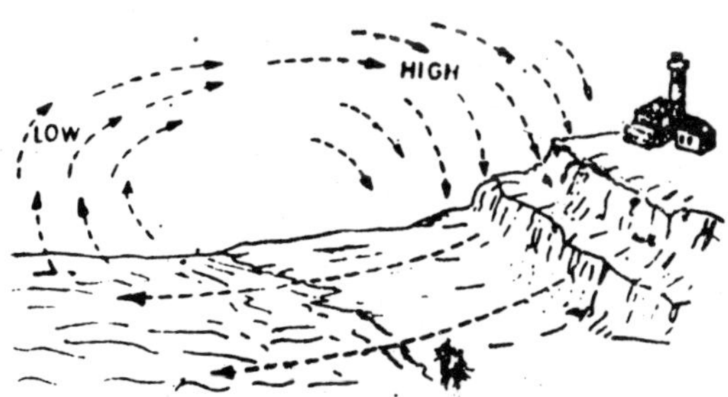

NIGHT — Land breezes blow off shore

Land and Sea Breezes.

At night, because of the high level of radiation from the earth's surface, the air immediately enveloping the surface of the slope cools more rapidly than at higher levels, and the heavy cold air begins to flow into the valley under the force of gravity, giving rise again to a cold downcurrent. Both currents flow across the direction of the valley, following the down line of the slope.

As the upcurrent rises along a slope during the day, air masses flow out laterally towards the valley at higher levels. Similarly higher level movement towards the mountain occurs during the night. The mountain valley breezes are, however, not a simple circulation, being complicated by air flow over lateral ridges of the valley.

SOLAR RADIATION

Radiation

Radiation has different meanings. In its most general sense, it refers to the energy waves found in the electromagnetic spectrum

'Radiation means:

1. energy (or particles) emitted by radioactive materials as they decay into more stable elements;
2. the energy waves found in the electromagnetic spectrum (a system for classifying energy sent out—'radiated'—from various sources): these include radiowaves, microwaves, visible light, X-rays and gamma rays;

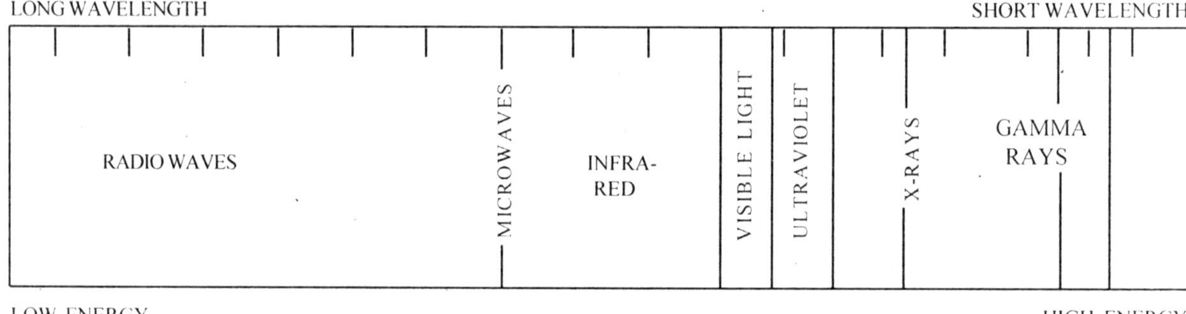

Visible light waves are the most familiar part of this range, but they are a very small part of the spectrum. Other waves vary from, radio waves (long wavelength, low energy) to high-energy (short wavelength) gamma rays and X-rays. Infra-red and ultraviolet light are also located in the spectrum—so sunlight is a source of radiation as are microwaves.

Longwave Radiation

The solar radiation reaching the earth constitutes short wave radiation, seldom longer than 3 microns in wavelength. When the ground or water are warmed under the impact of insolation, the molecules of ground become excited and cause release of infrared radiation of a wavelength exceeding 4 microns. These constitute the longwave radiation, also known by other names such as **ground radiation, terrestrial radiation** or **reradiation.** This radiation differs from reflected rays, which are simply sent back without alteration in wave length. The atmosphere also absorbs a part of the insolation and sends longwave radiation to the ground as well as to outer space. The ground radiation absorbed by the atmosphere is often again radiated back to the ground and termed **counter radiation.** Whereas the short wave solar radiation is transmitted only during the day, the long wave radiation continues even during the night. Also although the solar radiation largely penetrates through the water vapour and carbon dioxide in the lower atmosphere, the longwave radiation cannot escape and are sent back by these, in the sameway as the glass sheet sends back ground radiation in a green house. The resultant warming of the atmosphere below has appropriately given it the name **greenhouse effect**, a matter of considerable concern since carbon dioxide content of our atmosphere is on the increase and may be responsible for global warming. This, some fear may result in melting of polar ice caps, rise in ocean level and submergence of many low lying areas like those along the Bay of Bengal.

The longwave radiation from the ground and the atmosphere is lost to the outer space largely in the wave length between 8 to 13 microns, often referred to as a window. There are also lesser windows: one in the range 4 to 6 microns and another 17 to 21 microns.

Solar Radiation

The solar energy is emitted in the form of electromagnetic radiation, travelling at the uniform velocity of 3 X 10^8 metres per second. About 9 per cent of radiation constitutes short wave X-rays, gamma rays and ultraviolet rays. Nearly 41 per cent constitutes radiation of wave-length between 0.4 to 0.7 microns, termed visible light rays. Rest 50 per cent constitutes the infrared rays of longer wave length. All incoming solar radiation is generally called short wave radiation, with wave length less than 3 microns. X-rays and gamma rays, although constituting a very small portion of short wave length radiation, are intense energy beams capable of tearing electrons from the atoms they strike, and can destroy living cells. This ionising radiation, fortunately are absorbed high in the atmosphere, and do not reach the earth. Ultraviolet rays cause damage to the skin cells and kill microorganisms and can lead to skin cancer in humans. Fortunately, the ozone layer in the stratosphere screens and does not allow the ultraviolet rays to reach the surface of the earth.

The visible light rays, the ones we can see, range from violet to red of the spectrum produce a variety of colour to the objects, which absorb certain wave lengths and reflect others. Light rays are the source of energy in our biosphere, being fixed and converted into chemical energy by green plants. These rays also produce heat.

The Infrared radiation, another form of invisible radiation of wavelength longer than 0.7 microns are responsible mainly for heating of our land, water and atmosphere.

In addition to the electromagnetic radiation, the solar energy is also issued in the form of clouds of electrified particles, called **magnetic storm** about once a month in the form of tiny molecules of helium, hydrogen, oxygen and nitrogen. The magnetic field of our earth, which extends to about 80,000 km in space (magnetosphere) catches these particles above each pole. These particles on friction with nitrogen and oxygen in our atmosphere create green and red luminescence across polar skies, visible as **aurora borealis** at the North pole and **aurora australis** at the South pole. These particles are, however, not allowed to reach the earth by two bands of radiation belts, termed the **Van Allen radiation belts,** which encircle the earth at a distance of 40000 km and 16,000 km.

ATMOSPHERIC COOLING

The atmospheric cooling results from the fact that air has maximum density near the ground surface, and loses its density with increasing altitude. Also significant is the fact that, the solar radiation is non-existent at night and the heated ground surfaces becomes colder much faster. The cooling of the atmosphere is the result of expansion, evaporation and advection.

Expansion is the cause of altitudinal decrease of temperature in our atmosphere much in the same way as compression of a descending air results in adiabatic heating. A rising air mass expands and becomes cooler, with no net loss in heat energy of the air mass. This is termed as **adiabatic cooling**. As the atmosphere experiences a regular decrease in pressure with increase in altitude, the temperature decline, leaving aside the lower temperature inversion zone, is also constant, showing an environmental or **normal lapse rate** of 6.4°C per km. However, when a warm dry air rises up, the lapse, called **dry adiabatic lapse rate** averages 10°C per km as long as no condensation is taking place. But as the rising air normally containing water vapour undergoes condensation, some latent heat is released, offsetting the temperature lapse by adiabatic cooling. The resultant lowered lapse rate, 6°C per km is termed **wet adiabatic lapse rate**. The point at which the condensation begins is termed **dew point**.

As the water vapour on condensation releases latent heat, similarly for evaporation of water the heat needed for conversion of water into water vapour, is drawn from the surrounding atmosphere, bringing down its temperature. The atmospheric cooling by **evaporation** largely occurring from water bodies and moisture containing ground as also the phenomenon of transpiration is largely dependent on atmospheric relative humidity. Cooling by evaporation is maximum at low levels of relative humidity, but declines with rise in relative humidity.

Whereas the rising air mass experiences adiabatic cooling, the horizontal movement of an air mass, can bring about cooling effects, especially when a cold air is sliding under a mass of warm air. The phenomenon known as **advection**, is prominent when a cool breeze from lake or sea blows towards the land during the day time. The trade winds of low latitudes and the polar easterlies are the global manifestation of advection.

ATMOSPHERIC WARMING

The atmosphere absorbs a very small fraction of short wave solar radiation, to account for rise in its temperature. The atmospheric warming, overrule results from a variety of processes such as reradiation, compression, convection and condensation. The **ground radiations** or reradiation, in the form of longwave radiation is the major source of atmospheric warming, primarily because of greenhouse effect of atmospheric gases, mainly carbon dioxide, which easily allow shortwave solar radiation to reach ground surface, but do not permit the long wave radiation from ground to escape, in the same way as glass sheet in a greenhouse.

The temperature of atmosphere also increases, when a mass of cool air descends down from a higher to lower altitude. Such an air mass does not lose or gain any heat from surroundings but gets heated due to increase in air pressure, at lower altitude. This, warming of the descending air due to compression is commonly known as **adiabatic heating**. The Afghanet winds blowing down from the Pamirs Afghanistan, and Santa Ana winds in California are the result of this warming process.

The water vapour also contributes to heating especially when it cools down due to rising up of warm humid air or encountering a cold surface such as a mountain slope. Because of a high latent heat of water, the condensation of water vapour, releases considerable heat into the surroundings.

Although, **conduction** has very prominent role to play in ground warming at lower depths, it contributes little in atmospheric warming, chiefly because air is a poor conductor. A very thin layer of air is heated up near ground surface or ocean surface, by direct contact. This heated air, however, sets in tempo for convection currents wherein the heated air rises up and cold air descend down to take its place. The atmosphere with lot of dust particles which rise up with heated air, may in part contribute to heating by conduction when coming in contact with a cool air.

TEMPERATURE INVERSION

A temperature gradient is formed in the air layer on the earth's surface in which the temperature decreases steadily by day as the altitude above the earth's surface increases, whereas at night the temperature increases as the altitude increases. This process is known as thermal or temperature inversion. If, in the course of the following day the ground warms up again, the night's inversion will be cancelled out by the supply of energy from the ground to the cold air above it, and the normal temperature gradient will be restored. The form of such inversions depends to a large extent on the structure of the terrain and the climatic conditions and is not for the most part very stable. Under some circumstances the usual temperature profile near the ground is altered so that temperature increases with altitude. In this situation vertical mixing is suppressed over the altitude range where the temperature is increasing.

The significance of the inversion for the exchange of gas in the air layer near the ground lies in the difference in specific gravity of colder and warmer gases in the same combination. Cold air heated up by the ground rises and cools. If it reaches the layer of warm air, because of its lower specific gravity any further rise of the cold air as it warms up is now checked. Thus, above the boundary layer no exchange of gas can occur.

Such thresholds create special problems for the layer of air near the ground in industrial towns. The low-lying cold air is consumed in combustion processes in household, transport and industry, and the end product is further contaminated by combustion gases and dust. If no boundary layer existed, these noxious elements would be carried away into the upper atmosphere by normal dispersal processes and thereby dispersed to such an extent that there would be no immediate danger of injury to the biosphere. If an inversion occurs at a lower level (30—100 metres above an industrial and housing area) the combustion products emitted are unable to penetrate the boundary layer and are retained in the lower layer of cold air. The concentrations of noxious gases and dust then increase rapidly and cause considerable damage to people, animals or plants.

In unfavourable valley regions and hollows the nightly inversion can also make itself felt in this way. That the inversion situation in valley basins is very stable is clear from the often immobile fog banks. Columns of smoke from chimneys stop at the boundary layers as if cut off; the smoke itself gathers below the boundary layer and forms great billows. If the chimney rises above the boundary layer the smoke can disperse freely. If the smoke is held down by high atmospheric moisture it will spread right across the boundary layer without penetrating it.

The frequency of such inversion situations depends on the season of the year. In summer, because of sustained and powerful solar radiation, there is very little inclination to inversion. As the days shorten ground radiation may exceed solar radiation for several hours before sunset and several hours after sunrise. Solar radiation on the short winter days, thus, hardly has any effect even in fine weather, so that the frequency and duration of the inversion situation are greater in winter. This results in ground frost and fog conditions, often causing damage to the crops. The farmers counter such frost damage by using straw enclosures for sensitive crops like cueurbits, installing wind fans to ensure air mising, or by growing frost resistant crops like alfalfa, cotton or sugar beets. On hill slopes, the orchards are commonly located where the boundary layer, also called thermal belt or inversion lid intersects a mountain slope, to ensure high temperature even in cool winter nights.

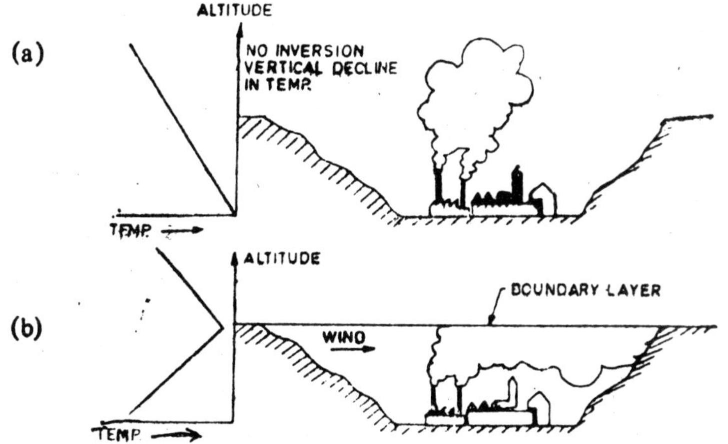

(a) Weather conditions without Invesion; (b) Inversion Weather Conditions.

OZONE LAYER

Ozone (O_3) is a gas, and a variant of oxygen (O_2) (see Ozone). Although it is present in very small amounts in the atmosphere, significant reductions or increases in this gas can have important environmental consequences.

Ozone occurring in that part of the earth's atmosphere nearest the ground (the troposphere) is frequently industrially—caused and can damage human and plant health (see Acid Rain).

A proportion of the Sun's energy is emitted as ultraviolet (UV) radiation or light. Ultraviolet light is divided into three types according to wavelength:

UVA 320 — 400 nm
UVB 290 — 320 nm
UVC < 290 nm

DNA, the genetic code present in all living cells, is damaged by ultraviolet radiation. UVC is the most damaging radiation followed by UVB, and UVA. Damage to DNA can either kill the cell or cause it to mutate and become cancerous.

The depletion of the ozone layer has recently become the subject of much concern. If this continues, the consequences could be disastrous for the planet.

An international treaty signed in 1987 has restricted the production of some of the substances that have been eroding the ozone layer. Because of the long life of the pollutants involved, however, those already released will continue to be there for the next century. A further pollution load will be deposited in the stratosphere by substances already manufactured, which will be released when the products they inhabit are scrapped. Additional loading will be created by substances permitted to be manufactured until the end of the century, and by substances not yet scheduled for abolition.

Ozone in the Stratosphere

Ozone is a minor constituent of the earth's atmosphere, found in varying concentrations between sea level and a height of some 60 km. Most of the atmosphere's ozone is found in its two lowest layers:

the troposphere, which extends up to 12 km above the earth's surface,
and the stratosphere above it which extends up to about 50 km.

The majority of ozone if found between 20 and 50 km above the ground, with the highest ozone concentrations occurring between 20 and 25 km.

Ozone occurring higher up in the atmosphere, between 15 and 50 km above ground level (*i.e.*, the stratosphere) forms a protective barrier called the ozone layer, shielding the earth from extreme heat radiated by the Sun.

The ozone layer plays a critical role in protecting living organisms from the harmful effects of ultraviolet radiation. Ozone is broken down by sunlight, and this process absorbs shorter wavelength ultraviolet light (ultra-violet radiation below 315 nm). Stratospheric ozone, therefore, acts as a shield around the Earth, which virtually eliminates UVC and greatly reduces the amount of UVB reaching the surface of the planet. Without the protective effect of ozone, terrestrial ecosystems could not have evolved in the manner that they have.

Ozone is simultaneously formed and destroyed in the stratosphere by naturally occurring chemical reactions. When oxygen molecules absorb sunlight (light energy), they dissociate into free oxygen atoms which then combine with O_2 to form ozone:

$$O_2 + hv \longrightarrow 2O$$
$$O + O_2 \longrightarrow O_3$$

Sunlight is also responsible for the destruction of ozone. It dissociates the ozone molecule into oxygen atoms and molecules. The free oxygen combines with ozone to form more oxygen molecules.

$$O_3 + hv \longrightarrow O + O_2$$
$$O + O_3 \longrightarrow 2O_2$$

In the absence of any interference from man, the result is a constant concentration of ozone in the stratosphere.

THE OZONE HOLE

If it were not for a diffuse gaseous layer 14-45 kilometres (9-28 miles) up, life on Earth could not exist. The ozone layer is all that protects us from the Sun's harmful ultraviolet radiation. Without the ozone layer, terrestrial life could not exist. But recent observations have shown that pollution of the atmosphere could be destroying this vital layer. Scientists have detected an "ozone hole" over Antarctica, which appears seasonally, extending over an area the size of the United States. Were this hole to spread, the consequences would be disastrous for both animal and plant life.

Without sunlight, ours would be a barren planet. Solar radiation warms the Earth's surface and is harnessed by plants in photosynthesis. But some wavelengths of light emitted by the Sun are less beneficial to life: indeed, ultraviolet light is an effective sterilizing agent and is used in hospitals to kill micro-organisms. Much of this radiation is filtered out in the stratosphere before it can wreak its destructive effects. Under normal circumstances, the concentration of the ozone that absorbs the ultraviolet light is constant.

Ozone is a form of oxygen with three atoms instead of two in its molecules. Ozone molecules can become oxygen molecules by losing one oxygen atom. The process is reversible. Chlorofluorocarbons are made of one carbon atom, one fluorine atom and three atoms of chlorine.

Each October since 1979 the thickness of the ozone layer over the Antartic has decreased by up to 50 per cent or more. The resulting ozone "hole" is constantly expanding. The thickness of the ozone layer is measured in Dobson units. A hundred Dobson units correspond to 1 mm (0.04 in) of compressed ozone. In the southern summer the thickness of the layer is about 350 Dobson units. The ozone hole is repaired in this way because the polar stratospheric clouds evaporate in the warmth of the summer, and fresh air returns to the Antarctic from other latitudes, allowing ozone levels to recover.

Punching a Hole in the Heavens

In the 1970s and 1980s, it became clear that human activity was disturbing this delicate equilibrium: gases —most notably chlorine — emitted into the atmosphere had depleted the levels of stratospheric ozone. Chlorine atoms are released in the form of *chlorofuorocarbons* (CFC_s), which are widely used as collant fluids in refrigerators, to make bubbles in foamed plastic and in aerosols.

After their release CFCs take many years to rise into the stratosphere. At these altitudes, ultraviolet radiation is sufficiently intense to splite CFC molecules and liberate chlorine atoms, which can then attack ozone. Chlorine can attack many thousands of ozone molecules before it is finally removed from the stratosphere. Small increases in the amounts of chlorine released can therefore cause enormous changes in the chemistry of the upper atmósphere.

Emissions of CFCs have been rising rapidly in recent decades. Because CFCs remain in the stratosphere for anywhere between 65 and 130 years, immediate reductions in their release are essential. International agreements may succeed in phasing out CFC production, and alternative ozone-harmless chemicals are being developed. Even so, a vast reservoir of ozone-depleting chlorine will remain in the stratosphere for many decades.

In 1985 scientists discovered that the ozone layer was being depleted not gradually all around the globe, but massively and seasonally in one particular area: the Antarctic. The amplified impact of CFCs in Antarctica is a result of the unique meteorology of that region. During the southern winter the cold, heavy air high over Antarctica is prevented from mixing with the warmer, strong winds that whirl around the continent. Within this thin, cold air, at temperatures around —90°C (—130°F), icy particles from clouds known as polar stratospheric clouds (PSCs). In the dark polar winter atmospheric chemistry is at a virtual standstill. But some reactions still take place on the surface of the cloud particles, slowly preparing the chlorine held in "reservoirs" to make a rapid escape when the sun begins to shine in the southern spring. This sudden springtime "flush" chlorine rapidly depletes the Antarctic ozone layer.

Increased levels of ultraviolet radiation, especially ultraviolet-B (290-320 nanometre wavelengths), are expected to cause marked increases in sunburn, skin cancer and in eye problems such as cataracts. For every 1 per cent decrease in ozone there may be a 5 per cent increase in the number of non-malignant skin cancers in humans each year.

The Role of CFCs

Ozone one oxygen absorb ultraviolet light by using up the energy of light to fuel chemical reaction. When ultraviolet light strikes an oxygen molecule (O_2), it splits it into two oxygen atom (1). Few are highly reactive atom and in turn combine violet other oxygen molecules to form ozone, O_3 (2).

$$O_2 + h_v \rightarrow 2\,O \qquad (1)$$
$$O + O_2 \rightarrow O_3 \qquad (2)$$

Sunlight is also responsible for ozone decomposition. Ozone absorbs sunlight and decompose to oxygen molecule and a reactive oxygen atom (3) the free oxygen atom may combine with another oxygen atom to form are oxygen molecule (4), or it may colloid with an ozone to form two oxygen molecules (5) and the whole process can begun again. In this way a chain reaction is established in which ozone formation keeps up with ozone loss. In the absence of any interference from man, the regular is a constant concentration of ozone is the stratosphere

$$O_3 + h_v \rightarrow O + O_2 \qquad (3)$$
$$O + O \rightarrow O_2 \qquad (4)$$
$$O + O_2 \rightarrow O_3 \qquad (5)$$

When a $CFCl_3$ molecule is struck by ultraviolet light, it loses a chlorine atom (6.) The free chlorine can steal an oxygen atom from an ozone to form a chlorine monoxide (ClO) molecule, leaving behind an ordinary molecule of oxygen (7). When the ClO collides with a free oxygen atom, the oxygens combine into a molecule (8) and leave the chlorine free to go and destroy another ozone. This is known as a "catalytic" process because the chlorine can destroy many ozones without itself being consumed in the process.

$$CFCl_3 + h_v \rightarrow CFCl_2 + Cl \quad (6)$$
$$Cl + O_3 \rightarrow ClO + O_2 \quad (7)$$
$$ClO + O \rightarrow Cl + O_2 \quad (8)$$

As stratospheric ozone decreases, more of the Sun's energy reaches the Earth. Ultraviolet-B may kill marine phytoplankton, which absorb large quantities of carbon dioxide from the air. As atmospheric CO_2 increases, less of the heat radiated from the Earth's surface escapes. Thus the ozone hole may also contribute to global warming.

WEATHER AND CLIMATE

The real engine of the Earth's weather is the Sun. Its heat generates powerful movements of air within the Earths atmosphere that circulate for tens of thousands of kilometres back to their starting point, providing the world's winds with a total kinetic energy greater than a million million jumbo jets flying at top speed.

All the world's weather is driven by the Sun's intense heating of equatorial regions, which sends air currents surging high into the atmosphere. The rising air cools and the moisture it carries condenses to form clouds and heavy rainfall, creating the characteristic wet climate zone of Africa, Asia and South America.

But the air currents themselves continue moving polewards. Eventually, this warm tropical air sinks at around 30° latitude and encounters increased pressure at the lower altitudes. It is compressed and heated, and moisture is squeezed away, so creating the subtropical climate zone of permanent deserts in places like Sustralia and the Saharan region of Africa.

On reaching the surface, the sinking air is pushed outwards, with some returning to the equator as the north-east trade winds (south-east trades in the Southern Hemisphere). This circulation of air between the equator and 30° latitude is called the ***Hadley Cell.***

Mixing up a Storm

The rest of the descending air is sent polewards to form the warm winds of the mid-latitudes. At 50°-70° latitude these winds, now moist from their journey over the oceans, encounter cold dry winds spilling from the bitterly cold polar and taiga climatic regions. A battle takes place where these winds meet, resulting in the creation of the large swirling storm called frontal depressions that are typical of mid-latitude, temperate climates.

Classifying Climates

When they occur again and again over a long period, these geographical patterns in the weather — the balmy doldrums near the equator; the steady trade winds of tropical latitudes; the calm, cloud-fee regions of sub-tropical latitude; the stormy winds of the mid-latitudes and the light, bitterly cold easterly winds blowing from the poles — can all be used to define the climate of a region. In addition to their precipitation and temperature, climates have been classified into zones by moisture index, vegetation and even estimates of human discomfort. But all classifications consider only the broad similarities between climates at various places around the world. Local differences are ignored and boundaries are approximate, and shift over time with the Earth's changing temperature.

Regional climates change enormously with seemingly small changes in global mean temperature. This happens because polar latitudes are highly sensitive to alterations in global temperature compared with the tropics. As the global temperature falls, the area of snow and ice in higher latitudes expands rapidly causing greater amounts of the Sun's energy to be reflected back into space, so chilling these areas even more. In contrast, the tropics change very little. As the temperature difference increases it forces a rearrangement of atmospheric circulation.

Photographed from space, the clouds and oceans and the various climatic zones with their characteristic landscapes make a spectacular global patchwork of colour. This is because not all the energy the Earth receives from the sun is absorbed. A lot is reflected and the proportion of solar energy the Earth bounces back into space is its *albedo.* Although the Earth's average albedo is about 30 per cent, a wide range of values exist throughout the world. The albedo of forests is 9-18 per cent, varying according to tree type and foliage density. The albedo of grasslands is 25 per cent, of cities 14-28 per cent, and of deserts 25-30 per cent. Fresh snow may reflect as much as 85 per cent of solar energy. The albedo of oceans rises from 2-3 per cent when the Sun is high in the sky to 50 per cent, when the Sun drops below 15° as it descends towards the horizon.

The Earth is spheroidal in shape. As a consequence, the equatorial latitudes receive for more heating from the Sun's rays per unit area than do the poles. Heat is also lost at the poles because the rays must travel through a greater thickness of air before reaching the ground. Since lower latitudes are clearly not overheating and higher latitudes not cooling, an annual balance between the energy received from the Sun and the heat lost by the Earth is achieved. It is the atmosphere and ocean currents that accomplish the task of transferring any excess energy between the equator and the poles.

Jet Streams

Pilots flying high over the Pacific Ocean in World War II were often astounded to encounter high-speed winds that held their aircraft stationary. They were among the first people to discover jet streams — narrow bonds of air travelling at speeds up to 500 km/h (300 mph). There are many jet streams in the atmosphere, and their presence is often revealed by ribbons of cirrus (high-level clouds of ice crystal) many hundreds of kilometres long.

CLOUD

Solar radiation cause the atmosphere to become warmer in its layers close to the earth than in its higher reaches. As molecules possess more kinetic energy in warmer air than in colder, they pick up more water molecules from the liquid zone and transmit them in the form of vapour. Warm air can accordingly absorb more water vapour than cold air. Warm air being lighter than cold, rises in the atmosphere while cold air sinks. As they rise, the warm air masses gradually cool down, with the result that the water molecules lose energy again and combine again into bigger units. If a position is reached in which no more water molecules pass from the liquid to the vapour state, this is known as water-vapour saturation. Atmosphere at this point has 100 per cent relative humidity.

Any further cooling of the air results in surplus water vapour getting converted into small droplets of water, around **condensation nuclei.** The formation of condensation nuclei stems essentially from combustion processes, volcanic eruptions, and wind effects. The combustion processes release sulphates and nitrogen compounds into the atmosphere, and these combine with ammonium sulphate and act as condensation nuclei. Nuclei of salts, primarily sodium chloride, which pass into the atmosphere in the form of spray, are of less importance than was originally assumed.

A cloud is a dense mass of suspended water or ice particles generally in the size range 20 to 50 microns, each condensation nucleus being 0.1 to 1 micron in size. Whereas the liquid water turn into ice at a temperature of 0°C, the minute water particles in cloud remain in liquid state upto a temperature of —12°C and called **supercooled water**. Between —12°C and —30°C half the particles in cloud are water and half ice, below —30°C mainly ice crystals and below

—40°C all ice crystals.

Two main types of clouds are recognised: Stratiform, which occur in layers and cumuliform, which occur like huge puffs or globular masses. The typical forms of these are called **stratus** and **cumulus**, respectively. A broad thick layer of puffy clouds is called stratocumulus, a tall dense form is called cumulonimbus, dense stratiform cloud as nimbostratus. Prefix nimbo is used for rain producing clouds. The high altitude clouds, generally between 3000 m to 6000m are thinner and called altostratus and altocumulus clouds. Higher up the clouds generally consist of ice crystals and are termed **cirrus** clouds.

A cloud is differentiated from a fog in that whereas the fog occurs at ground surface, a cloud is separated from the ground by a portion of the atmosphere having no visible moisture.

Smaller masses of stratus and cumulus clouds generally do not produce rain. A technique of **cloud seeding** has been developed in which dry ice (solid CO_2) pellets or silver iodide smoke is released into cumulus clouds. These provide nuclei for intensive condensation to form dense cumulonimbus clouds that can yield rain. Rainfall stimulated in this way is called artificial rain.

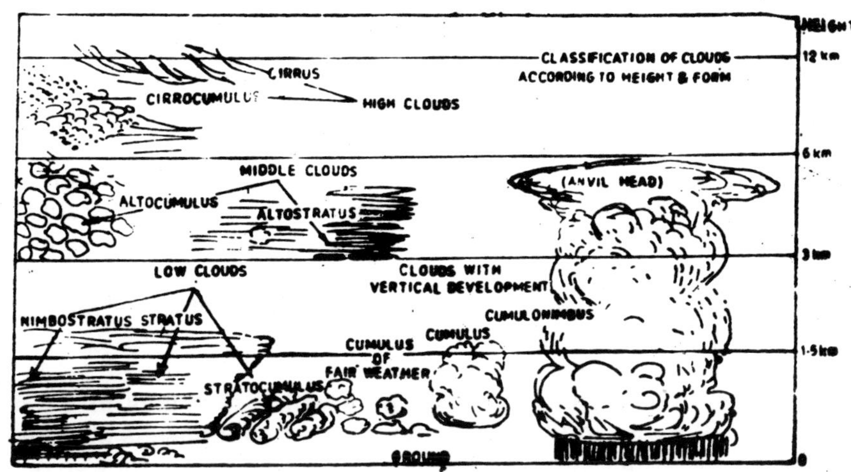

Cloud types.

MONSOON

Monsoon is a major wind system, providing distinctive monsoon climate to larger tropical and subtropical areas of Asia. The phenomenon results primarily from the unequal distribution of land masses in the two hemisphere. Whereas the three heat belts from equator to poles work very well in Southern hemisphere where land surface is much smaller in comparison to oceans, the same is not true for Northern hemisphere where vast continents of North America, Asia and Europe are located. These huge land masses in the Northern hemisphere, exert a powerful control over pressure conditions, so that typical latitudinal heat belts are lacking.

In winter with poor solar insolation, land masses are much colder than oceans and as a result cold land areas develop high pressure centers. At the same time, oceans which have retained much of the warmth from last summer, develop low pressure centers. There is, thus an air flow from continental Asia to equatorial oceans, called winter monsoon. As it originates from cold land masses, the winter monsoon is associated with dry weather for several months.

In summer with strong solar insolation land masses heat up faster and develop low pressure zones which pull in the moisture laden air from Indian ocean and Southwestern pacific ocean. These air masses move northward and northwestward over India, Indo-Chine and China constituting summer monsoon, and result in heavy rainfall, especially in coastal regions and proximal mountain ranges under conditions of intense solar insolation on land. The low pressure centers in summer often develop into cyclones resulting in forceful sucking in of oceanic moist air, resulting in devastating cyclonic storms in certain coastal areas.

PRECIPITATION (RAIN FORMATION)

Precipitation occurs when cloud bearing air is further cooled and water droplets or ice crystals are heavy enough, so that they cannot be held in the clouds and eventually fall down: Different forms of precipitation include rain, snow, hail and sleet.

The water droplets (mist) present in clouds are so small that they cannot leave the clouds without evaporating. For raindrops to be formed that will reach the earth before they evaporate, hundreds of cloud droplets must coagulate. For the aggregation of the cloud droplets, adhesion and condensation are not sufficient in themselves to explain the phenomenon.

There are two types of mechanisms of **raindrop** formation, each of which is valid under different climatic conditions. Not all the cloud droplets present in a cloud are the same size. The larger droplets fall faster than smaller ones, so that a bigger drop will overtake and absorb all the smaller ones lying in its downward path, unless an air current forms around the falling drop that causes part of the smaller droplets suspended in the air to be bypassed, so that the actual agglomeration is smaller than it might be. Nevertheless, a big cloud whose droplets have a long way to fall and are sufficient in number will still produce raindrops in this way. In addition, in a rising cloud there will be considerable upcurrents by which big drops will be held in suspension while smaller ones are caught up and merge with the big ones. If the individual drop gets too big it will lose its stability and burst, and the fragments will again be caught up and merged until they eventually fall from the cloud in big drops. This kind of raindrop formation, which is sometimes referred to as the **Langmuir chain reaction** is most likely to occur in the hot, humid atmosphere of the tropics.

In the temperate zone all the clouds from which rain falls in big drops consist, at the higher levels, of ice crystals. In the temperate zone the ice phase is obviously a precondition for formation of rain in big drops. Upon formation of the ice crystals they go on growing rapidly and in consequence fall faster and absorb additional supercooled droplets on their way down, which contribute in turn to the growth of the crystals. The more droplets that adhere to it, the faster a crystal falls and the less time remains for the formation of the crystal form, so that out of the original hexagonal crystal a shapeless, unwidely lump of ice is formed, which melts at less than the atmospheric 0°C limit (and so at temperatures above 0°C) and becomes a raindrop.

The adhesion of supercooled water droplets to ice crystals is considerably more effective than the process described by Langmuir. **Hail** is formed when as a result of strong upcurrents the grains of sleet are held in suspension longer and the ice particles become so large, owing to the adhesion of supercooled droplets, that they remain in the form of ice at temperatures of over 0°C.

Average rain drops have a diameter of 1000 to 2000 microns, but they can reach maximum diameter of 7000 microns. Above this size, they become unstable and break into smaller drops while falling.

Snow is produced in clouds that are a mixture of ice crystals and supercooled water droplets. The falling crystals serve as nuclei for formation of shapeless lump as described above. When the underlying air is below freezing temperature, the lumps fall to the ground in solid form. If the raindrops formed encounter a cold layer during descent, they freeze into pellets of ice that fall down as sleet. Sleet is also sometimes a mixture of snow and rain.

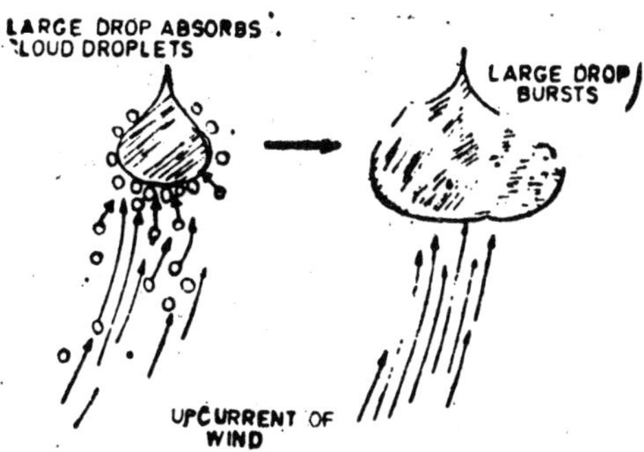

Rain Formation in Tropics.

There are three major causes of precipitation, which primarily results from further rising up of water saturated clouds. **Orographic precipitation** results when cool air mass encounters a mountain and the air is forced upward to undergo further condensation and final precipitation. **Convectional precipitation** occurs during summer when warm humid air rises up suddenly from ground level and cool air is forced down. The rising humid air on eventual cooling brings precipitation. **Frontal precipitation** results when a warm moist air encounters a mass of cold air (cold front) and rises up, being highter than cold air, when the air masses converge around a low pressure centre in anticlockwise direction (in northern hemisphere) or clockwise direction in southern hemisphere. The warm air is forced up spirally and ultimately develops into precipitation. This is a common phenomenon in tropical climates and is called **cyclonic precipitation.** When large in magnitude, it results in tropical cyclones which bring down very heavy precipitation.

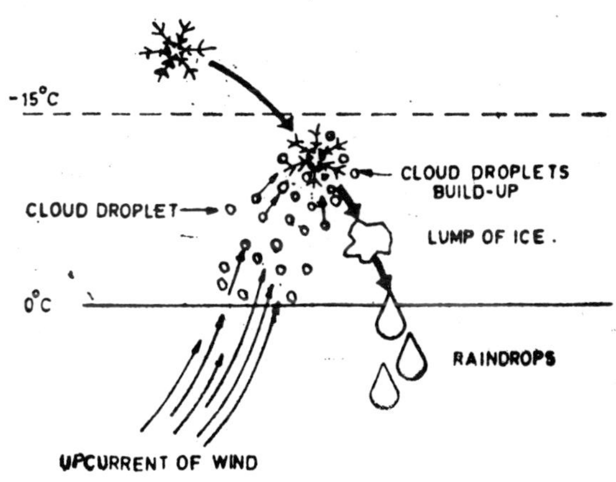

Rain Formation in Temperate Zone.

LIGHTNING AND THUNDER

The power of a lightning strike is awe inspiring: 100,00-40,000 amps of current surge to Earth down a narrow air channel which reaches temperatures of 30,000°C — over five times the temperature at the surface of the Sun. But lightning doesn't just strike downwards. A lightning flash consists of alternating upwards and downwards strokes. Each stroke may be over 30 kilometres long under certain conditions. Although 70-80 per cent of the people who are struck survive, around 100 people are killed directly by forked lightning every year in the United States alone.

Lightning is a massive electrical discharge which neutralizes the charges that build up in a thunder cloud. The discharge either occurs within the cloud as sheet lightning, or between the cloud and the ground as forked lightning. The flash is the same in both cases, and sheet lightning only appears to be different because of the intervening cloud. At any time there are 1,500-2,000 thunderstorms active on our planet, triggering some 6,000 lightning flashes a minute, the majority of which are sheet lightning. Nearly 250 years after Benjamin Franklin, the American scientist and statesman, proved that lightning was a form of electricity, scientist still lack a complete understanding of how it works.

To initiate the avalanche of electrons that marks the start of a lightning stroke, a build-up of electrical charge (potential difference) of about a million volts/metre is needed in the thunder cloud. Incredibly, the thunderstorm generates this voltage potential in only half an hour, as a result of strong adjacent rising and falling air currents.

Large falling raindrops, hailstones and ice pellets frequently touch and collide with smaller water droplets and ice crystals being swept aloft, and electrical charge is transferred. The falling stream gains a negative charge, while the rising stream gains a positive charge, resulting in an accumulation of negative charges at the bottom of the cloud and positive charges at the top.

When the lightning strikes it travels along an air channel 1-5 centimetres (0.4-2 inches) in diameter. The evidence for this comes from lighting strikes on sand, which fuses into a hollow tube of silica glass known as *fulgurite*. When lightning strikes a moist surface such as a tree, road or wall, the heating effect is so great that the moisture boils explosively and may blow the object to pieces.

Once an electrical potential of a million volts/m has been created in a thunder cloud, the lightning process begins. A stream of electrons flows down, colliding with air molecules and freeing more electrons, and in the process giving the air molecules a positive charge (ionizing the air). This intermittent low current discharge forms a highly branched pathway and is called the stepped leader. As the leading branches of the stepped leader, carrying large negative charges, near the ground they induce short upward streamers of positive electrical charges from good conducting points on the ground. When a branch of the stepped leader contacts an upward positive streamer, a complete channel of ionized air has been created. This allows a huge positive current called the return stroke to flow upward into the cloud in the form of a bright lightning stroke. The different strokes have been given different colours to distinguish between them. In nature, all lightning is colourless. The return stroke causes the first of the shock waves we hear as thunder. The flash effectively reaches the eye instantaneously, whereas the sound of thunder travels at approximately 330 m/s. So multiply the number of seconds between the flash and the thunder by 330 to find the distance in metres.

Lightning conductors generate a strong positive streamer which encourages electrical contact with an approaching stepped leader. Consequently, potential lightning strikes within 50-100 m of a building are attracted to a lightning conductor. Return strokes and dart leaders are safely routed to the earth along a wide copper strip which has one end buried in the ground.

Thunder is caused by the narrow lightning stroke heating the column of air surrounding it to around 30,000°C, expanding it explosively at supersonic speeds under a force of 10 —100 times normal atmospheric pressure. The immense shock wave becomes a sound wave within about a metre, producing the sound of thunder.

The Dangers of Lightning

The downward flowing pathway of electrons, or stepped leader, of lightning always seeks out the best conducting route to the ground. Tall buildings, television aerials and trees are struck most frequently, but people outdoors also face a serious risk, and the danger is increased if they are holding a metal object such as a golf club or umbrella. One of the safest places to be if lightning strikes is inside a vehicle. Since the current flows safely through the metal of the bodywork, around the occupants, before earthing to the ground across the wet, often steel-wired tyres. Taking shelter beneath a tall, isolated tree is making a mistake: the lightning is initially attracted to the tree because of its height, but may side-flash through or over the body of anyone underneath because the human body offers a better path. About a quarter of all lightning victims are struck while misguidedly standing under a tree.

There are sometimes signs that lightning is about to strike. Positive electrical charges streaming upwards from pointed objects such as trees and church spires may be visible as faint, luminous glows and may produce buzzing noises. People's hair may stand on end because of the same effect, and if this happens if is probably a good idea to leave the area immediately.

GREENHOUSE EFFECT

A warming effect produced by the presence of carbon dioxide, CFCs, methane, nitrous oxide and water vapour in the lower atmosphere.

How the Greenhouse Effect Works

The sun is the most important source of energy for the earth, sending a spectrum of radiation to the planet, as X-rays ultra-violet and visible light, infra-red radiation, microwaves and radiowaves.

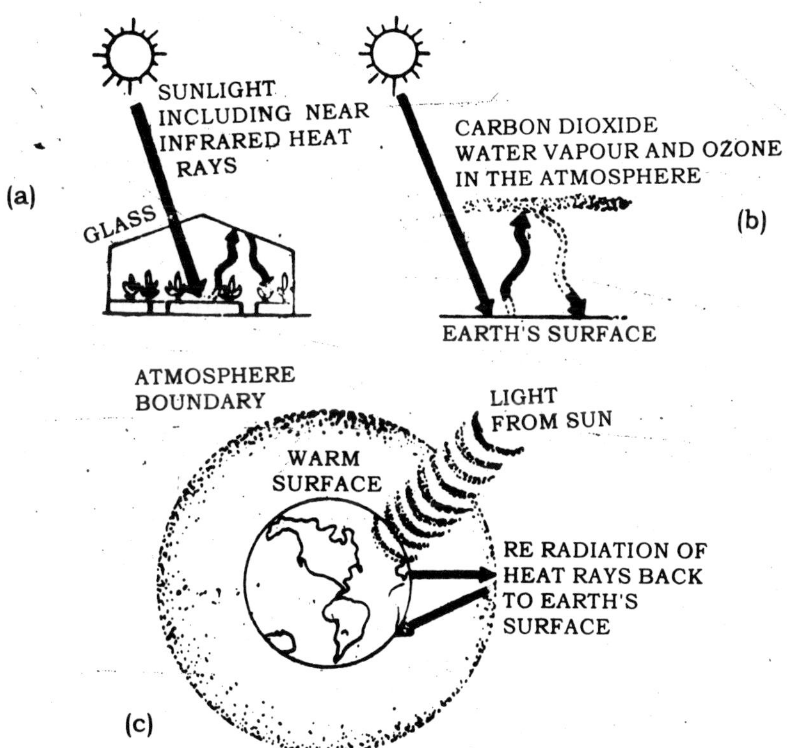

Greenhouse effect. (a) Heat from the sun readily penetrates the glass roof of a greenhouse in the form of near-infrared rays. These are absorbed by soil, plants, and other objects and reradiated as longer wavelength heat rays. However, the glass is not as transparent to these radiations of longer wavelength; thus much of the heat is held inside the greenhouse. (b) In a similar way, the atmospheric carbon dioxide, water, and ozone are transparent to the near-infrared rays of the sun but absorb the reflected longer wavelength heat rays, thus (c) holding the heat within the boundaries of the earth's atmosphere and having a warming effect.

Carbon dioxide, which is present in small quantities in the earth's atmosphere, allows higher energy (visible light, short wave length) to pass through it. The light waves are absorbed by the earth, which in turn emits radiation, but mostly of a longer wavelength—as infra-red radiation. Carbon dioxide (CO_2) absorbs infra-red radiation, or reflects it back to the earth.

CO_2 thus works in the same way as glass in a greenhouse, allowing solar radiation to pass through it, but then holding back the longer wavelength radiation that attempts to return. The result is a warmer temperature in the greenhouse than outside it.

This is a natural process, and an essential system for maintaining the earth's temperature. Without carbon dioxide in the atmosphere, the earth's surface would cool very rapidly at night, much more rapidly than at present. On the other hand, increasing or reducing the amount of carbon dioxide in the atmosphere could increase or slow down the speed at which this greenhouse effect occurs.

The Greenhouse Gases

It is not just carbon dioxide that contributes to the greenhouse effect: there are four other pollutants that contribute to global warming.

Carbon Dioxide

CO_2 is released into the air by burning coal, oil or gas, with coal-burning being by far the largest contributor. About 13 per cent of the gas emitted from coal-burning power stations, for instance, is CO_2. Equipment exists to remove CO_2 from these emissions. If the whole world used the technology, emissions would drop by 30 per cent.

Chlorofluorocarbons (CFCs)

CFCs' contribution to the greenhouse effect could be the second largest (after CO_2) in the next 50 years. Restriction in production of CFCs seems likely in the future, as this is the best way of preventing their build-up in the atmosphere (see Ozone Layer).

Ozone

The problem with this gas occur in two areas: in the troposphere, near the earth's surface, and in the stratosphere, higher up. The higher concentrations form the Ozone Layer, which shields the earth from harmful ultraviolet rays. Depletion of ozone at this height would result in greater entry of ultraviolet rays to earth, increasing the energy reaching the earth's surface, and assisting the greenhouse effect. Prevention of this occurrence would entail reduction of CFC emission.

Build-up of ozone in the troposphere is the result of emissions of nitrogen oxides and hydrocarbons (see Acid Rain), primarily from power stations, vehicles and the use of solvents. There is evidence that tropospheric ozone is increasing over North America and Europe: it has been calculated that a doubling in this concentration throughout the entire troposphere would lead to a 1°C increase in surface air temperature.

Nitrous Oxide

Tropospheric nitrous oxide (N_2O) has increased over the last few years. Caused by extensive use of nitrogen fertilizer, by combustion of wood and fossil fuels, and partly by the chemical industry. Emissions of N_2O are likely to increase less rapidly as a result of fuel consumption trends and limits on the amount of land available for cultivation. On the other hand, N_2O can stay in the atmosphere for about 170 years, so concentrations of the gas are expected to increase by a significant proportion during the next few decades.

Methane

Released in the production of coal and natural gas, by burning vegetation and as a result of intensive cattle farming, methane, concentrations began to increase 300 years ago but have accelerated over the last 20 years, increasing in the atmosphere by 1 to 2 per cent per year. There is considerable uncertainty over the reasons for this increase.

Carbon Dioxide Build-up

Carbon dioxide forms a very small part of the earth's atmosphere, which consists roughly of 78 per cent nitrogen, 21 per cent oxygen, 0.9 per cent argon, and 0.03 per cent carbon dioxide. Although small in percentage terms, this is still an enormous amount, approximately 2,600 billion tonnes of CO_2.

The gas emitted into the atmosphere partly by burning wood but primarily by burning fossil fuels (*i.e.*, coal and oil). At the moment, about 5 billion tonnes of carbon a year are sent into the atmosphere as a result of this process, equivalent to 18 billion tonnes of carbon dioxide. A large proportion is absorbed by the oceans, but about 8 billion tonnes of CO_2 are not assimilated in this way, and remain in the air. This is a significant annual addition to the CO_2 already in the atmosphere.

Concentrations of carbon dioxide are usually measured by checking how much CO_2 there is in a million molecules of air. Beginning in 1957, the Mauna Loa Climate Observatory at Hawaii, USA collected data which revealed a systematic increase in atmospheric carbon dioxide. To date, the change from 290 parts per million (ppm) in 1880 to 352 parts per million in 1989 represents more than a 20 per cent increase over the course of the past century.

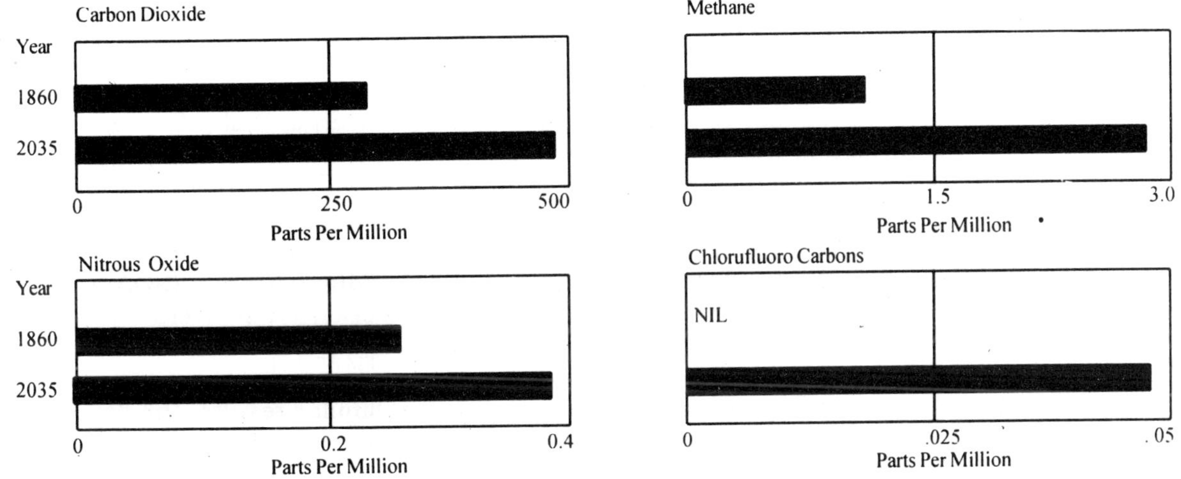

The Projected Growth of Greenhouse Gases in the Atmosphere.

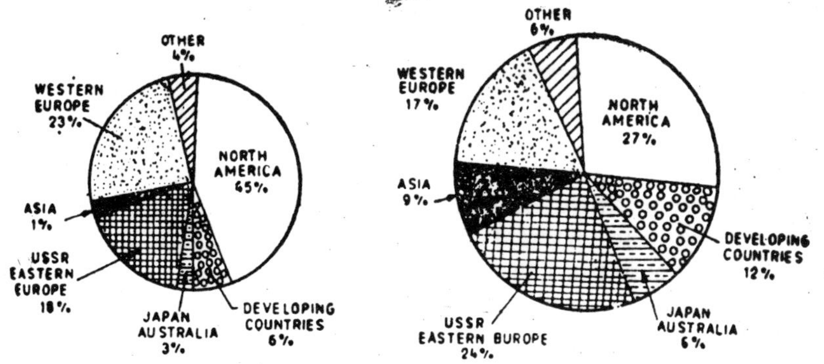

The Changing Pattern of Global Carbon Dioxide Emissions.

Global Warming Trends

In this final decade of the 20th century the greenhouse effect and the prospects of global warming is the subject of scientific and political controversy and a cause for widespread concern.

Efforts to unravel the climate consequences of increasing carbon dioxide emissions have been going on since long. The world famous mathematician John Von Neumann at the Institute for Advanced study in Princeton, N.J. (USA) made the first attempt to represent the atmosphere mathematically on digital computers in 1950. In 1963, an unusual laboratory of the National Oceanic and Atmospheric Administration (NOAA) was established at Princeton University under the leadership of Joseph Smagorinsky. The laboratory was devoted to the mathematical modeling of the atmosphere using the largest and the fastest digital computers available. Called the Geophysical Fluid Dynamics Laboratory, the center kept researchers from many nations interested in this new approach to the study of the atmosphere. Among them was a young Japanese scientist, Syukuro Manabe who developed the first climate model in collaboration with his colleague Richard T. Wetherald in the 1960's. In 1975 they calculated that a doubling of the carbon dioxide content of the atmosphere would produce a global climate warming of about 3°C averaged over the surface of the earth. This calculation has been verified in many different laboratories and has not changed substantially. Studies have since been undertaken in many parts of the world, including Europe and the Soviet Union.

A quantitative assessment of global warming trends can be made on the basis of historical temperature records. Unfortunately, such records are difficult to come by. Today global temperature measurements are compiled through the World Weather Watch System, a global cooperative network of national meteorological services. Observations in previous centuries were largely compiled by individual observes working without coordination.

About 10 years ago, in the face of growing global climate concerns, Philip D. Jones and Tom M.L. Wigley and coworkers at the Climatic Research Unit of the University of East Anglia, Norwich, England, initiated a project to collect and analyse, once and for all, every available historical temperature record. This was sponsored by the U.S. Department of Energy. Raymond S. Bradley of the University of Massachusetts at Amherst and Henry F. Diaz of the Environmental Resources Laboratory of the National Oceanic and Atmospheric Administration (NOAA), USA were their collaborators. Their assessment is that the earth has experienced an overall warming trend of half a degree celcius since the late 19th century.

Although the mathematical models of all the groups yielded similar results, the details of the geographic distribution of climate changes differed from one model to the other. All projected that an increase in carbon dioxide would bring about a gradual warming, but the timing of this warming would depend on the rate of global energy use. They all agreed that if reasonable assumptions were made about future global energy consumption, it would be around the middle of the next century that the carbon dioxide content of the atmosphere might double. Just how much this doubling of carbon dioxide would increase temperatures, however, varied greatly from model to model. Some predicted as little as 1°C increase, whereas others predicted increases of as much as 5°C.

These projected temperature changes may appear innocuous because variations of this magnitude are experienced in the normal course of daily and seasonal weather. Their full implications can be appreciated by noting that it took only 1°C average decrease in temperatures in Europe to cause the run of several frigid centuries (from the 1400's to the 1800's) known as the Little Ice Age. Five degrees C is believed to be the difference in temperature that separates the end of the last great ice age 12,000 years ago from the present. Further, the projections indicate that the Northern Hemisphere would experience in just a half century an unprecedented temperature change, 10 to 50 times faster than the change since the last ice age.

Since the carbon dioxide content of the atmosphere has increased by more than 20 per cent over the past century, we ought to be able to detect the climate warming in the global temperature record during the same period. Researchers have sought to do this, but it is a much more difficult task than it first appears. The problem is that climate is always in a state of natural fluctuation. Separating out the changes that are caused by increasing carbon dioxide from the natural changes is difficult. Moreover, the climatic temperature record is based on scattered and irregular observations not taken specifically for the purposes of determining climatic conditions.

Even so, careful analysis of these temperature records by scientists in the U.S. and in the U.K. sought to detect whether a climate warming has occurred and whether such warming is consistent with the prediction of the models. The prevailing view is that the climatic record over the past century for the entire globe reveals a net increase in temperature ranging from 0.3 to 0.8°C. But set against this conclusion is the disturbing result that similar increases in temperature cannot be detected over the past century in the U.S., where observations are numerous and accurate.

Even if the temperature rise is real, a puzzle remains that workers have not been able to solve: Is the rise in global temperatures a natural fluctuation or a result of the increase in greenhouse gases? All that can be said is that the observed increase is consistent with the lower end of the temperature increases predicted by the computer models. Consequently, the temperature records, as well as the predictions of mathematical models provides support to both viewpoints: Those who believe the evidence warrants action now and to those who believe the evidence is too weak.

The Consequences

Serious Climate Changes

If climate change in a historical fact and a future inevitability, why worry? The argument is often put forward that people have survived climate change in the past and will continue to do so. Moreover, if civilization adds its own influences to natural climatic trends, how do we know the result won't be an improvement?

The response to this has several parts. First, to talk merely of survival as a species is to evade the issue of the degree of social disruption and human suffering that have accompanied regional climatic change in recent history and, presumably, at times in the more distant past.

Today, moreover, the absolute magnitude of the human misery that could be caused by climate change is much large than it ever was because the human populations in virtuality all regions of the world are much large than they were before. Far more people would be affected by any given regional climatic change, and there is not likely to be any room in adjacent regions into which they could flee.

The part of the ecosphere most sensitive to climatic change —that is, the element most easily disrupted to the point of directly influencing human well-being—is agriculture. Farmers know very well that a year's crop can be ruined by rain that comes too early or too late, or in too great quantity or too little, by too much hot weather or too much cold, by early frosts, and by other vagaries of weather.

The consequences of changed climatic conditions can be seen in historical records. One of the recent climatic events was the disastrous drought of the summer of 1988 in USA. During this drought, one of the worst on record, the water in the Mississippi River fell so low that navigation was impossible over long stretches, urban water supplies were threatened and crops throughout the grain belt were devastated. Both officials and the public wondered whether this was the greenhouse effect manifest. Records show that in the U.S. five of the years of the 1980's were among the hottest on record, and the average temperature for the decade as a whole was the warmest since instrumental records have been kept. The temperature went up to 32°C. The worst effected was the corn belt. Leaves of corn plants curled up to conserve water. Pollens missed the flowers and consequently no corn was produced.

In fact we do not know enough to predict the severity of the consequences, because the warming would not be uniform over the surface of the earth. Some parts of the earth would become warmer, some wetter and some drier. On the basis of the evidence at hand climatologists are not able to predict who would benefit and who would lose in such a global redistribution of so-called climatic resources.

Some aspects of global climate warming would be greatly beneficial according to agricultural researchers. Increased carbon dioxide will enhance more active photosynthesis and improve crop growth. About the lowered plant requirements for water in a CO_2—enhanced atmosphere nothing can be said with certainity. Models do, however, agree that the polar regions of the world would undergo greater increases in temperature than would the tropics. Some of the projections of temperature increases in polar areas are startling in their magnitude, predicting as much as 10°C (by 2050) on the average in the Northern Hemisphere and only slight increases in tropical regions. What are the general consequences of such a change in the temperature difference between equatorial and polar regions? If arctic regions were to undergo significantly greater warming than equatorial regions and if precipitation belts were to move farther north, countries in the north temperate and polar zones would probably benefit greatly. Their growing season would lengthen, and their precipitation would increase. With suitable soils, agriculture might thrive.

Seal Level Rise

The polar ice caps would melt resulting in rise in sea level. Estimates of the resulting rise in sea level vary between an increase of 30 and 150 centimetres by the year 2050. If the latter increase occurred, many major cities would be inundated: the projected rises would occur 100 times faster than has occurred globally before.

When the sea rises, coastal areas are eroded or flooded. Populations could be defended by sea defences, but at huge cost. The US government estimates that, if current trends continue, it will have to spend $ 3 billion to protect its coastline.

Other countries more at risk do not have this kind of financial solution at their disposal. A one metre sea-level rise would flood over 10 per cent of Bangladesh. Other countries like Egypt and India would suffer on a widespread basis. Half the Commonwealth countries in the Pacific—all small island nations—are at risk. Some would disappear.

As it is the fertile, food-growing delta areas which are most at risk, large food-growing regions would be lost. Drinking water would be poisoned because of salination. Low-lying island nations such as the Maldives could be completely wiped out creating the world' first **"Greenhouse Refugees"**. Fires, floods, and prolonged droughts may be more common. These are however, speculations.

Another, somewhat more speculative respect in which climate change could lead to great increases in human misery is by altering the abundance and the geographical distribution of various disease-producing organisms. As is the case with crops, the degree to with such organisms and the other organisms that transport them (vectors) thrive is governed by such environmental conditions as temperature and moisture, in terms of both averages and extremes. Changes in climatic patterns therefore might give certain of those organisms access to human populations that have no prior evolutionary experience with them and hence little or no resistance to them. Alternatively, such changes might remove checks on the abundance of organisms pre-existing in area, to the extent that a previously minor hazard becomes a plague. This is true of pathogens that attack crops and trees, as well as those that attack people.

The Alternatives

If the climate, changes the expectation is that it will do so gradually. We should be able to see the initial evidence of coastal inundation in an increasing frequency of high tides and in the undercutting of seacoasts. Climate warming itself should be evident in a rising frequency of heat waves or in other weather anomalies. The effects of a global climate warming are likely to take 30 to 50 years to become serious, and that is a long enough span in which actions to adapt to these changes should be possible.

Where might we start? According to experts energy conservation and efficiency should be the first priority. Achieving greater energy efficiency justifies itself in economic terms, increased energy efficiency would also reduce urban air pollution and acid precipitation. Shifts in the fossil-fuel mix from coal and oil to natural gas could significantly reduce carbon dioxide emissions per thermal unit. Technology is also available for more efficient power generation and for increased gasoline mileage.

Major investments in non fossil-energy sources are desirable. The circumstances favour significant new investments in possibly safe, publicly acceptable nuclear power. Further development of forms of solar energy—photovoltaics or biomass, for example—makes good sense. Reforestation and forest preservation yields many ecological and climatic benefits. Research aimed at producing stress and disease-resistant crops would also be recommended.

It is likely that humanity will have to adapt to some climate changes. Modes of adaptation by society have not been well studied. Individuals, corporations and communities can adapt to climatic changes in numerous ways. Farmers can change crops, water use can be regulated and management practices can be altered.

Other modes of adaptation would be needed if climate changes were severe. Sea-level rise, which is one of the predicted consequence of a climate warming, might inundate low-lying coastal areas and cause salt water to intrude into freshwater bodies. If this were to occur, society would have to decide whether to invest in protective structures along coasts or adapt by changing land-use patterns. The North Sea dikes in the Netherlands are an outstanding example of adaptation to relative rise in sea level.

The developing nations are not severely affected so far. They can escape the severity of such climatic changes by not adapting the same technology as the Western countris. the cjoice may be harder to make such as finding alternatives to freons which might be very expensive. However, a simple check could be to control the population growth.

Our global environment is under attack on many fronts. Climate warming is but one, perhaps the most complex, of these issues. If the changes occurring in our atmosphere are likely to cause consequences, we must understand the problems and promote sensible policies to remedy them.

HURRICANES AND TORNADOES

A cyclone that struck Bangladesh in 1970 caused half a million deaths. "Cyclones" and "hurricanes" are, in fact, exactly the same phenomena—the most violent and destructive storm on Earth. Although tornadoes are less than one hundredth the size of hurricanes their winds can reach similarly devastating speeds. Two-thirds of the world's tornadoes last less than 3 minutes, however, and under 2 per cent are true killer tornadoes, creating swathes of destruction hundreds of metres wide and over 160 kilometres long.

Most tornadoes have wind speeds of about 180 km/h, which can damage roofs, uproot trees and fill the air with lethal debris. Some tornadoes exhibit wind speeds up to 470 km/h (290 mph) and can crush sturdy buildings and toss vehicles, people and animals through the air.

Tornadoes are mainly found in the mid latitudes. They are produced by thunderstorms, such as the cyclones of Asia and Japan. The USA suffers many tornadoes, especially at its centre, where humid air streaming north from the Gulf of Mexico meets the cold, dry air moving westwards from the Rocky Mountains. The basic reason for a torando's spin is the Coriolis force caused by the Earth's rotation. The thermal does not begin to spin at ground level, but within the cloud. The rotation gradually extends to the surface as a funnel stretching from the base of the parent cloud.

The storm that are collectively known to science as tropical cyclones are given many different popular names in different areas of the world. In the Atlantic and eastern Pacific Oceans. For example, they are called hurricanes. In the western Pacific they are known as typhoons. Around the Philippines they are called baguios. And in the Indian Ocean they are cyclones. Only when wind speeds in a tropical storm exceed 120 km/h, should it be called any of these names.

Tropical cyclones are powered by the latent heat released when vast quantities of water—evaporated from the oceans—condense to form towering clouds. A seasonal variation in cyclone formation occurs because high sea-surface temperatures are needed to encourage strong evaporation. Tropical cyclones do not form close to the equator because the Coriolis force (due to the planet's rotation) is not powerful enough to cause the air to spiral wildly. The light shaded areas suffer on average 0.1-1 storm of force 8 or stronger/year. The darker, ringed, areas have over 1 such storm/year.

Hurricane Force

Hurricanes develop over ocean water that is at or above a temperature of 27°C. A hurricane is composed of bands of thunder-storms and cumulus which spiral around the storm centre, the calm, cloud- free eye, which is typically 30-40 kilometres across. In the cloud-wall around the eye exceptional wind speeds occur: the smaller the eye, the higher the wind speeds. Hurricanes vary greatly in their intensity, and in meteorological terms are classified according to their "damage potential" on a five-point scale, ranging from minimal to catastrophic. Such measurement is based on air pressure in a storm's eye, because the lower the pressure, the greater the wind speeds in the hurricane. A hurricane that measures 5 on the scale is characterized by air pressure below 920 millibars, producing wind speeds of over 250 kilometres/hour and a coastal storm surge in the sea of over 5.5 metres above the normal level. Fortunately, fewer than 1 per cent of all hurricanes fall within this category.

One hurricane that did, however, was Hurricane Gilbert, which swept through the Gulf of Mexico in September 1988. As this hurricane, 1,500 kilometres in diameter, passed over Jamaica it generated as much energy as the country would need for the next 1,000 years, at current rates of consumption. Although Hurricane Gilbert travelled at a mere 18-25 kilometres/hour, the wind around the eye at the centre, where air pressure was a record low of 885 millibars, reached speeds of over 320 kilometres/hour.

A tidal surge of 6 metres, flash floods from torrential rainfall that totalled 250-380 millimetres in a few hours, and the creation of 24 tornadoes as the storm reached land all together contributed to the devastation caused: 318 people lost their lives, 100,000 people were evacuated from the Mexican coast, and 500,000 people were rendered homeless in Jamaica.

Forecasting the power and path of a hurricane remains problematical, although satellite monitoring since the 1970s has allowed far more of a warming to be given to those in the path than was once the case.

Taming a Hurricane

When a hurricane travels across the ocean and finally reaches the coast it begins to die as its supply of energy from the warmer ocean is cut off, and as the increased surface friction upsets the storm's circulation and reduces the wind speeds. Even the terrible Hurricane Gilbert was eventually laid low.

An American attempt to weaken hurricane winds by seeding a storm with silver iodide or dry ice some distance outside the eye-wall clouds — an enterprise called Project Stormfury—met with limited success. The intention was to produce rainfall, so releasing latent heat that would otherwise have sustained the high wind speeds in the eye-wall. The decrease in wind speeds should in theory have caused the hurricane simply to fizzle out. Instead, a new eye-wall was created farther out, with lower wind speeds.

ANTARCTICA

Antarctica is a centre of great attraction for the last few decades, and not just a seventh continent, merely covered with ice. Antarctica is the last great wilderness in the world—an enormous land mass which, because of its inaccessibility and hostile climate, has remained virtually unchange for millions of years. The world's only example of a large pristine environment, Antarctica has attracted a number of territorial claims this century, with reasons varying from national pride to Antarctica's possible mineral reserves.

In the last 30 years, treaties and conventions have established a workable system which allows scientific research while limiting the types of national and international activities which may take place in the area. The evidence of increasing environmental disturbance has encouraged those who wish to strengthen the conservation of Antarctica. The environmental movement has put forward a suggestion that Antarctica should become a World Park: a proposal that has received support from several countries. 'Antarctica' is not one land mass. The over lying huge sheet of ice upto 4.5 km thick binds together ***Greater Antarctica*** to the smaller land bodies known collectively as ***Lesser Antarctica*** to give the appearance of a single continent.

Antarctica is 99 per cent covered by ice, and contains most of the planet's fresh water reserves. The volume of ice is estimated at roughly 30 million cubic km formed by the compacted accumulation of 100,000 years of snow. If this ice sheet were to melt, the levels of the world's oceans would rise as much as 65 metres.

It is the coldest place in the world, with average annual temperatures of minus 49 degrees C near the South Pole. Central region of Antarctica is technically a desert as only 15 centimetres of snow falls each year, equivalent to 70 mm of rain. It has an average height three times greater than that of other continents.

Colder than the Arctic because of its height, the size of the ice sheet and its southerly position which shelters it from the sun. Antarctica has a significant impact on climate in the southern hemisphere, cooling the surrounding air and ocean and thus contributing to the climatic balance maintained by the equatorial and polar regions.

Abundant wildlife abounds in the region, based not on the continent but in the surrounding Southern Ocean, which accounts for about 10 per cent of the world's seas. The northern boundary of this ocean sinks below the warmer surrounding oceans, with an abrupt change in temperature of 2 to 3 degrees C, resulting in conditions which encourage the formation of rich supplies of oxygen and chemical nutrients. Thus the cold Southern Ocean is able to support an abundance of plants and animals.

Antarctica has remained untouched over the centuries, and provides an important monitoring zone to determine the spread of world air pollution and the extent of pollutants such as PCBs. It is also, for other reasons, the ideal site for measuring ozone levels in the upper air.

The harshness of the environment, however, means that scientific research, and any exploitation of the continent at all, could disturb the delicate ecosystems that have struggled into existence. Antarctica supports a limited variety of land species, and a wealth of marine species. Over-exploitation carries, as always, the risk of ecosystem destruction; but the possibilities of regeneration are much slimmer than in other parts of the world.

Current geological thought holds that Antarctica, South America, Africa, India and Australia were originally joined together in a super-continent called Gondwana. About 1800 million years ago, the continents started to separate and move apart, a process that is still continuing. It is thus logical to assume that the geological features of the different continents would be continue in Antarctica, and mineral resources found in India, Africa and South America would also be located in Antarctica.

Substantial deposits of iron in the Prince Charles mountains, and possibly the world's largest reserves of coal in the Trans-Antarctic mountains, are expected. The Dufck Massif, which covers 50,000 sq. km of Antarctica, is a geological formation known as a layered intrusion. Some of the layered intrusions found on other continents, such as the Bushveld complex in South Africa, have proved to be among the richest mineral sites in the world. A number of countries the UK, France, Norway, Australia, New Zealand, Argentina and Chile all made formal claims to possession of sectors of Antarctica, with the UK, Argentina and Chile claiming overlapping territories. The efforts of the international scientific community followed by a diplomatic initiative resulted in ***Antarctic Treaty***, which was signed in 1959 and came into effect in 1961, established a number of provisions: Antarctica is to be used only for peaceful purposes, though military personnel may be used for scientific tasks, nuclear explosions and disposal of radioactive waste are banned; there is to be freedom of scientific investigation, but with all stations and equipment open to inspection by any Treaty member; territorial claims are not recognized, disputed, or established by the Treaty.

Responsibility for managing Antarctica rests with the twenty nations called consultative parties with full decision-making status under the Treaty.

Two conventions have been concluded under the Antarctic Treaty : the **Convention for the conservation of Antarctic Seals** (1972) and the **Convention on the Conservation of Antarctic Marine Living Resources** (1980). The later convention, known as CA MLR, holds annual sessions and is charged with considering the ecosystem as a whole, rather than just individual species. CCAMLR has been criticized by the environment movement for its slowness in moving to protect fish stocks. Antarctic cod has been used as an illustration of the dangers of overfishing, as stock have been heavily depleted since 1969, and now declined to the extent of becoming a protected species. The Antarctic icefish, the finfish have also shown a continuing decline in stocks in recent years.

CCAMLR's inability to protect these species has led to doubts about whether it will effectively manage the krill fishery. The effect of krill depletion in the Antarctic foodchain is impossible to guess; some scientists believe that the declining population of elephant seals in the Southern Ocean is linked to the industrial fishing of krill. Until the fishing nations, change their philosophy within CCAMLR, adequate protection of commercially—targeted marine species seems impossible.

There have also been problems on land. The USA operated a pressurized-water nuclear reactor at its McMurdo base for ten years: when it was demolished in the early 1970s, over ten thousand cubic metres of contaminated soil and rock had to be removed and shipped to the USA for dumping. It took six years for the site to be cleaned up.

Australia's Casey station dumped huge quantities of rubbish. Rubbish was dumped there twice daily, with no attempt to separate combustible, toxic and hazardous materials. The lichens and mosses on the hill around Casey are the most extensive and complex examples of plant life in continental Antarctica, which may suffer from the activity.

The environmental movement supports the notion, first put forward at the Second Conference on National Parks in 1972, that Antartica should become the first World Park, under the auspices of the United Nations. It was suggested that: the wilderness values of Antarctica should be protected; there should be complete protection for Antarctic wildlife (though limited fishing would be permissible); Antarctica should remain a zone of limited scientific activity, with cooperation and coordination between scientists of all nations; Antarctica should remain a zone of peace, free of nuclear and other weapons.

Although initially supported by New Zealand and Chile, the idea of Antarctica as a World Park has never been discussed by the Treaty States.

HOW SEASONS CHANGE

Parts of Siberia can vary in temperature from -78°C in winter to over 36°C in summer. This is the world's most extreme temperature variation, and the farther away from the equator and the closer to the poles, the more life is influenced by seasonal change. Plants reproduce on a seasonal cycle. Some animals change fur colour in winter, others migrate to escape winter's ravages. Many Arctic and temperate animal, however, reduce all bodily functions to the absolute minimum to sustain life, when, to the casual observer, a hibernating animal appears dead.

We have seasons mainly because the Earth's rotational axis tilts 23.5° from the perpendicular to its solar orbital plane. Since the tilt is constant as the Earth orbits the Sun, the angle at which the Sun's rays strike various latitudes changes through the year. Summer daylight hours are longer and there is a stronger warming effect, for the Sun's rays strike the atmosphere less obliquely than in winter, when much of their heat dissipates.

Polar zones experience the greatest annual variation in attitude relative to the Sun and thus the greatest seasonal change, from long, dark and cold winters to short, lukewarm summers. The contrast between seasons gradually becomes less marked closer to the equator, where tropical zones, which experience minimal change in attitude to the Sun, have little annual temperature change, although they display great seasonal rainfall variation. Between the equatorial and polar extremes, mid-latitude temperate zones exhibit four well-defined seasons.

Each of the four seasons begins and ends at one of the two **solstices** (the shortest and longest days of the year), which are separated by six months, or one of the **equinoxes** (when day and night are equal), also separated by six months. The solstices mark the days when one pole is at its closest position to the Sun and the other is furthest away from it. Equinoxes mark the days when each pole is the same distance from the Sun. For example, spring in the Northern Hemisphere at the spot marked by a cross on the globe commences on the equinox of 20 March, when day and night are equal length, and is followed by the summer solstice on 20 or 21 June, (the dates vary because of leap years), when the North Pole is at its closest to the Sun and the day is the longest of the year, Autumn equinov commences on 22 or 23 September, when day and night are once again equal in length, Winter in the Northern Hemisphere begins on 21 or 22 December, when the day is shortest in the year, This seasonal sequence is reversed in the Southern Hemisphere, where, for example, winter starts on 20 or 21 June when the South Pole is farthest away from the Sun but the North Pole is closest and the Northern Hemisphere summer begins.

The start and finish of each season can thus be defined in very precise astronomical terms by the Earth's position relative to the Sun.

Obviously, the climate does not suddenly change on a specific astronomically determined date — we usually mark seasonal change by factors such as changes in animal behaviour and leaf and flower growth or decay.

In fact, because of the greenhouse effect seasonal temperature extremes occur two months later than purely astronomical factors would suggest. Many other trrestrial factors modify astronomical effects.Temperature and dryness or wetness are substantially modified by such factors as altitude, while the fact that thc sea warms up slowly and loses heat slowly modifies purely astronomical factors and delays the onset of warm or cold periods. Warm currents, such as the Atlantic Gulf Stream, have a general modifying effect on adjacent land massess, resulting in far warmer climates in, for example, parts of northern Scotland, where purely latitudinal factors would suggest much colder winters than is actually the case. In general, the farther away from the sea, the more a climate is determined by purely astronomical factors.

The Earth's solar orbit is nearly circular, but its slight ellipticity means that the Sun is not exactly at the centre of the orbit; thus, at present, the Earth is closest to the Sun in January, reinforcing the Antarctic summer and weakening the effect of winter in the Arctic. The effects of solar radiation on air circulation also affects the weather, producing, for example, phases of high pressure and long spells of hot weather in summer but unpredictable bouts of wind and rain in spring.

The Earth's tilted rotational axis causes the phenomenon of "The Land of the Midnight Sun". Each polar extreme has six months of continuous day and then six months of continuous night. Inside the polar circles the effects is less extreme, the regions alternating between six-month periods of very long nights and short days and very long days and short nights. Regions inside the polar circles experience one day when the Sun never sets.

Sleeping the Winter Away

Extreme seasonal variation demands greater animal adaptability to cope with the changes. Many animals simply migrate to avoid cold and lack of food. Small mammals use insulating snow-trapped air to survive; snow-covered soil may be 10°C warmer than air temperatures: having eaten as much as they can in times of plenty, mice or voles can find enough food beneath the snow to sustain them through the long winter.

One of the most fascinating ways of coping with extreme cold, however, is by sleeping to conserve energy. Large animals, like badgers, sleep much of the time, readily awakening if disturbed. But this winter "doze" uses considerable energy, which must be replenished by hunting. True hibernation is a state of suspended animation when a mammal slows its metabolism, lowering its temperature from perhaps 32°C (90°F)) to 4°C (39°F) and its pulse and breathing rates until barely discernible; a hibernating hedgehog, for example, breathes once every six minutes. As well as minimizing energy needs, a mammal stores energy by gorging itself before winter—dormice resemble furry balls in autumn, and some sheep store food in fatty tail deposits. In fact, a herbivore's fat deposits, not the cold, may trigger hibernation; an artificially warmed animal still hibernates at the same time as wild animals.

A hibernator can be picked up without awakening, but it wakes up if it is so cold it risks freezing solid, shivering violently and even running around to raise its body temperature. Usually, however, it simply stretches occasionally in hibernation (which may help prevent cramp in its muscles), slowly beginning to stir or twitch its tail as spring nears. Fully awake, the mammal is ravenous after losing perhaps half its body weight during its sleep.

The Long, Dry Siesta

Animals such as kangaroo rats reduce water loss by becoming nocturnal in droughts. Others avoid fatal dehydration by *aestivating*, using similar techniques to hibernators to "sleep" through droughts. Male Californian ground squirrels aestivate but the females stay swake to feed and tend the young. The Jersey tiger moth aestivates in Greece and, having hibernated as a caterpillar, spends over half its life asleep. Some lungfish aestivate in dried-up ponds, using a mucus-lined mud cocoon to reduce skin evaporation.

Part III

Noxious Substances in the Atmosphere

- Arsenic
- Asbestos
- Acid Rain
- Aerosol
- Air Pollution
- Carbon Monoxide
- Chlorofluoro Carbon, CFCs
- Hydrogen Sulphide
- Nitrogen Oxides
- Natural Gas
- Ozone
- Sulphur Oxides
- Particulates
- Fly Ash
- Photochemical Smog
- Pollution
- Smog
- Pans
- Metal Toxicity
- Barium
- Beryllium
- Caesium
- Cadmium
- Chromium
- Copper
- Iron
- Lead
- Manganese
- Mercury
- Nickel
- Organomercurial Compund
- Plutonium
- Radon
- Selenium
- Tin
- Tritium
- Uranium
- Vanadium
- Zinc

ARSENIC

A grey, shiny, metallic — looking brittle element. Highly toxic by ingestion and inhalation. A known carcinogen. Tolerance 0.2 mg per cubic metre of air.

Arsenic occurs widely in nature. It is used in alloys, pesticides, wood preservatives and some medical preparations. It was formerly used in paint pigments, but this use discontinued when it was found that, under damp conditions, moulds converted the arsenic to the highly toxic gases, arsine, AsH_3, and trimethyl arsine, $As(CH_3)_3$. Arsenic is a cumulative poison, causing vomiting and abdominal pains prior to death. It may also cause dermatitis and bronchitis, and may be carcinogenic to tissues of the mouth, oesophagus, larynx and bladder. At the cell level, it can uncouple oxidative phosphorylation and compete with phosphorous in metabolic reactions.

Arsenic is concentrated by organisms exposed to it and accumulates along food chains. Accumulation in fish seems to be favoured by increasing salinity. Crabs and lobsters have been especially noted for accumulating high concentrations, but no cases of human poisoning seem to have arisen from this.

ASBESTOS

Asbestos has been known for more than 2000 years. It is a material that has common applications in households, industrial premises and public buildings. Until recently the substance was regarded as harmless: now, however, the environmental hazards of asbestos are well recognised. Most uses of asbestos are now banned or are being phased out, although some continue. Asbestos remains an environmental threat because it can be found in many locations in houses, schools, hospitals and factories.

Types of Asbestos

Asbestos is a generic term describing a number of hydrated silicate minerals. This mineral rock is mined in many parts of the world, especially the Soviet Union (43%), Canada (26%), and Southern Africa (13%). Its attraction from the technical point of view is that it can be divided into millions of fine fibres that are often soft and silky to the touch, but are also strong and resistant to heat and chemical attack. The fibres can be separated into bundles of fibres or woven into fabrics used as reinforcement for cement and plastics, or sprayed on to surfaces.

The danger from asbestos is to be found in the fibres themselves, especially the very fine ones which, invisible to the naked eye, can be fatal when inhaled. For that reason, processes which produce very small airborne fibres are generally the most hazardous.

There are six common types of asbestos, separated into two groups (Table-1). Out of these three are in common industrial use:

crocidolite—'blue' asbestos

amosite—'brown' asbestos

chrysolite—'white' asbestos

The names are misleading, as the colour refers to appearance when the substances are freshly mined. Ageing and heat turn all asbestos a similar colour and only scientific tests can identify a particular type.

Nowadays, about 95 per cent of all asbestos mined is chrystolite, or white asbestos.

TABLE 1. Asbestos Minerals and Nonasbestiform Analogs

Asbestiform	*Non asbestiform*
Serpentine	
Chrysolite	Antigorite
Amphibole	
Anthophyllite asbestos	Anthophylhite
Amosite	Cummingtonite-grunerite
Actinolite asbestos	Actinolite
Tremolite asbestos	Tremolite
Crocidolite	Riebeckite

Source: Rajhans and Sullivan (1981).

Asbestos has been in industrial use since the last century. There are currently about 3000 uses of asbestos in commercial, public and industrial facilities. The most important application is in the manufacture of cement products, such as pipes, shingles, electrical panels, and corrugated or flat sheets. Other uses include floor tiles, gaskets and packings, friction products, paints and sealants, textiles, paper products and asbestos reinforced plastics. The uses to which the different forms of asbestos are put depend mainly on their physico-chemical properties. For example, amosite is used mainly in building products because of its heat-resistant characteristics, whereas the acid resistance of crocidolite makes it useful in the manufacture of batteries and packings for acid pumps. Chrysolite is used in friction material such as brake linings because of its reinforcing qualities and heat stability.

Generally, these products are considered to be safe in everyday use, provided that they are not broken open, drilled or allowed to deteriorate in such a way that fibres are released. Substitutes for asbestos have been developed from many industrial applications.

Dangers of Different Asbestos Types

The situation concerning the relative danger of the three asbestos types is confusing. It is agreed that blue and brown asbestos should be treated as extremely dangerous substances, but the health warnings about white asbestos are not as clear as they could be.

The importing of blue asbestos has been banned in the UK since 1970.

The importation, supply, use in manufacture and marketing of blue or brown asbestos, or products containing them, was prohibited on 1 January 1986.

From 20 March 1986, in response to an EC directive, products containing asbestos have to be labelled 'Warning; contains asbestos—Breathing asbestos dust can be dangerous to health—Follow safety instructions'.

This chain of events appears to carry the implication that white asbestos is safe if used carefully, or is at least a tolerable risk. *Asbestos Materials in Buildings* makes the point that asbestos—cement products, which contain only 10 per cent asbestos, will continue in use as slates, corrugated sheet and cavity-roof decking.

Today most industrialized countries maintain standards of 2 fibers cm^{-3} or less. There is a proposed limit in the UK of 0.2 fibers cm^{-3} for crocidolite and an existing standard of 1.0 fibers cm^{-3}.

Asbestosis is most commonly associated with fibers more than 5 μm in length. Since it is impractical to determine the concentration of fibers shorter than 5 μm, the current recommendations for occupational exposure standards is 0.1 fibers cm^{-3}. This is based not on epidemiologic data but on the presumption that any level of exposure will produce cancer. In order to achieve the 0.1 fiber cm^{-3} limit, plants would need to enclose processing totally, with advanced exhaust ventilation.

This assumes that crocidolite is relatively hazardous because of its rapid uptake and slow clearance.

Asbestos Killer Dust (1979), published by the British Society for Social Responsibility in Science (BSSRS), points our that Denmark, The Netherlands, Finland and Sweden think as the UK does, that blue and brown asbestos are more dangerous than white. On the other hand, again according to BSSRS, West Germany, East Germany, Italy, France, the USA, USSR, Canada, France and Norway all believe the three types of asbestos to be equally dangerous.

It does not appear that UK authorities have acted very quickly to counter the health effects of asbestos. The first cases of asbestos-linked death were reported to the British government in 1906. Regulations concerning the 'safe' use of asbestos were introduced in 1931. The Yorkshire Television documentary about an asbestos sufferer 'Alice—A Fight for Life' was a stimulus to government decision-making when it was broadcast in July 1982. Blue and brown asbestos were formally banned in 1986.

It is not impossible that white asbestos will also eventually be banned.

Asbestos Diseases

There are three main asbestos-related diseases:

Asbestosis, a chronic, progressive pneumoconiosis (Nicholson, 1984). The disease is characterized by fibrosis of the lung parenchyma and is found in 50-88% of occupationally exposed individuals with more than 20 years of service. Damage to lungs is irreversible and may progress following cessation of exposure. All forms of asbestos are apparently capable of inducing asbetsosis in workers.

Lung Cancer, all three types of asbestos appears to act as a promoting agent in the onset of lung cancer. However, because of the lengthy residence time of fibers in the lung the promotional effect does not decrease with time after cessation of exposure, as it may with some chemical agents. In addition, deposition of fibers in the lung can easily precede the onset of other carcinogenic processes because the fibers remain continuously available in the lungs. Moreover, the chance of contracting lung cancer is greatly increased for asbestos workers who smoke.

Mesothelioma, a previously rare cancer of the inner lining of the chest or abdominal wall. Very painful and always fatal, mesothelioma is almost wholly linked to asbestos exposure: 85 per cent of all mesothelioma cases are believed to be asbestos caused. Asbestos fibers are known to reach the chest cavity of exposed individuals and scar the membranes (pleura) that line the chest cavity. The most common effect is to produce a thickening in the exposed area. Pleural plaques may also be formed when calcium salts are deposited in the target area. Although pleural thickening does not necessarily lead to mesothelioma, individuals with pleural disorders have an elevated risk of developing cancer.

All three diseases take a long time to develop-usually ten to twenty years, though this may be up to 40 years in the case of cancer. This may partly explain why it took so long to ban what are particularly dangerous substances. If it had not been for the outbreak of mesothelioma—a rare disease, but of an unusual form, and therefore easy to discern and record—it is arguable that the true dangers of asbestos would not yet have been fully appreciated.

Asbestos Deaths

Recent improvements in dust control have reduced mortality but in high-risk areas, such as textile mills and mines, 5-40% deaths may be attributed to asbestos in workers with more than 20 years exposure. Some who died, however were relatives who had been affected by fibres brought home in the clothes of parents or spouses. Children are likely to be more liable to mesothelioma than adults, given the same exposure to the fibres.

These people are at risk from any activity that disturbs the asbestos in their homes, whether it be putting up shelves or taking out an old cooker. If they have to deal with the asbestos because it is disintegrating, or needs to be cut or drilled, they should avoid breathing any asbestos dust, and should consult Health and Safety Guidance Notes on how to handle the substance. Waste materials are classified as controlled wastes, and must be disposed of in sealed, clearly marked plastic bags at sites licensed by a Waste Disposal Authority.

This is not the end of the asbestos industry, however. In January 1987, worldwide production of asbestos began to climb, as manufacturers moved their output to developing countries. Growing use of asbestos—especially for construction— in the USSR, and in much of Asia, has cancelled out the drop in orders from North America and Western Europe. Overall world asbestos consumption, according to the *New Scientist* (29 January 1987) is expected to rise steadily for the foreseeable future.

ACID RAIN

Normal, unpolluted rain would contain almost pure water (H_2O) in which some carbon dioxide (CO_2), some ammonia (NH_3) originating from organic matter and existing in water as NH_4^+ ion, and varying but small amounts of cations (Ca^{2+}, Mg^{2+}, K^+, and Na^+) and anions (Cl^-, SO_4^{2-}) would be dissolved.

The component of acid rainfall is not an acid in the strict sense of the term, but it is more acid than it should be. Acidity is measured on the pH scale, which runs from pH 14 (very alkaline) through pH 7 (neutral) to PH 1 (very acid). The scale is logarithmic, so pH 3 is ten times more acid than pH4, a hundred times more acid than pH5, thousand times more acid than pH6, etc.

Although the pH of pure water is neutral, 7.0, the pH of normal, "unpolluted" rain is usually 5.6; in other words, it is already acidic (Fig.). Such rain, however, is considered "normal" and only when the pH of rain or snow is below 5.6 it is considered acidic ("**acid rain**").

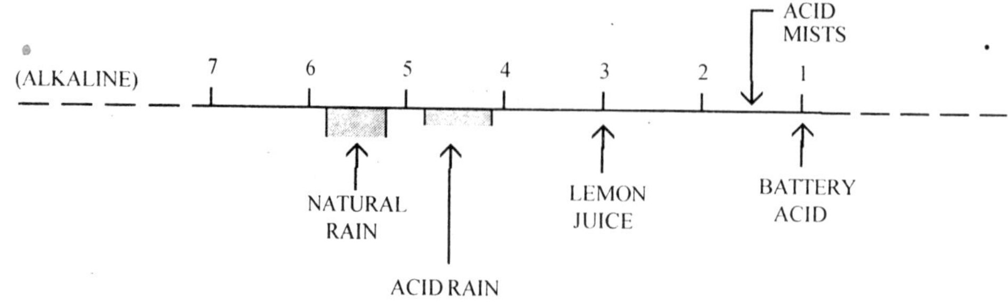

Relative Acidity of Rain.

The Causes of Acid Rain

Acid rain is the result of human activities, primarily the combustion of fossil fuels (oil, coal, natural gas) and the smelting of sulphide ores. These activities release in the atmosphere large quantities of sulphur and nitrogen oxides, which when in contact with atmospheric moisture are converted into two of the strongest acids known (sulphuric and nitric)) and fall to the ground as localised rain, snow or mist. The pH of rain and snow over large regions of the world ranges from 4.0 to 4.5, which is from 5 to 30 times more acid than the lowest pH (5.6) expected for unpolluted areas. The lowest rain pHs reported so far (2.4 in Scotland, 1.5 in West Virginia and 1.7 in Los Angeles) are more acidic than vinegar (pH 2.4) and of lemon juice (pH 3.0). It is estimated that about 70 per cent of the acid in acid rain is **sulphuric acid**, with **nitric acid** contributing about 30 per cent. In addition to sulphur contained in the acids carried in the rain, it is believed that an approximately equal amount of sulphur reaches all surfaces through dry disposition of particulate sulphur. In humid or wet weather, this sulphur too is oxidized to sulphuric acid.

The Chemistry of Acid Rain

Acid rain is a direct consequence of the atmosphere's self-cleansing nature. The tiny droplets of water that make up clouds continuously capture suspended particles and soluble trace gases. When precipitation coalesces from cloud water, it washes the impurities out of the atmosphere. All trace gases cannot be removed by precipitation. However, sulphur dioxide (SO_2) and oxides of nitrogen emitted into the atmosphere an chemically converted into forms that are readily incorporated into cloud droplets; sulphuric and nitric acids.

The processes that convert the gases into acid and wash them from the atmosphere began operating long before human beings started to burn large quantities of fossil fuels; sulphur and nitrogen compounds are also released by natural processes such as volcanism and the activity of soil bacteria. But human economic activity has made the reactions vastly more important. They are triggered by sunlight and depend on the atmosphere's abundant supply of oxygen and water.

The reaction cycle occurs in the troposphere, the lowest 10 or 12 kilometres of the atmosphere. It begins as a photon of sunlight strikes a molecule of ozone (O_3) which may have mixed downward from the ozone layer in the stratosphere or may have been formed in the troposphere by the action of nitrogen, and carbon-containing pollutants. The result is a molecule of oxygen (O_2) and a lone, highly reactive oxygen atom, which then combines with a water molecule (H_2O) to form two hydroxyl radicals (OH). This scarce but active species transforms **nitrogen dioxide** (NO_2) into nitric acid (HNO_2) and initiates the reactions that transform **sulphur dioxide** into sulphuric acid (H_2SO_4).

The concentration of the hydroxyl radical in the atmosphere is less than one part per trillion, but it is practically inexhaustible; several of the oxidation processes it triggers end up by regenerating it. For example, one byproduct of the initial oxidation of surphur dioxide is the hydroperoxyl radical (HO_2) which reacts with nitric oxide (NO) to produce nitrogen dioxide and a new hydroxyl radical. In effect each hydroxyl radical can oxidize thousands of sulphur-containing molecules. As a result only the amount of pollutant in the air determines how much acid is ultimately produced.

Fig. on page 90 Illustrates the formation of acid rain from suplhur dioxide and oxides of nitrogen given off by industry and vehicles.

The sulphuric and nitric acids formed from gaseous pollutants can easily make their way into clouds. (Some sulphuric acid is also formed directly in cloud droplets, from dissolved sulphur dioxide and hydrogen peroxide). Nitric acid gas readily dissolves in existing cloud droplets.

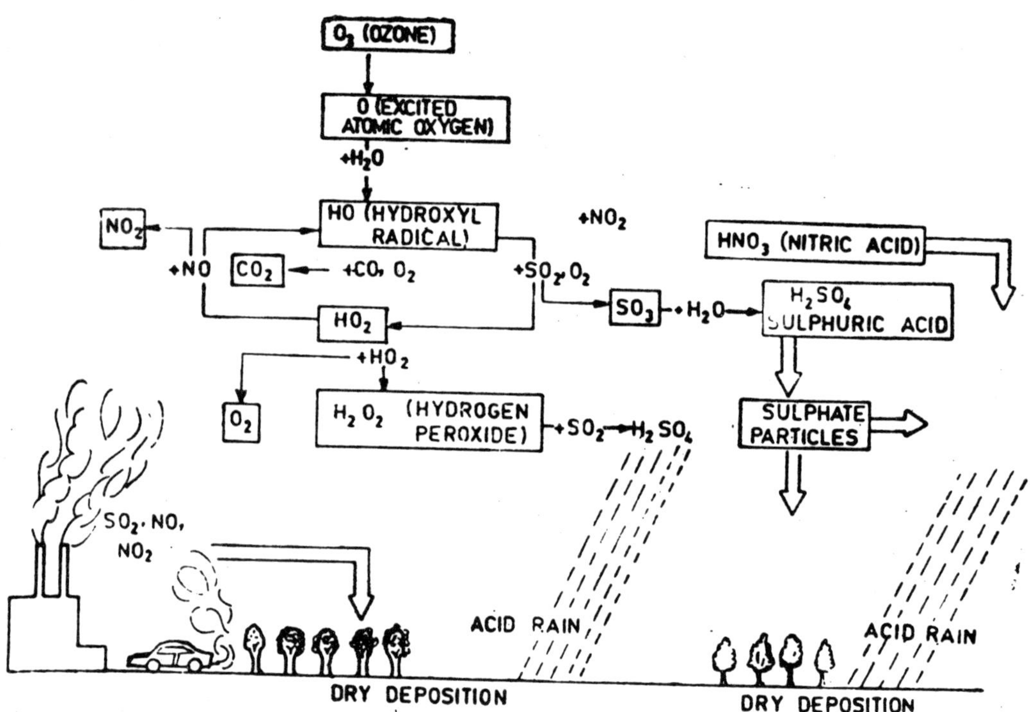

Chemistry of Acid Rain.

Atmospheric chemistry generates sulphuric and nitric acids from sulphur dioxide and oxides of nitrogen given off by industry and vehicles. The hydroxyl radical, formed when a molecule of ozone breaks apart and releases an oxygen atom that can react with water is the major actor. It converts nitrogen dioxide (NO_2) into nitric acid and initiates the conversion of gaseous sulphur dioxide (SO_2) into sulphuric acid. A different reaction sequence forms sulphuric acid from sulphur dioxide and hydrogen peroxide dissolved in cloud water. The hydroxyl radical is regenerated by reactions involving nitric oxide (NO) and the acids come to the earth as dry particles and in rain and other forms of precipitation.

Sulphuric acid formed through gas-phase reactions condenses to form microscopic droplets, from roughly 1 to 2 micrometres (10—6m) diameter. Some of these sulphate particles settle to the ground in a process known as dry deposition. (Dry deposition also refers to the capture of sulphur dioxide gas by vegetation)). Most of them, however, are incorporated in clouds. Moisture readily condenses on an existing surface—a condensation nucleus—and sulphate particles are ideal condensation nuclei. They grow into cloud droplets containing dilute sulphuric acid.

The sulphuric and nitric acids in cloud droplets can give them an extremely low pH. Water collected near the base of clouds in the eastern U.S. during the summer typically has a pH of about 3.6, but values as low as 2.6 have been recorded. In some parts of the world the pH of fog has fallen as low as 2.

The acid rain may fall hundreds of miles from the pollution source. Wherever it lands, it undergoes a new round of physical and chemical changes, which can reduce the acidity and alter the chemical characteristics of the water that eventually reaches lakes and streams. Alkaline soils, such as soils rich in limestone, can neutralize the acid directly. In the slightly acidic soils typical of the evergreen forests exposed to acid rain in the U.S. Canada and Europe two other processes can counter the effects of acid deposition. The acid can be immobilized as the soil or vegetation retains sulphate and nitrate ions (from sulphuric and nitric acids respectively). It can also be buffered through a process that is known as cation (positive ion) exchange.

In cation exchange the ions of calcium, magnesium and other metals found in many soils take the place of the acid's hydrogen ions. The source of the metal ions is rock weathering: the dissolution of minerals by precipitation and groundwater containing dissolved carbon dioxide, which releases the positive metals ions into the soil together with anions, or negative ions, of bicarbonate (HCO_3-). Then, when sulphuric acid is added to the soil, the sulphate (SO_4^{2-}) of the acid can displace the calcium or magnesium ions. As the sulphate solution washes the metal cations from the soil, the hydrogen ions responsible for the acidity are left behind.

The Effects of Acid Rain

'Natural' rainfall is already slightly acidic, because of dissolved carbon dioxide from the atmosphere. It is usually assumed to have an acidity range of pH 5 to pH 5.6. The rain that falls over Britain and southern Scandinavia has an average range of pH 4 to pH 4.7. The most acid rainfall recorded in the UK was at Pitlochry in 1974—pH 2.4.

Polluted mists and fogs are frequently more acid than this: **acid fogs** in California have been recorded with a pH of 1.6, and it is estimated that the great London fog of 1952, which killed 4,000 people, had a pH of 1.7 (almost the strength of battery acid). The acidity arises from a transformation of the sulphur dioxide into sulphuric acid, and of the nitrogen oxides into nitric acid. This increases the acidity of the rain or mist in which the pollutant finds itself.

The pollutants occur in their original state, as SO_2 or NO_2 in their acid state, as sulphuric or nitric acid; or in another form, when the acids break down into their component parts, as hydrogen, sulphates and nitrates. Each has a particular effect on the environment, but all share the same consequence—that of destabilizing natural cycles.

The situation is complicated by the interaction between the sulphur and nitrogen pollutants, and also with other air pollution. Nitrogen oxides can form ozone, given the right weather conditions and the presence of hydrocarbons in the air. Ozone, which is essential for the protection of the planet when it is present in the upper atmosphere (see Ozone Layer), is a pollutant when it is industrially created and occurs nearer the surface of the earth.

The combined effect of acid pollution and ozone is believed to be a central factor contributing to the decline and death of forests across Central Europe.

Acid rain exerts a variety of influences by greatly increasing the solubility of all kinds of molecules and by directly (through the low pH and the toxicity of the SO_4^{2-} and NO_3- ions) or indirectly (through the dissolved molecules) affecting many forms of life. The adverse effects of acid rain on the microorganisms, plants, and fishes of rivers and lakes have been well documented. The effects of acid rain on crop plants have been more difficult to document. Experiments in which acidic rain (pH 3.0) was applied to plants showed that under some conditions, treated leaves developed pits, spots, and curling and that treated plants, with or without symptoms, sometimes showed reductions in dry weight. Also, more seed of some plant species germinated when the soil in which they were planted received acid rain than when it did not, while the opposite was observed for other species. Experiments conducted to determine the effect of acid rain on the initiation and development of plant diseases have shown that in some diseases, such as *Cronartium fusiforme* rust of oak, only 14 per cent as many telia formed under acidic (pH 3.0) rain treatment than under a pH 6.0 rain treatment and the beans treated with acidic rain (pH 3.2) had only 34 per cent as many nematode egg masses than they did under a pH 6.0 rain treatment. On the other hand, a bacterial disease (halo blight) and the rust disease of bean were sometimes more severe and other milder with the acidic rain than with the pH 6.0 rain. In general, although some evidence exists that acid rain causes variable amounts of damage to at least some plants, consistent quantitative data are still insufficient to determine the extent of such damage on various crops in the areas where they occur.

Acid rain degrades lakes and streams if it falls long enough to overcome natural defence mechanisms. This has happened in Scandinavia and in certain parts of Britain. The lakes are based on rock or thin soils, and cannot buffer the acidic input. They become more acidic, and all the fish die.

There is circumstantial evidence that acid air pollution kills forests on an enormous scale. There is disagreement over the specific causes of the forest deaths, with over 180 theories as to the mechanisms involved.

There is agreement as to the widespread damage caused by acid pollution to materials. It corrodes sandstone, limestone, leather, paper, certain metals, historic monuments and stained glass.

Certain birds are known to be affected by acid changes in the environment, where their food supply is disrupted. Others have suffered health damage as a result of aluminium contamination indirectly caused by acid rain. Damage to other animals *e.g.* seals, through deterioration of their food supply, is suspected but difficult to prove. Changes in the flora of areas previously considered to be unaffected, such as Cornwall, have been discovered.

In a report published in November 1986, the World Wildlife Fund identified damage in the areas listed below.

Forests : West Germany (52% of forest damaged); Switzerland (33%); Austria (16%); Netherlands (50%); Luxembourg (19%); Czechoslovakia (27%); Poland (7% to 40%); Hungary (11%); France (20% in Vosges), Belgium (18% on German border);Yugoslavia (450,000 hectares); Rumania (170,000 ha); Spain (23,000 ha); Denmark (1,000 ha).

Lakes : Norway; 35,000 square kilometres of lakeland affected, with a loss of more than half the trout populations; Sweden; 18,000 of 85,000 lakes acidified, 4,000 severely (*i.e.,* unable to support fish life); Scotland; at least 22 major lakes acidified. Wales; most small rivers in Central and West Wales acidified, Upper Tywi and Wye strongly so; England; Duddon and Usk in Cumbria, and parts of the Lake District acidified.

Control

The most direct way of controlling the pollutants that cause acid rain would be to burn less fossil fuel for transportation and energy generation. Rapid mass transport system and fuel efficient cars can reduce oil consumption in the transportation sector, but energy generation is less manageable. In spite of worthy strategies for conserving energy, consumption is likely to increase in the long run, and current alternatives to fossil fueled power plants do not look promising. Hydroelectric power is limited by a variety of problems, including scarcity of appropriate sites, resettlement of people, deforestation etc. Nuclear power is beset by economic problems and lack of public confidence in its safety.

The key to controlling acid rain, then, must be the reduction of emissions from fossil-fueled power plants, particularly coal burning power plants. The approach that has already led to reductions in sulphur emissions in the U.S., West Germany and Japan combines the use of coal that is naturally low in sulphur, or has been washed to remove sulphur and other contaminants, with **Flue-gas Desulphurization** (FGD). In FGD wet limestone is sprayed into the plant's hot exhaust gases, where it removed as much as 90 per cent of the sulphur dioxide. The suplhur-containing waste can be difficult to dispose off, however and FGD reduces the efficiency of a power plant, causing it to consume several per cent more coal for a given output. Furthermore, the process does nothing to reduce nitrogen oxide emissions.

The new power plant technologies developed in USA, jointly by the Government and industry under the Clean Coal Demonstration Programme, enacted in 1984, offers a more comprehensive solution. There clean-coal technologies are already being demonstrated in full-size plants.

In the system known as **atmospheric fludized—bed combustion**, a turbulent bed of pulverized coal and limestone is suspended by an upward blast of air; the combustion region is threaded with boiler tubes, which supply steam to the plant's turbines. The turbulent mixing of the coal and air allows combustion to take place at a lower and more even temperature than it does in a conventional boiler, which reduces the formation of oxides of nitrogen. Meanwhile, the limestone efficiently captures the sulphur dioxide. In a related technology known as **pressurized fluidized—bed combustion** the coal is burned in compressed air, which improves the plant's efficiently as well.

In the third technology, **gasification/combined-cycle**, coal is reacted with steam and air at high temperatures to produce a gas consisting mainly of hydrogen and carbon monoxide. The gas can then be burned, spinning a turbine; waste heat in the gas turbine's exhaust serves for generating steam, which drives a steam turbine to yield additional electricity. A gasification/combined-cycle plant operates much more efficiently than a conventional plant and gives off considerably less sulphur dioxide and nitrogen oxides.

AEROSOL

In atmosphere science, an aerosol is a dispersion or suspension of very small particles or liquid droplets, or both, in air. Their diameters are in the range of 6 X 10^{-4} to 0.5 microns. The microparticles in aerosols, together with larger ones, make up the particulates of polluted air.

In consumer products, an aerosol or aerosol can is a metal can fitted with a valve. The can contains some product — hair spray, paint, cheese, toothpaste, insecticide, deodorant, deodorizer, shaving lotion, cologne, for just a few examples — together with a propellant, usually a Freon fluorocarbon or a volatile hydrocarbon (see. Ozone; Freon)

Dust, regardless of the individual particles sizes, is called a **dispersion aerosol** if it is formed from a grinding and powdering action on a solid followed by dispersal in air.

Smoke is an aerosol of the **condensation type** with a considerable amount of solid particles dispersed in air; smoke may be composed entirely of solid particles, but liquid particles are usually present.

Mist, is a dispersion aserosol if it forms by the atomizing of a liquid in air, and a condensation aerosol if it forms by the condensation of a liquid's vapour (as in the formation of fog).

A cloud is a naturally occurring, **visible aerosol** well above ground level. A fog is a cloud at ground level.

AIR POLLUTION

The air is one of the most used and abused parts of the biosphere. Industrial, automotive, and domestic activities have resulted in increasingly outrageous insults to the atmosphere. Yet the air is finite; it cannot be manufactured and replenished as the need for it increases. It is at the same time our most precious and most fragile resource. Even animals and plants that live in the sea depend upon dissolved gases from the atmosphere. The enormous abuse of the atmosphere has become a health hazard and a threat to life, damaging both plants and animals in areas polluted with poisonous fumes, dust, and smoke.

Air pollution (also termed **atmosphere pollution**) sometimes has natural causes, such as form frost, and grass fires caused by lightning. Volcanoes are a source of air pollution with ash, dust, sulphur compounds, and other gases that can be irritating or even fatal. The ash from smoking volcanoes often becomes distributed over wide areas and causes distress due to irritation of the respiratory tract and possibly aggravating the respiratory inflammations. The eruption in 1883 of the volcanoes on Krakatoa, an island between Sumarta and Java, blew most of the island away and started a tidal wave that killed 36,000 people. Volcanic dust from the explosion spread around the earth and was visible for many months.

Air pollution is caused by particulate matter as well as certain gases. The particulate matter includes solid particles such as smoke, dust, salt crystals, pollen, flyash, bacterial cells spores, etc. and liquid droplets such as terpens, benzene and fog. The chief gases causing air pollution include sulphur dioxide, hydrogen sulphide, nitrogen oxides, ozone, Freons (chlorofluorocarbons or CFCs) carbon monoxide and ammonia.

One of the serious irritating situation of air pollution occurs when a smoke from automobile exhausts and industries combines with fog and certain pollutant to form **smog**. The smog occasionally persists for several days in urban climate because of the barrier (lid or boundary layer) created by temperature inversion. The pollutants in the smog are commonly the result of **photo-oxidation** by sunlight: hydrocarbons and nitrogen oxides on photooxidation form peroxyacetyl nitrate (PAN) and hydrogen peroxide; NO_2 on photooxidations forms ozone. Such secondary pollutants in the smog form **Photochemical smog** and are highly injurious for plants as well as animals. SO_2 and NO_x in the smog can prove very harmful when sufficient fog is present. These gases combine with water droplets to form sulphuric acid and nitric acid which on precipitation may produce acid rains. In all these photochemical reactions, sunlight helps combining various gases with oxygen to produce secondary pollutants.

Although nature has the power to recover from her own ravages, following the industrial revolution, humans have caused an upsetting of nature's air ecology. If our rush for increased comforts continues as in the past, we shall not only suffocate ourselves, but also destroy all life around us. Air pollution, basically the presence of foreign substances in the air, has become a serious problem because the concentration and qualities of these substances are injurious not only to property, but also to vegetation and animal life.

Motor vehicles are by far our worst polluters, followed by industry and power plants. Carbon monoxide constitutes the greatest mass of any air pollution, followed by sulphur dioxide, hydrocarbons, nitrogen oxides and particulates. Most air pollutant are toxic to plant and animal life. The effects are discussed separately.

Among the pollutants of serious concern for our atmosphere are increased concentration of carbon dioxide and its influence in global warming, the role of chlorofluorocarbons and nitrogen oxides in depletion of Ozone layer are creating Ozone holes.

The balance of pollutants in the atmosphere is illustrated in Fig. below, the direction of pollution are marked by dark arrows; white arrows indicate the process of purification.

Other Relevant Entries

Air pollution and Plants; Air Pollution and Human Health Carbon monoxide; Nitrogen oxide; Oxides of Sulphur; Pollution; Smog; Temperature Inversion.

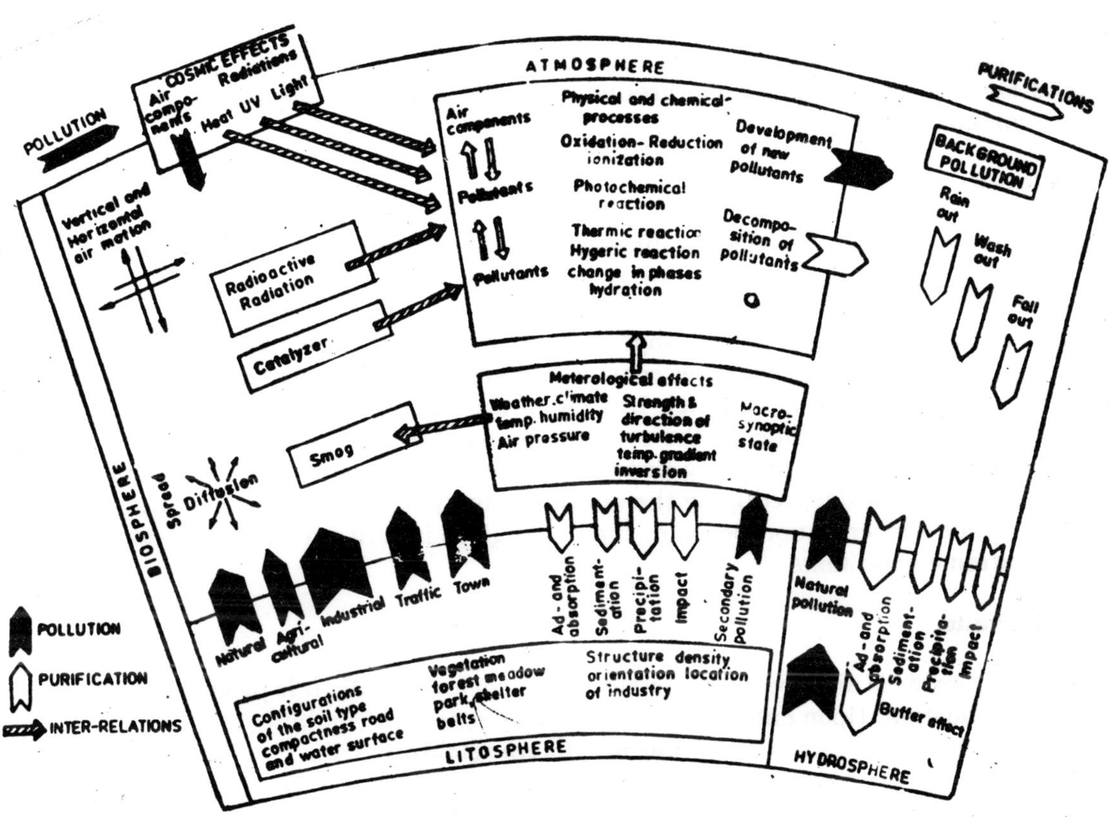

The Balance of Pollutants in the Atmosphere.

CARBON MONOXIDE, CO

Carbon monoxide, a colourless, odourless poisonous gas, is the product of incomplete combustion of carbon and its compounds. Created largely by the automobile, CO constitutes more than half the pollution caused by humans. Carbon monoxide is very dangerous to humans because it combines with haemoglobin in the blood to form carboxyhaemoglobin . If taken in sufficient quantities, CO prevents the haemoglobin from carrying the needed oxygen to the tissues.

On the annual basis, about ten times as much as carbon monoxide enters the atmosphere from natural sources as from transportation and other human activities. Most of the this (77.6%) comes from atmospheric oxidation of methane arising from the breakdown of organic residues, especially in tropical regions. Smaller contribution come from the oceans (3.9%) and from the growth and breakdown of chlorophyll in plants (2.6%). This naturally generated carbon monoxide does not pose any serious problem because the rate of emission at any given place is relatively low. However, man-made carbon monoxide tends to be related rapidly in urban areas. This results in localized concentration 50 to 100 times higher than the global average.

The concentration carbon monoxide in the atmosphere at any time depend not only on the rate of production but also on the rate of removal. This occurs mainly in the soil, though a small amount is oxidized to carbon dioxide in the lower atmosphere. In the soil, fourteen species of fungi have been identified as the active agents oxidizing carbon monoxide to carbon dioxide. The activity of these fungi varies with the type of soil, being least active in desert soils and most active in the tropics. Cultivated soils are less satisfactory to these fungi than are those covered with natural vegetation. The total capacity of soil fungi for carbon monoxide oxidation seems to be more than adequate to deal with the carbon monoxide produced on a global basis. Unfortunately, the urban areas, where most anthropogenic carbon monoxide is released, have among the poorest soil reservoirs of these fungi which, therefore, cannot make an effective contribution to lowering the localized high concentrations in these areas. One study showed that in big cities, the air in heavy traffic contained more than 30 ppm carbon monoxide, a concentration that approaches the threshold for acute poisoning. Underground garages have been found to contain 100-200 ppm carbon monoxide in air.

Effects on Plants

Carbon monoxide has the effect of inhibiting nitrogen fixation. At concentrations of 0.01 to 1 per cent (100 to 100,00 ppm), visible effects are abscission (leaf drop), premature aging, and the inititation of roots on stems. These effects resemble those of ethylene, a gas produced naturally by ripening fruit and some pathogenic fungi. Carbon monoxide is believed to inhibit cellular respiration in plants by reacting with the enzyme system cytochrome-oxidase which is vital to the utilization of sugars as metabolic fuel. In view of this, there has also been interest in the possible effects of carbon monoxide on the cytochrome-oxidase system of animals. Cytochrome oxidase is the principle oxygen-reducing enzyme of the body. It is a haeme protein that will combine with carbon monoxide, reducing the enzyme activity. However, it appears unlikely that the carbon monoxide concentration becomes high enough to inhibit cytochrome oxidase activity except under extreme conditions.

Effects on Humans

Carbon monoxide is an odourless, tasteless, and colourless gas. It is produced as a product of incomplete combustion. Upon entering the respiratory system it combines in the lung with the haemoglobin in the bloodstream to form **carboxyhaemoglobin** COHb. This reduces the ability of the haemoglobin to carry oxygen to the body tissues. The affinity of CO is some 200 times that of O_2 for attaching itself to the haemoglobin so that low levels of CO can still result in high levels of COHb.

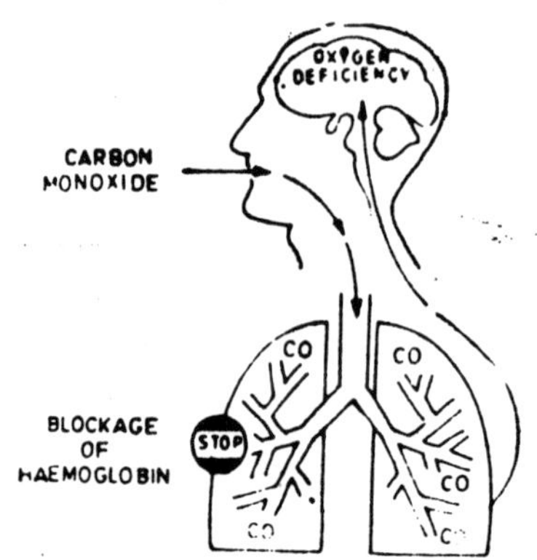

Damaging Effect of Carbon Monoxide.

After exposure the CO is slowly released from the blood with a clearance half life of 3 to 4 hours. Cigarette smokers commonly have COHb levels of about 5 per cent with heavy smokers reaching values to 10 per cent. This is due to high CO concentration, about 400 ppm, in cigarette smoke.

At COHb levels of 2 to 5 per cent effects are found to affect the central nervous system such as impairment of time interval discrimination and visual acuity. At levels greater than 5 per cent there are cardiac and pulmonary function changes. There is evidence that CO concentrations encountered in the Los Angeles area (10 to 15 ppm ambient Los Angeles *basin* 24-hour *average* values) are associated with excess martality. Evidence has also been presented that a relationship exists between ambient CO levels and myocardial infraction (heart attacks). There is ample evidence that ambient air concentration of CO in major traffic areas are high enough to produce significant effects on health.

CHLOROFLUOROCARBONS, CFCs

CFCs are simple compounds produced from hydrocarbons by replacing one or more of the hydrogen atoms by both chlorine and fluorine atoms. When all the hydrogen atoms have been replaced, the CFC is said to be fully halogenated; partially halogenated compounds (HCFCs) still contain one or more hydrogen atoms. Six grades are produced in significant quantities; the fully halogenated CFCs 11, 12, 113, 114 and 115, and the partially halogenated HCFC 22.

CFC 11	Trichlorofluoromethance	CCl_3F
CFC 12	Dichlorodifluoromethane	CCl_2F_2
CFC 113	1, 1, 2, Trichloro-1, 2, 2-tri fluoroethane	CCl_2FCClF_2
CFC 114	1, 2-Dichloro 1, 1, 2, 2-tetrafluoroethane	$CClF_2CClF_2$
CFC 115	Chloropentafluoroethane	$CClF_2CF_3$

Chlorofluorocarbons (CFCs) are man-made gases which are relatively inexpensive to produce, non-flammable, non-toxic, non-reactive and very stable. Since their invention in the early 1930s they have been developed for use as propellants in aerosol cans, as blowing agents to aid cell production in plastic foams (particularly polyurethane and Polystyrene), as refrigerants in refrigerating and air conditioning equipment and as solvent cleaners for use particularly in the micro-electronics industry.

CFC 11 has been basic propellant in 50-60 per cent of all aerosol cans sold. Because it boils at 24°C its vapours are easily liquefied, and its liquid form is easily vaporized to create the pressure in the aerosol can. CFCs are exceptionally stable chemicals toward almost anything normally encountered in the environment or put into aerosol cans; they are popular among aerosol manufacturers because they have essentially no odour and impact no flavor. CFC 12 is widely used as a **refrigerant**. Earlier, the common refrigerants were smelly, dangerous chemicals (*e.g.*, ammonia or sulphur dioxide) that caused corrosion of equipment. Others were flammable. The CFCs are completely noncorrosive and do not burn.

At present, worldwide annual demand for all types of CFCs exceeds 1 million metric tonnes. The majority of CFC production is located in the European countries, USA and Japan although in recent years manufacturing plants have been built in India, South Africa, Australia and several countries in Central and South America, Eastern Europe and the Far East.

Use of CFCs was relatively limited for the first 30 years of their existence. The two best-known CFCs — CFC 11 and 12 — have always dominated the market, and now account for almost 80 per cent of CFC manufacture. From 1960 to now, production of CFC 11 and 12 has mushroomed.

CFCs last a long time (CFC 11 has a lifetime of 78 years, with CFC 12 lasting 139 years, and CFC 115 even longer at 380) and are also cheap to make.

Not surprisingly, they began to be used in other industrial products. Aerosols using CFCs as propellants first went on the market in 1950; by 1974, 1.4 billion CFC aerosol were being made every year in North America alone.

CFCs are emitted into the atmosphere either directly from aerosols; through leakage from refrigeration units or air conditioning systems; by escape from factories during the foam blowing process; and at the end of the waste stream, for example when refrigerators are dumped. Concern first developed in the early 1970s when it was realised that CFCs emitted into the atmosphere would not break down rapidly in the way that most chemicals do, but could persist for several decades.

Scientists Sherry Rowland Mario Molina worked out that CFC releases of 800,000 tonnes a year (the rate of release in 1972) would result in half a million tonnes of chlorine being deposited in the atmosphere over 30 years, destroying around 20 to 40 per cent of the **ozone layer**. The US government banned the use of CFCs in aerosols in 1978.

HYDROGEN SULPHIDE, H_2S

Hydrogen sulphide is a colourless gas, having about the same toxicity as cyanide. The gas has a penetrating odour resembling rotten eggs, and is therefore detectable by the human nose at very low concentrations, about 2 ppb (parts per billion). However, the olfactory sensory organs become fatigued quickly and can no longer respond. The sense of smell for hydrogen sulphide is lost within 2 to 15 minutes at 100 ppm or higher. This characteristics makes hydrogen sulphide extremely dangerous in cases of excessive or prolonged exposure.

Effect on Health

Low concentration of hydrogen sulphide cause headache, nausea, lassitude, (Lassitude is a state of tiredness, laziness or lack of interest), collapse, coma, and death. Hydrogen sulphide readily passes through the alveolar membrane of the lungs and penetrates to the blood stream. Death may come within a few seconds from even one or two inhalations. At 1,000 ppm it is almost instantneously fatal. Death is from respiratory failure. The California standards for ambient air quality specify an "adverse" level of 0.1 ppm for 1 hour as sufficient to produce sensory irritation. The maximum allowable concentration for an 8-hour day is 20 ppm.

Natural Sources

The principle natural source of hydrogen sulphide are decaying vegetation and animal material especially in shallow aquatic and marine environments. Anaerobic bacteria attack sulphates in deep sediments, reducing the sulphate to sulphide. Anaerobic decomposition sometimes takes place in underground organic deposits, causing hydrogen sulphide to be present in domestic water supplies. Water from such sources has an objectionable odour but in some parts of Europe water containing hydrogen sulphide is considered to have medicinal properties. A concentration of 0.05 ppm is detectable by taste. Hydrogen sulphide is given off by sulphur springs, volcanic discharges, gas wells, coal pits, and sewers. It has been estimated that about 30 million tons of hydrogen sulphide is released annually by ocean areas and 60 to 80 million tons by land areas.

Industrial Sources

Industrial pollution is probably not often a significant factor in public health, although it can be a threat in localized area and form exposure during industrial operations. The important industrial sources of hydrogen sulphide are those which also give rise to sulphur oxides and sulphuric acid, the users of high-sulphur fuels. About 80 per-cent of the sulphur compounds in fuels are converted to hydrogen sulphide which is soon oxidized in the atmosphere to other sulphur compounds. Notable emitters of hydrogen sulphide are petroleum refineries, kraft paper mills, tannery wastes, and mining operations.

Most of the hydrogen sulphide released to the atmosphere does not remain in that form for long. It is oxidized to elemental sulphur and sulphur dioxide. The latter is further oxidized to sulphur trioxide and subsequently to sulphuric acid and sulphates. An important reaction in the atmosphere is the oxidation of hydrogen sulphide with ozone:

$$\underset{\text{hydrogen sulphide}}{H_2S} + \underset{\text{ozone}}{O_3} \rightarrow \underset{\text{sulphur dioxide}}{SO_2} + \underset{\text{water}}{H_2O}$$

The reaction takes place fairly rapidly, with a half-life of atmospheric hydrogen sulphide of 2 to 28 hours, depending on location and conditions. The hydrogen sulphide does not remain as a pollutant for a long period of time, but is a transient toxicant. It can be highly injurious, however, if concentrations are high or if exposure is prolonged.

NITROGEN OXIDES, NO_x

Nitrogen combines with oxygen to form a family of oxides collectively called nitrogen oxides (indicated by the symbol NO_x), which play a multiple role in the air pollution. They enter into chemical reactions with other substances in the atmosphere with the production of additional irritating and toxic ingredients of smog.

The fixation of nitrogen is a vital biological process certain plants and microorganisms and industrial nitrogen fixation by the chemical processing of air is an important commercial enterprise. The automobile and other combustion-power devices are also nitrogen-fixing machines, although in combustion the immediate end product is a contaminant rather than a useful product. In petrol and diesel engines, the high temperature and pressure in the combustion chamber provide suitable conditions for the production of nitrogen oxides, which unfortunately, enter into a number of reactions in the atmosphere responsible for a number of air pollutants.

The nitrogen oxides are among the most toxic substances found in the atmosphere. One of the lesser toxic members, **nitrous oxide,** called "**laughing gas**", is useful in medicine as a mild anesthetic. Nitrogen oxide, and nitrogen dioxide, because of their abundance and toxicity, are the most dangerous in terms of public health. A third highly toxic members, **nitrogen pentaoxide,** is a potential threat.

Nitrogen Oxide, NO

A colourless gas, is the primary product of combustion of nitrogen. It is produced when the temperature is high enough to cause a reaction to take place between the nitrogen and oxygen in the air that is sucked into the cylinder with the air fuel mixture. Thus the form of the nitrogen oxide that is emitted to the atmosphere in the exhaust is almost all nitric oxide. However, a large portion of it is converted to the more toxic nitrogen dioxide by a complex series of reactions in the atmosphere are shown in fig. on p. 100.

On contact with air, nitric oxide combines with oxygen, or even more rapidly with ozone, to form the poisonous nitrogen dioxide:

$$\underset{\text{nitric oxide}}{2NO} + \underset{\text{oxygen}}{O_2} \rightleftharpoons \underset{\text{nitrogen dioxide}}{2NO_2}$$

At high temperatures the equilibrium is reversed, converting nitrogen dioxide back to nitric oxide.

Nitrogen Dioxide NO_2

NO_2 is only the widely prevalent pollutant gas that is coloured. Pure nitrogen dioxide is deep reddish brown and causes much of the atmospheric discolouration on bad smog days in metropolitan areas. Power plants and other energy conversion systems as well as a variety of chemical process industries emit nitrogen dioxide into the atmosphere, but in most metropolitan areas the most important source is vehicular exhaust.

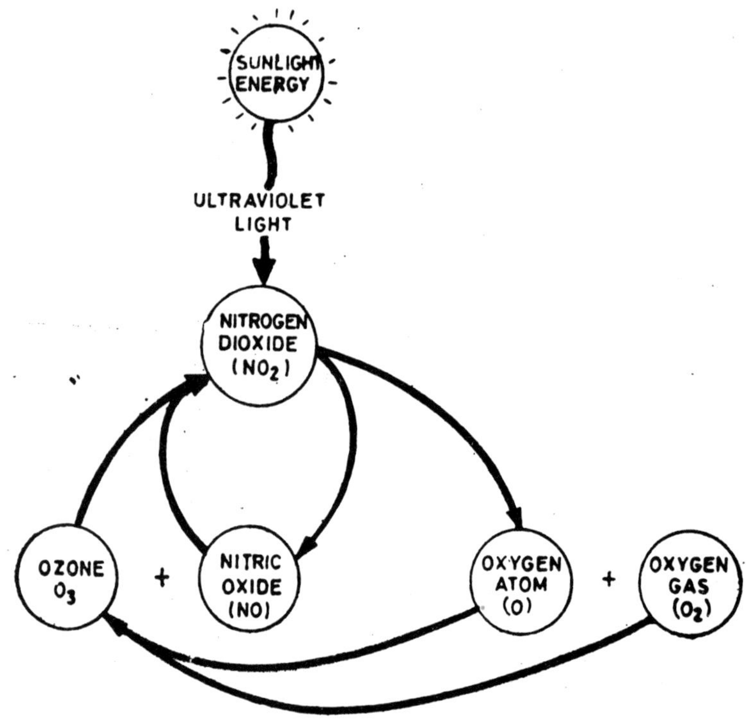

Atmospheric nitrogen dioxide photolytic cycle. The ultraviolet light acts upon nitrogen dioxide, breaking its bond and yielding two substances, nitric oxide (NO) and oxygen atom(O). The oxygen atom then reacts with oxygen (O_2) in the atmosphere, forming ozone (O_3). The ozone reacts with the nitric oxide to reform nitrogen dioxide. Thus the energy of ultraviolet light acts as a "pump" in the rapid destruction and reformation of nitrogen dioxide.

Nitrogen dioxide is responsible for much of the ozone on polluted atmosphere due to its decomposition by the action of ultraviolet light;

NO_2	+	UV	→	NO	+	O
nitrogen dioxide		ultraviolet light		nitric oxide		atomic oxygen

O	+	O_2	→	O_3
atomic oxygen		molecular oxygen		ozone

These reactions, in conjunction with others of a complex nature, account for daily cyclical fluctuations in the concentration of ozone in polluted areas of high sunlight intensity. One such area is the Los Angeles basic where the amount of ozone is characteristically low in the morning but rises sharply by afternoon.

Nitrogen dioxide itself is a deadly poison and is one of the most treacherous gases known. Exposure may go unnoticed until it is too late for the victim to recover. Inflammation of the lungs is immediate, but it may not cause enough pain to be noticed. Oedema (accumulation of fluid) may occur several days later, resulting in death. In some cases the effects may be prolonged for weeks or even months. Apollo astronauts Thomas P. Stafford, Donald K. Slayton, and Vance D. Brand narrowly escaped death while making a landing after their historic rendezvous in space with a Soviet spacecraft in July 1975. They were hospitalized after inhaling a poisonous gas, believed to be nitrogen tetraoxide, N_2O_4, leaked into the spacecraft from the fuel system. Much of the nitrogen dioxide in the atmosphere exist in the form of nitrogen tetraoxide, with which it is in equilibrium.

$$2NO_2 \rightleftharpoons N_2O_4$$

Even short exposure to nitrogen dioxide is dangerous. Breathing air containing 20 ppm of the gas for a brief time may be fatal with prolonged exposure, much smaller amounts in the air are dangerous. The MAC (maximum allowable concentration) for occupational exposure has been set at 5 ppm (9 milligrams per cubic metre) for an 8-hour period.

Experiments with rats showed that the acute toxicity of this gas is less than that of ozone, but greater than that of carbon monoxide. However, delayed, rather than immediate, death is the characteristics response of nitrogen dioxide exposure. It is predicted that the nitrogen dioxide concentration will continue to get worse for several years until adequate controls can be established. Nitrogen dioxide remains one of the most potentially dangerous of modern day pollutants.

Tobacco Smoke

Cigarette and cigar smoke contains 330 to 1,500 ppm nitrogen oxides which are removed completely by inhalation, presumably by absorption in the lungs.

Nitrogen dioxide reacts with water to form nitric acid and nitric oxide:

$3NO_2$	+	H_2	→	$2HNO_3$	+	NO
nitrogen dioxide		water		nitric acid		nitric oxide

Much of the nitric acid in the atmosphere ultimately combines with ammonia to form ammonium nitrate, and probably all of the nitrogen oxides that do not react photochemically become nitrate salts that form aerosols (suspensions of small particles) in the atmosphere in sizes larger than 1 micron.

Effects on Plants

Nitrogen dioxide is also highly injurious to plants. It suppres growth when plants are exposed to 0.3 to 0.5 ppm for 10 to 22 days. A concentration of 1 ppm for 8 hours produces significant growth reduction with no visible damage. Exposure to 4 to 8 ppm for 1 to 4 hours causes visible leaf injury to sensitive plants. Low-light intensity increases the plant's sensitivity.

Effect on Human Health

Both oxides of nitrogen, NO and NO_2 are potential health hazards. Animal mortality studies indicate that NO_2 is about four times more toxic than NO. No cases of human death from NO poisoning have been reported, at the concentrations found in the atmosphere. NO is not an irrant and is not considered a health hazard. The greatest toxic potential of NO is its ability to undergo oxidation to the more toxic NO_2.

The proven effects of NO_2 on humans and animals are confined almost entirely to the respiratory tract and occur only with NO_2 levels higher than those now prevailing in the atmosphere. A higher NO_2 dose results in the following effect sequence: nasal irritation, breathing discomfort, acute respiratory distress, pulmonary oedema, and finally death. Even the mildest effects, such as mucous membrane irritation, do not occur at the prevailing atmospheric NO_2 concentrations. The odour threshold range for NO_2 in human is reported to be 1-3 ppm. Concentration above 12 ppm resulted in eye and nasal irritation in a smaller number of volunteers. In these cases, the nasal irritation was more intense than that of the eyes. Concentrations of NO_2 greater than 100 ppm are lethal to most animal species, and 90 per cent of the resulting deaths are due to pulmonary oedema.

NATURAL GAS

Natural gas is composed mainly of **methane (CH_4)**, is the simplest organic hydrocarbon. Methane is found in underground rock layers both by itself and lying above natural deposits of petroleum. It is generally believed that natural gas is our least abundant fuel.

Natural gas was once dismissed as a nuisance by crude oil producers, and billions of cubic feet have been wasted by burning it off in the oil fields. But the advantages of natural gas over other forms of fuel stimulated the building of a network of high—pressure pipelines in the United States that made it possible to distribute the fuel at reasonable cost from the gas fields to major parts of the country. Gas is now preferred by both residential and commercial users because it is convenient, clean-burning, efficient, and reasonably priced. Natural gas is favoured by industries that must control heat.

The ability of natural gas to meet air pollution control standards without costly treatment or alteration, such as desulphurization, has greatly increased the interest in gas as a clean-burning fuel. The demand for natural gas has been growing.

In India, natural gas is emerging as an important source of commercial energy in view of large reserves that have been established in the country particularly in South Basin off the West Coast. As on 1 January 1985, India had 47,900 crore cubic metre of recoverable of gas reserves which had increased to 64,755 crore cubic metre by January 1989. There has been substantial increase in supply of natural gas. Against the supply of 807 crore cubic metre in 1988-89, supply in 1989-90 was estimated to the tune of 940 crore cubic metre. Oil strikes at Kaveri Offshore and at Nada in Cambay Basin as also gas found at Tanot in Jaisalmer Basin in Rajesthan were major discoveries during 1988-89. Another noteworthy feature was commencement of production from South Basin Gas Field since September 1988. During 1989-90 oil gas structures had been discovered in Adiyakkamangalam in Tamil Nadu, Andada in Gujarat, Khovaghat in Assam, Lingala in Andhra Pradesh, Bombay Offshore and Kuchchh Offshore. Natural gas is also making significant contribution to household sector by way of Liquid Petroleum Gas, (LPG) extracted from associated gas. A'bout 30 per cent of the counrty's output of LPG comes from this source.

Gas authority of India Limited (GAIL) was incorporated in August 1984 with an immediate objective of construction of the cross-country HBJ gas pipeline. GAIL could complete lying of HBJ gas pipeline including its extension from Babrala of Delhi. However, associated systems of the pipeline such as telecommunication, compressor stations, etc., are yet to be fully commissioned. During 1989-90, GAIL started supply of gas to GTPC's power plants at Anta and Auraiya and Delhi Electric Supply Undertaking. GAIL is implemented LPG Recovery Project at Bijapur in Madhya Pradesh. Government has also in principle agreed to assign GAIL as one of the marketing companies for LPG in some specified areas.

OZONE

Ozone, O_3, is a gas present in significant amounts in both the stratosphere and the troposphere, the combined amount being known as **total column ozone. Stratosphere ozone** is generated as follows:

Molecular oxygen (O_2) separates into two oxygen atoms under the effect of ultraviolet sunlight (with a wavelength of 240 nanometres). An oxygen atom formed in this way can combine with an oxygen molecule (O_2)to form a triatomic oxygen molecule (O_3 or ozone). The O_3 molecule, for its part, rapidly absorbs ultraviolet light in the region of a wavelength of 300 nm and again separates into O_2 molecules and oxygen atoms. Some of the oxygen atoms combine again with O_2 molecules. In this way a photochemcial balance is established. Other gases also participate in these process by purely chemical reactions.

$$O_2 \xrightarrow{h\nu} O + O$$
$$O + O_2 \rightarrow O_3$$

The ozone that is formed at altitudes of over 30 km finds its way into lower levels of the atmosphere only as a result of air turbulence. For this reason, the high O_3 concentrations are found in the atmosphere at altitudes of between 20 and 30 km, while in the layer of air close to the earth only very low proportions are recorded. Enrichment of the ozone in lower air layer is not possible, since on coming into contact with the organic substances to firm ground, or with the sharply articulated upper surfaces of woodlands, ozone gives off the third oxygen atom to the organic substances and so returns to the state of molecular oxygen.

Tropospheric ozone is generated by the action of ultraviolet light on molecules of nitrogen dioxide as follows:

In the atmosphere of big cities polluted by nitric oxide and other noxious gases, a reaction between ultraviolet radiation and these noxious gases can result in the formation of ozone. The nitrogen dioxide contained in vehicle exhaust fumes, for instance, decomposes under solar radiation through photo-dissociation into nitrogen monoxide and atomic oxygen. Atomic oxygen combines with molecular oxygen, as already described, to form ozone:

$$NO_2 \xrightarrow{h\nu} NO + O$$
$$O + O_2 \rightarrow O_3$$

This explains why, in a smog situation in the daytime, the ozone concentration in the air layer close to the ground in big cities can rise to five times the normal content and in extreme cases to a thousand times the normal content.

Effect on Plants

Ozone has long been known, to be toxic to vegetation, but injury to crops from atmospheric pollution has only recently been recognized. In 1958 grape leaves in California vineyards were described as having the upper surfaces stippled from ozone toxicity. Shortly afterward similar injury tobacco leaves, called **weather fleck**, was identified as ozone injury. Since then, ozone toxicity has appeared as stipple, fleck, spots, streaks, tipburn, or premature yellowing on at least twenty crop plants. Ozone affects the mature leaves first and spots are usually visible only on the upper surface. White pine needles are also affected with tipburn caused by ozone.

Plants are extremely sensitive to ozone. Concentrations as low as 0.08 ppm for period of exposure as short as 1 hour cause injury to sensitive plant tissues. The sensitivity of plants to ozone is dependent upon the species, variety, and environmental conditions both before and during exposure. Among the edible plants that are susceptible to ozone, some varities of tomatoes and radishes are injured by 0.15 to 0.25 ppm when exposed for 1 to 2 hours .

Ozone is taken into leaves through stomata and injures primarily palisade but also other cells by disrupting the cell membrane. Affected cells near stomata collapse and die and appear as white (bleached) necrotic flecking or stippling on broad leaves, as streaks on the leaves of cereal crops, and as browning on the tips of pine needles.

Many crop plants, such as alfalfa, bean, citrus, grape, potato, soybean, tobacco, and wheat, and many ornamentals and trees, such as ash, lilac, several pines, and popular, are quite sensitive to ozone, while some other crops, such as cabbage, peas, peanuts and pepper are of intermediate sensitivity, and some, such as beets, cotton, lettuce, strawberry, and apricot, are tolerant. Flecking of tobacco leaves reduces their value, particularly as cigar wrappers.

Certain tobacco varities are the most sensitive organisms known for indicating the presence of ozone and will show symptoms when the atmospheric concentration is above 0.05 ppm. The presence of sulphur dioxide enhances the effect of ozone. For example, symptom appeared after 2 hours of exposure of 0.037 ppm ozone plus 0.24 ppm sulphur dioxide but when the plants were exposed to the same concentrations of the gases separately, they did not develop visible signs of injury. This effect is described technically as reducing the threshold concentration by synergistic action.

Effect on Human Health

At high concentrations, troposphere ozone has deleterious effects on human health (WHO recommend an 8 h maximum in air of 60 parts per billion). Tropospheric ozone also reduces crop yields in sensitive species and damages forests. Tropospheric ozone is a potent greenhouse gas and may contribute significantly to the problem of global warming.

Ozone is an important component of so-called oxidant **smog**. There is very little emission of ozone from automobile exhaust. However, ozone plays a key role in the photochemical formation of atmosphere pollutants (see Photochemical Smog). Ozone is the most reactive form of molecular oxygen and the fourth most powerful oxidizing agent known, exceeded only by fluorine (F_2), oxygen difluoride (OF_2), and atomic oxygen. Ozone has a distinctively pleasant odour at concentrations of about 2 ppm or slightly less, but it is irritating at higher concentrations. It is used as a disinfectant for air and water, and in industry for bleaching waxes and oils, as well as for organic synthesis. The use of ozone as an air disinfectant for closed spaces is dangerous.

Ozone can cause irritation of the respiratory passages. It penetrates much deeper into the lungs than sulphur oxide. Rats exposed to 1 ppm for an 8-hour day develop bronchitis, fibrosis (formation of fibrous tissue), and bronchiolitis. A concentration of 1.25 ppm for an our causes an increase in the residual lung volume (reduced expiration) and a decrease in the breathing capacity. Higher levels cause pulmonary oedema (accumulation of fluid in the lungs), haemorrhaging, and impairment of gas exchange through the alveolar membrane. The maximum allowable concentration for occupational exposure for an 8-hour work day has been set at 0.1 ppm.

A curious characteristic of ozone toxicity is that tolerance develops in laboratory animals from exposure to subinjurious concentrations. Experiments with animals have shown, that repeated inhalations of ozone at low concentrations can promote a certain resistance to its irritant effects. The tolerance develops quickly, within 24 hours. It may persists for 4 to 6 weeks in rats and 100 days in mice, as shown by their survival from exposure to concentrations that would ordinarily be lethal. After tolerance develops, the pulmonary oedema and haemorrhaging that usually follow strong dose exposures are absent. Moreover, cross-tolerance can be demonstrated between ozone and the lethal gases phosgene and nitrogen oxides.

Ozone increases susceptibility to infection. Experiments with mice showed that ozone increases the survival of bacteria inside the lungs. In rabbit, cells washed from the lungs after exposure to ozone showed that there was a reduction in the ability of the alveolar macrophages to engulf bacteria. Also, laboratory animals that have been subjected to prolonged exposure to ozone show signs of premature aging.

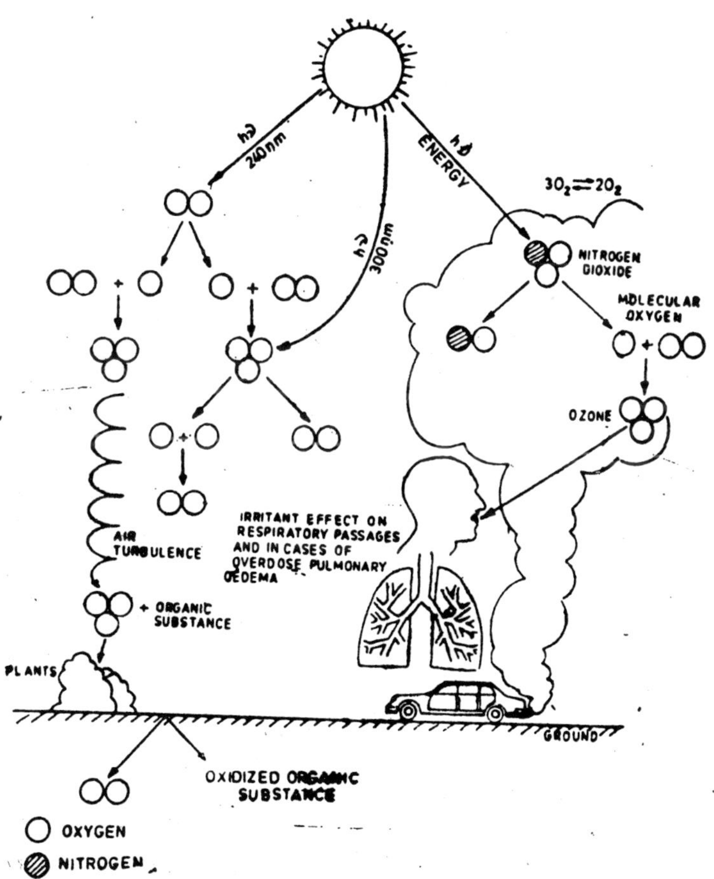

Formation and Decomposition of Ozone and its Harmful Effect on the Organism

The injurious effect of ozone may be partly due to its ability to oxidize lipids (fats). A deficiency of tocopherol (member of the vitamin E complex containing several tocopherols) is known to make tissues more susceptible to oxidation. Rats with a deficiency of vitamin E are more susceptible to lethal concentrations of ozone than those not deficient.

SULPHUR OXIDES, SO_x

Sulphur pollutants are belched into the air in enormous quantities as part of the industrial effluent. Atmospheric sulphur comes mainly from the sulphur content of fuels. Nearly 80 per cent of the sulphur in sulphur dioxide is initially emitted as hydrogen sulphide and is converted to sulphur dioxide in the atmosphere. Sulphur dioxide and the products of its reactions are highly damaging to structural and other materials, as well as being physiologically harmful and injurious to health. Sulphur pollutants were probably the predominant pollutants in most of the "killer smogs."

Sulphur dioxide is a colourless gas with a penetrating and pungent odour. It produces respiratory irritation at low concentrations. As an industrial air pollutants, it is formed by the oxidation of the sulphur compounds in fuel. The gas is moderately soluble, and in the presence of moisture it forms sulphurous acid, a fair oxidizing agent :

$$\underset{\text{sulphur dioxide}}{SO_2} + \underset{\text{Water}}{H_2O} \rightarrow \underset{\text{sulphurous acid}}{H_2SO_3}$$

Sulphurous acid is converted slowly to sulphuric acid (H_2SO_4). Sulphur dioxide also reacts very slowly in oxygen to form sulphur trioxide gas (SO_3), which in turn reacts quickly with water, forming sulphuric acid. Aerosols (fogs) of sulphuric acid contribute to the haze from the combustion of high sulphur fuels in industrial areas. The bluish white plume often seen coming from the stacks of electric power plants is often largely a sulphuric acid aerosol.

Sources of Sulphur Oxides

The major natural source of sulphur oxides volcanic in origin. However, this contribution to pollution is believed to be small. Most of the total emission of sulphur oxides is from burning of fossil fuel by industries, and furnaces. Fossil-fuel power plants, for generating electricity, are the major source, and they are being built more rapidly that anticipated because nuclear power plants are being installed more slowly than planned. Most of the remaining sulphur oxide emissions is accounted for by nonferrous smelters and petroleum refineries. The major source are highly varied in size and design, and reduction of their emissions to acceptable levels requires a diversity of highly complex control systems. Emissions are increasing at an alarming rate.

In India sulphur dioxide emissions are maximum (10 tonnes per day) in Calcutta followed by Bombay, Delhi, Ahmedabad, Kanpur, Hyderabad, Madras, Nagpur and Jaipur in decreasing order.

Physical Damage

Sulphur dioxide and its end products have a severe corrosive and deteriorating effect on many materials. Sulphur dioxide and sulphur trioxide are especially damaging to building materials that contain carbonates, such as limestone, marble, mortar, and roofing slate. The reaction forms soluble calcium salts that are washed away by condensation and rainwater. Works of art in or near industrial areas and cities are in serious jeopardy. One of the two specimens of Cleopatra's Needle was moved to New York's Central Park in 1880. The belisk has deteriorated more since its arrival than during its previous 3,000 years in Egypt where it originally stood before the Temple of the Sun at Heliopolis. Some works of art are being moved indoors to preserve them.

The atmospheric deterioration of marble and limestone structures in Italy is referred to as **stone cancer**. The art works and irreplaceable structures of Venice are decaying from the emissions of nearby industries. Nearly all of the thousand or so piazzas and churches contain material of marble having a high carbonate content, highly susceptible to sulphur oxide corrosion. Venice has other environmental problems including floods humidity, salinity, industrial water pollution, changes in the populations of plants and animals in the lagoon, impaired tidal flow for removal of sewage, a rising water level, and sinkage of the foundations. These, combined with corrosive air pollution, make Venice appear to be a dying city.

Sulphur oxides even cause the destruction of extremely durable structural materials. Steel corrodes up to four times as fast in cities as in rural areas. Particulate matter accelerates the attack of sulphur oxides on metals, especially damaging when moisture is present. The oxide and the acid attack iron, steel, copper, nickel, zinc and at high concentrations, even aluminium.

Sulphur oxides attack some paint pigments and modify fresh paint films in a way that delays drying and increases the susceptibility to moisture. Sulphur pollutants even attack nylon. I makes paper brittle due to the presence metallic impurities that catalyze the conversion of sulphur dioxide to sulphuric acid. Sulphur dioxide also attacks leather and textiles, causing slow deterioration. Synthetic fibres, but particularly cotton and wool, are susceptible. Laboratory experiments under controlled conditions show that photo-chemically produced compounds in the atmosphere cause fading of dyes in fabrics, and the addition of sulphur dioxide causes a synergistic effect, greatly increasing the damage.

Effect on Plants

Sulphur dioxide has been known for many years to be a major cause of injury to higher plants. Some species are so sensitive that its effects are sometimes the first indication that the air is polluted. The cause of the symptoms, however, is not always easy to identify due to the fact that the injury often resembles other afflictions such as sunscald, frost, aging, or other disorders having obscure environmental or cultural causes.

Plants with high physiological activity are among the most sensitive species. Alfalfa, grains, squash, cotton, grapes, apples and white pine are susceptible. Species with thick, waxy leaves, such as citrus are more resistant. Trees with needles, such as most pines, are generally resistant except for those with newly formed needles. Conditions that are ordinarily beneficial to plant growth, such as high morning light intensity, high humidity, good soil moisture, and moderate temperature are apt to enhance injury.

Prolonged exposure of plants to sublethal levels of sulphur dioxide causes them to develop a chronic type of injury which first appears between the veins, leaving the vein areas green. Older leaves gradually becomes yellow, then white, and eventually die, However, in leaves that are only partially injured, the unaffected parts regain their function upon removal from exposure, and new foliage grows normally. The symptoms of such chronic exposure are sometimes described as "early aging" and are not always easy to identify. Concentrations that cause chronic injury are usually less than 0.4 ppm.

Since sulphur dioxide is absorbed through the leaf stomata, conditions that favour or inhibit the opening of stomata similarly affect the amount of sulphur dioxide absorbed. After absorption by the leaf, sulphur dioxide reacts with water and forms phytotoxic sulphite ions. The latter, however, are slowly oxidized in the cell to produce harmless sulphate ions. Thus, if the rate of sulphur dioxide absorption is slow enough, the plant may be able to protect itself from the buildup of phytotoxic sulphites.

The most serious cause of plant damage have been from high concentrations in the vicinity of smelters. A pollution episode occurred at Anaconda, Montana, during 1910 to 1911.

Large amounts of sulphur oxides were emitted, causing nearly all the important specied of trees within the radius of 8 to 13 kilometres to die. During another incident in 1929 in the area of a smelter at Trail, British Columbia, damage to plants occured as far as 85 kilometres from the smelter. Douglas fir, ponderosa pine, and other forest vegetation were 60 to 100 per cent damaged for a distance of 50 kilomètres. The emission rate from the smelter was as high as 20,000 tonnes and averaged 18,000 tonnes of sulphur dioxide per month. Inhibition of cone production by pine trees was still evident two years after installation of a control system. Acute injury to broad leaved plants causes the leaves, or areas of the leaves, to dry out and become bleached. Injured grasses show a similar pattern except that the tips are killed and the injury generally has a more streacked appearance. The response in plants is clearly the result of the oxidizing or reducing properties of sulphur oxide and not due to the action of acid. Sulphuric acid aerosols are not generally toxic to plants, although emissions of large droplets from factories have been known to damage vegetation.

Sulphur dioxide has a powerful toxic action on micro-organisms. It is used as a fungicidal fumigant, as a preservative for food and wine, and as a disinfectant in breweries. Its toxic action is believed to depend on the formation of sulphurous acid or sulphite ions in the tissues of the organisms.

Effect on Human Health

The dangerous nature of sulphur dioxide is due to the fact that it penetrates the respiratory tract and damages the numerous small delicate cilia of the bronchial mucous membrane, whose task it is to expel dust and aerosol particles. Owing to the effects of sulphur dioxide, inhaled dust and soot particles remain in the lungs and pursue their toxic activities there.

In small children, sulphur dioxide causes the Krupp syndrome, a sinister violent affliction which is characteristised by a raucous hacking cough, fever and increasing and sometimes fatal breathlessness. The mucous membrane in the larynx and the vocal cords is swollen and inflamed and the windpipe becomes seriously obstructed by phlegm. The Krupp syndrome is particularly serious in the evening and at night and is especially prevalent in the winter months.

Sulphur dioxide is intensely irritating to the eyes and the respiratory tract. A concentration of 0.3 to 1 ppm can be detected by the average person, and this may be taste as well as smell. Most people will show a response from exposure to concentrations in the range of 1 to 5 ppm. Exposure to the higher concentration for 1 hour cause a pronounced choking sensation, and 10 ppm for the same period produces severe distress. Exposure can cause serious inflammation of conjunctiva of the eyes. Concentrations of 6 to 12 ppm cause immediate irritation of the nose and throat. A few minutes of exposure to 1 per cent (10,000 ppm) produces irritation, even to moist areas of the skin.

Sulphur dioxide is so irritating that at acutely dangerous concentrations, it serves as its own warning agent. However, this safety does not always exist at low concentrations. Moderate resistance develops from continuous exposure so that concentrations as high as 5 ppm can no longer be smelled. Concentrations of 400 to 500 ppm may be lethal and are considered to be immediately dangerous to life, while 50 to 100 ppm is regarded as the maximum permissible range for 30 to 60 minutes exposure. The accepted maximum allowable concentration for an exposure of 8 hours is 10 ppm.

Moist air and fogs probably increase the danger of sulphur dioxide greatly, partly due to the formation of sulphurous acid and sulphite ions. A further explanation may be that sulphur dioxide becomes absorbed on finely divided aerosol particles, resulting in deposition deep in the air spaces of the lungs. The toxic end-product of sulphur dioxide is probably sulphuric acid, which is stronger irritant than sulphur dioxide. Sulphuric acid causes 4 to 20 times the effect on an equimolar basis (molecule for molecule).

PARTICULATES

Matter in the form of extremely small particles of either solids or liquids is particulate matter. The level of particulates in air are generally reported as total **suspended particulate matter** (SPM) per cubic metre. The amount and nature of suspended particulates in the atmosphere and the chemical and physiological effects are among the least understood aspects of air pollution.

Source of Particulates

Particulates are not necessarily visible. Suspended particles of microscopic dust, chemicals, liquids, sea-salt, pollen, aero-allergens, bacteria, and viruses are abundantly present in the atmosphere. Some of them are the cause of much human discomfort and diseases. Other injurious aerosol constituents include soot, lead particles from automobile exhaust, asbestos, fly ash, volcanic emissions, pesticides, sulphuric acid mists, and metallic dusts. Dozens of substances have been identified as minor particulate components of air, possibly including many toxic materials that have not yet been determined to be injurious.

A surprising amount of dust settles on the earth from outer space. Every day 8 billion solid particles penetrate the atmosphere. The average size is less than a grain of sand. There is enough dust and ash left over from the attrition of the larger bodies and the remains of the disintegrated material to leave an accumulation of about 2 million tons of debris that settle on the surface of the earth annually. It drifts down and is found everywhere from the high mountain peaks to the sediment at the bottom of the oceans.

Microscopic study of air samples taken in Australia showed that an enormous number of particles are dispersed in the air. The number varied widely from 5,000 per cubic centimetre in downtown Vienna to 800 particles per cubic centimetre in the residential area, as compared to 300 per cubic centimetre in natural aerosols.

Size of Particulates

Aerosol particulates in the atmosphere range in diameter from about 0.001 micron to more than 100 microns. Particles larger than 10 microns settle rapidly, although they may remain air borne for moderately long periods under turbulent conditions. Particles of 0.1 to 1.0 micron would normally remain air-borne for extended periods, except that collisions and electrostatic attractions, as well as other forces, result in coagulation and the formation of effectively larger particles. Particles of less than 0.1 micron can form extremely stable aerosols, but by random movement in the air, eventually collide with other particles and form larger particles by coagulation.

Most of the particles injected into the atmosphere fall out within minutes, hours, or days. But the smaller particles may remain air-borne for years. Many of the aerosol particles originating from man-made sources in the cities and industrial areas, as well as from volcanoes, forest fires, and nuclear explosions, travel across international borders.

Some of the aerosol particles, mainly sizes above 2 microns, are removed from the air by rainfall. Particles also disappear from the air by impinging on objects such as trees and buildings. Extremely small particles do not readily impinge on objects because of their aerodynamic properties. The size judged to be most liable to enter the respiratory tract and become lodged in the air spaces of the lung is about 1 micron.

Particles emitted from automobile exhaust are mostly less than 5 microns in diameter. They consist of carbon particles, lead compounds, and reaction products from the combustion of motor oil. Other particulates originating indirectly from automobile pollution of aerosols are formed as reaction products from nitrogen oxides and hydrocarbons in the atmosphere. Diesel engines emit more particulate material than gasoline engines. Diesel exhaust is estimated to contain 62.5 per cent particles less than 5 microns in size and 37.5 per cent in the 5 to 20-micron range.

Surface Area

The effects of small particles are best understood by considering the relationship between the size of the particles and their surface area. Assuming the shape of particles in an aerosol to be perfect spheres, the total surface area is inversely proportional to the diameter of the particles. In order to visualize this, let us take a droplet having a diameter of 100 microns and a volume of 523,600 cubic microns. Then split this large droplet into smaller droplets, each having a diameter of 10 microns. The volume will be the same, but instead of 1 droplet there will now be 1,000 droplets; and they will have a total of 10 times the surface area of the original droplet. Split these droplets further into smaller ones each having a diameter of 1 micron, and there will now be 1,000,000 droplets with a surface area 100 times the original droplet.

The enormous surface created with chemicals are divided into small droplets greatly increases the possibility that chemical reactions will take place. Large surface also bring about physical changes such as increased absorption of heat and light scattering.

Physical and Physiological Effects

Particulate matter has damaging physical effects, partly due to the acceleration of the corrosive effects of other pollutants such as sulphur oxides and sulphuric acid and partly due to the formation of grime. Dirt and soot are damaging to fabrics and other materials, which are then costly to clean, requiring the use of large quantities of solvents and detergents which further add to the pollution cycle.

The physiologically injurious effects of pollutants such as sulphur dioxide and its reaction products are made more severe by the presence of particulate matter. The synergistic effect was noted in the case of the London 'killer smog" of 1952; the evidence indicated that it was the combination of particulate and sulphur oxides that caused 4,000 deaths during that episode.

Cancer-inducing substances have long been known to be present in particulate pollutants. They induce cancer when painted on the backs of mice or when injected under the skin but do not consistently do so when injected into the lungs. However, lung tumours, resembling human lung cancer, are produced in experimental animals when benzpyrene is mixed with iron oxide or haematite and injected into the trachea. These materials are commonly found in mixture with other atmospheric pollutants. Toxicity experiments involving injections of extracts from polluted air into newly born mice shows that there is a broad spectrum of carcinogenic materials in the atmosphere.

Soots are physically complex structures, mostly highly porous agglomerates of great absorptive power. Since soots are produced in conjunction with other combustion products, many of them are potent toxicants, the potential for synergistic action is inherent. Polycyclic hydrocarbons, to which group the potent carcinogen benzpyrene belongs, are produced generally in combustion processes. In humans certain "inert" particulates, including those in cigarette smoke, cause resistance to the passage of air in the respiratory system.

FLY ASH

Very small, often powder like particles of glassy material carried away from a fire, usually a coal fire, by the warm air current, and consisting of noncombustible, mineral remains (ash) of the fuel, often together with small particles of the fuel. Fly ash is composed chiefly of the oxides of silicone, aluminium, iron, and calcium. Coal-fired furnaces and incinerators in particular tends to release fly ash. Unless trapped, it settles over the surrounding area. The finest particles in fly ash are the most difficult to remove, and on lodging in the lungs they create a health hazard (See Particulates).

Emission of fly ash is controlled largely by arrangements for settlement by gravity or by entrapment by baffles or sprays of water.

Delhi Electric Supply Undertaking's (DESU) fly ash dump is situated along the ring road behind the Pragati Maidan (Exhibition Grounds). People travelling along this route have to inhale the flyash. Ironically, on the world Environment Day—June 5th, 1991—a strong breeze early morning blew up the flyash from this dumping ground. Commuters passing through the area had a tough time with their eyes watering from this environmental hazard.

PHOTOCHEMICAL SMOG

The most irritating, and some of the most injurious, components of smog are the products of reactions in the atmosphere between oxygen, ozone, and emission pollutants. The mixture of undesirable products is sometimes called **photochemical smog** because some of the chemical reactions are initiated by the energy from ultraviolet light. The cycle of reactions is not completely understood; much of the work on air pollution is done is **smog chambers**, which may differ in their characteristics from conditions in the atmosphere. However, the important steps in the atmospheric production of photochemical smog have been determined. They consist of a series of reactions in which oxygen, ozone, nitrogen oxides, and hydrocarbons participate to produce compounds that are both irritating and toxic. Some of the principal reactions are as follows :

Step I

One of the most important steps in the formation of photo-chemical smog is the action of ultraviolet light on nitrogen dioxide, producing nitric oxide and atomic oxygen :

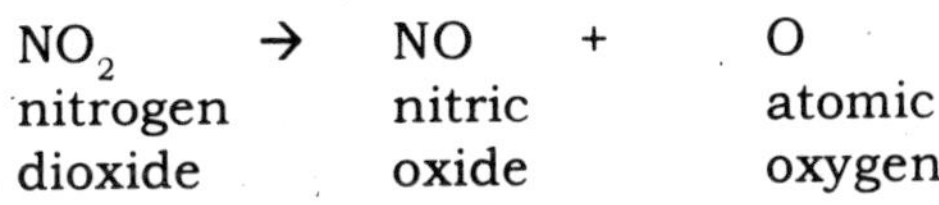

$$\underset{\text{nitrogen dioxide}}{NO_2} \rightarrow \underset{\text{nitric oxide}}{NO} + \underset{\text{atomic oxygen}}{O}$$

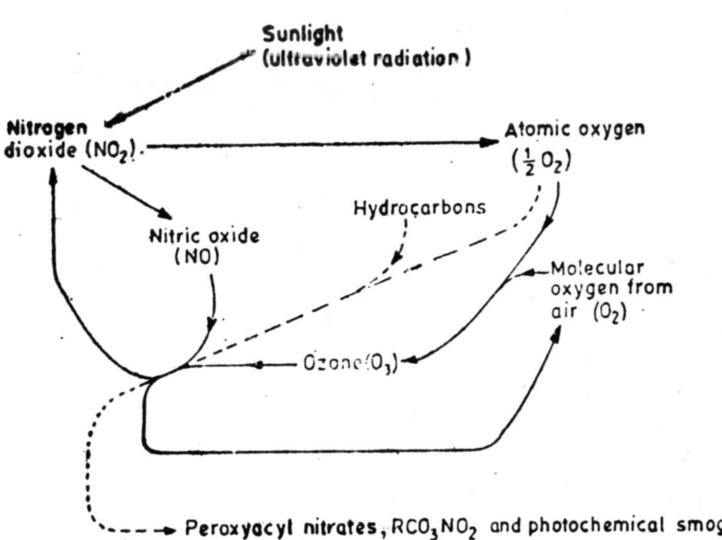

The nitrogen dioxide photolytic cycle, showing the interaction of hydrocarbons to produce photochemical smog.

Step 2

The atomic oxygen further reacts with a molecule of oxygen with the formation of ozone :

O	+	O_2	→	O_3
atomic oxygen		molecular oxygen		ozone

Some of the ozone is used up in oxidising nitric oxide to nitrogen dioxide is shown in Fig. on p. 111.

Step 3

If hydrocarbons are present, they can react with the ozone to form hydrocarbon free radicals which are very reactive because of the unpaired electrons which they contain. These free radicals can react with each other, with atmospheric toxicants and with other components of air to produce photochemical smog:

O_3	+	RCH_3	→	RCH_2
Ozone		hydro carbon		free radical

O	+	RCH_2	→	RCO
atomic oxygen		free radical		acyl radical

The acyl radicals undergo a complex series of reactions, some of which consists of a further reaction with oxygen to form more highly reactive substances called **peroxy radicals**, which in turn react with hydrocarbons to form aldehydes, ketones, and other compounds.

$RCO + O_2$	→	$RCOO_2$
		peroxyacyl radical

Step 4

A crucial stage in the photochemical smog complex is the reaction between peroxy radicals and nitrogen dioxide to form the highly toxic **PANs (peroxyacyl nitrates)**:

$RCOO_2$	+	NO_2	→	$RCOO_2NO_2$
peroxyacyl radical		nitrogen dioxide		PAN (peroxyacyl nitrate)

The foregoing scheme is a greatly abbreviated and simplified version of the scheme for the formation of the photochemical smog.

Hydrocarbons very widely in their potential to form free radicals. Therefore, the nature of a hydrocarbon mixture must be known before its contribution to photochemical smog can be assessed. In general, unsaturated compounds (—C=C—) are the most reactive. Following reactions with ozone, atomic oxygen, molecular oxygen and nitrogen dioxide, they give rise to the group of toxicants known as peroxyacyl nitrates. These toxicants, like ozone, can react further with hydrocarbons to generate free radicals.

Effect on Plants

Before the biological importance of air pollution was fully realized, it was noted that plants grown in bad smog areas suffered injury from what often appeared to be obscure causes. It is now apparent that many plants are more sensitive than humans to the acute effect of smog and, in fact, can be used as indicators of harmful levels of air contaminants.

Two ingredients of photochemical smog are recognized as being particularly injurious: ozone and PAN (peroxyacetyl nitrate). With PAN, injury appears on some plants as a characteristic silvering on the outer surfaces of the leaves. Ozone, on the other hand, kills cells on the upper surfaces of the leaves and appears as flecking or stippling.

Injury to plants from photochemical air pollution was noted in the mid-1940s when losses occurred to spinach and other sensitive plants in California. Since then crop damage has occurred in many parts of the United States, particularly California, New Jersey, and Florida, and in many other countries. The damage to agriculture in the United States in 1966 was estimated at 500 million-dollars. In California alone, the damage to agriculture crops during 1960 was estimated at 44.5 million dollars.

POLLUTION

Pollution of our environment is the problem of major concern for the human beings. Environmental pollution includes directly or indirectly through human activities, any undesirable or deleterious change in our environment through infusions of matter and energy at levels appreciably higher than natural levels. Any substance causing pollution is termed a pollutant. The pollution of the atmosphere (Air pollution or atmospheric pollution) is caused by gases as well as solid and liquid particles. Water gets polluted (water pollution) through disease producing microorganisms (biological pollution) and undesirable compounds (chemical pollution). Higher levels of sound energy causes Noise pollution. Groundwater also gets polluted (groundwater pollution) through seepage of sewage and other forms of liquid disposal.

A major part of the pollutants can be carried by air, sometimes over long distances, which can eventually reach the ground and be turned into soil and water pollution, reach the troposphere and stratosphere and be a grave cause for concern (in particular the possible depletion of the protective ozone layer), and also reach our eyes and lungs through smog or "ordinary" pollution. A quite unhealthy picture, which gives a poor idea of the way Man, the Lord of Creation, has dealt with his environment and created the conditions of his own discomfort or hazard. This hazard can sometimes be frightening: the memories of Bhopal are in all the minds.

The sources of pollution are many, but the general public thinks chiefly of three risks; (a) Oil spills, the most recent one occurred in the Persian Gulf on 26th January 1991 during the Gulf War between Iraq and U.S. led multinational forces. The oil slick was reported to be 140 km long and 40 km wide. (b) Nuclear hazards, because of Hiroshima and Nagasaki. The layman does not make the difference between a drug and a poison, between "the" bomb and a nuclear reactor, although the nuclear industry is abundantly proved to be the safest of all industries; (c) The chemical industry, the permanent villain judged to be the cause of all pollution, the source of bad smells and a danger to all.

Chemistry and the chemical industry have not been able to give of themselves a faithful image. It may bear the responsibility for some damage to our world, and we all have to be informed of the part of chemistry in the Bhopal tragedy. But chemistry is a science of creation, the chemical industry brings to mankind priceless contributions : drugs, pesticides, fertilizers, synthetic textiles, etc., and both chemical research and the chemical industry will probably be in charge of finding the proper solution to the environmental problems linked to the technical development of the contemporary world.

The chemists have to face a number of difficulties in their relationship to the public, a by-product of the emission of many pollutants by the chemical and related industries. Despite all that comes from other sources or causes, what comes to mind when chemistry is mentioned is bad smells, pollution and toxicity. The benefits we all draw from chemistry in our well being, comfort, health, safety and economy can be priceless, the layman will in most cases look at us and say "Bhopal". It may be unfair, that the assets of chemistry are just taken for granted, but life is often unfair. The Bhopal tragedy is the worst industrial disaster probably known. What happened was the result of an important leak of gaseous methyl isocyanate (MIC) from a number of equipments which were out of order due to the neglect of several safety requirements and operating instructions. It is very said to think that most of the victims would have been saved if they had protected their face with a damp cloth, and that this so simple fact was ignored.

It seems to be true that industrial production increases pollution but things are fortunately not so simple and not so bad. Many of the industrial wastes are raw materials or end products, and preventing their escape is a saving at the same time as an environmental protection. Other pollutants are really refuse, but in all cases the fight against pollution is on. In fact the pollution level has appreciably decreased in all developed countries in recent years. It is probable that the public is not really aware of the fact, and sees in pollution what remains, not what has been removed. And although industrial pollution has really decreased, the fact remains that a number of problems are as yet unsolved, among which the damage to natural vegetation figures prominently.

For a number of years damage to vegetation, and most of all to forests, have been evident in several countries. In Europe the extent of damage is very grave in Germany, Gzechoslovakia and Poland, slighter but not negligible in other countries. It seems quite certain that this damage is related to the presence of some of the best-known gaseous pollutants, sulphur dioxide, nitrogen oxides, ozone, and also perhaps hydrocarbons.

In the case of plant damage through air pollution no lives are at risk but other landscapes are menaced and the economic damage is very high. This is the type of problem we encounter in several cases. The origins of the forestry problem are far from being completely clear, so that even choice of remedies has to be very carefully weighed. Whatever the remedy the cost will be quite impressive. An analogous situation has been met in all cases of large-scale damage. Cleaning the Ganga water for instance, involve a huge investment. Several other examples could be given and, to mention one of the serious European problems, cleaning the Rhine will be an expensive task, In the case of the Rhine Switzerland, the Federal Republic of Germany, France and the Netherlands are the polluters. The heaviest damage is suffered by the Netherlands are all polluters. The heaviest damage is suffered by the Netherlands, for whom the Rhine is the only source of sweet water, and the heaviest single polluter are the Alsatian salt mines. All Mediterranean countries are damaging their sea, but it is evident that pollution comes chiefly from industrialized countries on the Northern shores. The USSR does probably more for the pollution of the Black Sea than Bulgaria etc.

Turning towards the sources of air pollution, we might think that the SO_2 emission from volcanoes cannot be checked but certainly from a power-station. However, the costs would be tremendous, whether the power-station burns coal or heavy fuel oil, both sulphur-rich, but emitting a plume in which SO_2 is quite diluted. The greater the dilution the higher the cost of recovery.

With nitrogen oxides the situation is comparable. Thunderstorms are great sources of nitrogen oxides but an abundant daily supply comes from car-exhausts among others. The problem has been studied now for a number of years. Car exhausts can be adapted and improved some day, at a higher costs of the car and its maintenance. Quite similar considerations would also apply in the case of other pollutants.

It is established that in many places the damage to plants continued to increase, whereas pollution had markedly decreased. That is the reason for the careful theories on the effects of accumulating pollutants and the mechanisms through which this effect is felt. Now nearly all theories take into account accumulation effects., but the parameters are many. Damage to forests has been shown to be altitude-dependent, raising the possibility that the chief cause of damage might be the nitrogen oxidesozone chemistry. Another view is the possibility of long-term interaction between soil nutrients and air pollutants having accumulated in the soil. It could also be argued that no single compound is responsible for the observed damage, because several countries are submitted to the same pollution level with the same pollutants but widely differing damage. The possible synergy between a well-known pollutant present everywhere at low concentrations, and a second, "activating" compound, which would trigger the synergy, cannot be ruled out.

Acid rains have certainly played a part in damage to trees but the existing theories are not in complete agreement. A new crop of facts, a new idea, and fresh hypotheses are coming up each day, all of which have to be considered.

Today we are faced with a grave and difficult situation and, in order to improve things, we need a better comprehension of the factors involved in environmental effects.

SMOG

The word smog is supposed to have originated in Great Britain as a short-form of "smoke-fog." The term may have been suggested by a report on smoke-fog deaths by H.A. Des Voeux in 1911. The report dealt with two ocassion during the autumn of 1909 in Glasgow, Scotland, when 1,063 deaths were attributed to smoke and fog. In recent years, atmospheric pollution has become increasingly critical. There have been several catastrophic instances of smog resulting in serious injury and multiple deaths.

Killer Smogs

From time to time a combination of high emission of air pollutants and adverse atmospheric conditions combine to cause the formation of "Killer smogs". During a disastrous week in the winter of 1930 in the Meuse Valley of Belgium, a large number of people were made sick and many were killed by smog (Firket, 1931). The Meuse Valley is a narrow river valley, 5 kilometres long with hills on each side. The area is highly industrialized with blast furnaces of steel mills, zinc smelting plants, glass factories, and sulphuric acid plants. On December 1, 1930, and for the rest of the week, a thermal inversion layer confined the increasingly heavy concentration of industrial emissions to the valley. By the third day many people complained of respiratory ailments, and before the atmospheric conditions changed to clear the smog from the valley, 60 people died. There were also fatalities in cattle. It could not be determined with certainty what toxicants caused the difficulty, but probably a combination of several pollutants was responsible. It was estimated that the sulphur dioxide content of air was 9.6 to 38.4 ppm which if true would be way high indeed (Firket, 1936). If the sulphur dioxide had been completely oxidized to sulphuric acid (although this is improbable), the concentration of sulphuric acid mist theoretically could have reached 14.6 to 58.5 ppm.

The London Smog

The December 5-9, 1952, London fog is the best known air pollution episode in England. It began on Thursday, December 4, as a high-pressure air mass created a temperature inversion over southern England. A white fog formed in the London area. As particulate and sulphur dioxide levels built up, because of extensive use of coal as fuel for space heating and electric production, the fog become a black fog. At the same time the high-pressure area stalled and became stationary. The buildup of pollutants combined with the fog resulted in essentially zero visibility. By Saturday pollutants were sufficiently concentrated to cause deaths, which were estimated to be about 4000. The fog cleared on December 9.

Since 1952, similar conditions have occurred in London in January 1955, December

1959. January 1956, December 1957, January 1959, December 1962. However, the smoke levels were markedly lower in 1962 because of the clean up effected by British Clean Air Act.

Table 2 : Survey of Selected Air Pollution Episodes

Date	*Location*	*Pollution*	*Effects*
December 1-5, 1930	Meuse Valley, Belgium	Particulates, SO_2. (9-38 ppm)	63 excess deaths, cough, chest pain, eye and nasal irritation—all age groups
December 5-9, 1952	London, UK	Particulates (4000 $\mu g\ m^{-2}$) SO_2 (1.3 ppm).	4000 excess deaths
December 5-10, 1961	London, UK	SO_2 (2.0 ppm)	700 excess deaths
Jan 29-Feb. 12, 1963	New York, USA	Particulates, SO_2 (0.5 ppm)	200-400excess deaths
November 24-30, 1966	New York, USA	Particulates SO_2 (1.02 ppm)	168 excess deaths
December 3rd, 1984	Bhopal, India	Methyl Isocyanate (MIC)	18,000 excess deaths, blindness all age groups

PANs

The PANs (peroxyacyl nitrates) are believed to be responsible for most of the eye irritation in photochemical smog, although other compounds are known to contribute. All of the substances responsible for irritation have not been identified. Formaldehyde and acrolein, both highly toxic substances, are known to be present along with various types of peroxides and free radicals.

The first known member of the group is **peroxyacetyl nitrate**, or simply **PAN**.

$$CH_3\text{—}\overset{\overset{\displaystyle O}{\|}}{C}\text{—}OONO_2$$

It is a patent eye irritant at concentrations of about 1 ppm of less. PANs and other eye irritants are known as persist for more than 24 hours in contrast to ozone, which is believed to have, in photochemical smog, an average half-life of about 1 hour. One distant cousin of the PAN family, **PBzN (per oxybenzoyl nitrate)**, causes eye irritation at a concentration only 1/200 that of formaldehyde. Since it takes approximately twice as much formaldehyde as PAN to produce irritation, the results indicate that PBzN is 100 times more powerful than the familiar PAN.

Effects on Plant

PAN is taken into leaves through stomata and causes injury at concentrations as low as 0.01 to 0.02 ppm. In large urban areas, concentrations of 0.02 to 0.03 ppm are common. Once inside leaves, PAN attacks preferentially the spongy parenchyma cells, which collapse and are replaced by air pockets that give the leaf a glazed or silvery appearance. The symptoms on broad-leaved plants appear on the lower leaf surface, while monocot leaves show symptoms on both sides. Young leaves and tissues are more sensitive to PAN, and periodic exposures of leaves to PAN often cause **banding** and in some plants even margin **pinching** of leaves because of discolouration and deaths of the most sensitive affected cells, respectively.

METAL TOXICITY

In geological terms, trace elements are defined as those occurring at 1000 ppm or less in the earth's crust. Only twelve elements do not come within this classification, *i.e.*, oxygen, silicon, aluminium, iron, calcium, sodium, potassium, magnesium, titanium, hydrogen, phosphorus and manganese. The trace metals may be divided into those that are 'heavy' with densities greater than 5 g cm^{-3}, and those that are 'light', with densities less than 5 g cm^{-3}. Many trace metals are essential at low concentrations for normal life. Since most organisms are not adapted to deal with high concentrations of trace metal, the problem of metal toxicity arises.

Excessive levels of trace metals may occur naturally as a result of normal geological phenomena such as ore formation. Weathering of rocks, leaching or, in the case of mercury, degassing may make these metals available to the biosphere. Man releases more of the metals by burning fossil fuels, mining, smelting, discharging industrial, agricultural and domestic waste, and by deliberate environmental application in pesticides. Once made available to the environment, metals are not usually removed rapidly, nor are they readily detoxified by metabolic activity. As a result they accumulate. Thus, their release into the environment must be carefully monitored and controlled.

The EPA has defined beryllium (a light trace metal) and mercury (a heavy trace metal) as hazardous, meaning that slight exposure can endanger human health. Nine other metals have been defined as potentially hazardous and they must be kept under review. They are barium, cadmium, copper, lead, manganese, nickel, tin, vanadium and zinc. Of these, all but manganese are trace metals and all but barium are heavy metals. Toxic effects of some metals are summarized in Table 3.

Table 3 : Toxic Effects of Metals

Metal	*Sources*	*Toxic Effects*
Antimony	Industry	Shortening of life spans in rats.
Beryllium	Coal, industry	Most toxic, accumulates in the lungs of produce berylliosis, a serious disease.
Bismth	Coal	Kidney and liver damage in large doses.
Cadmium	Coal, zinc mining, tobacco smoke	Cardiovascular disease and hypertension in humans suspected; inter feres with Zn and Cu metabolism.
Lead	Auto exhaust	Brain damage, convul-sions, behavioural disorders, death.
Mercury	Coal, industrial fungicides, electrical appliances	Nerve Damage, death
Nickel	Diesel oil coal, tobacco smoke, chemical catalysts, steel and non-ferrous alloys	Carcinogenic properties in animals, and in human when inhaled as the carbonyl, Ni $(CO)_4$.
Tin	coal, iron and steel production, tin plating	Low order of toxicity decreased life span in rats and mice.

Source: Schroeder, H.A. 1970. *Arch. Environment Health*, 20, 803

BARIUM

Silver-white metallic element. Toxic and flammable at room temperature in power form. Tolerance for soluble compounds in 0.5 mg per cubic metre in air.

The most common naturally occurring form of barium is barium sulphate, sometimes called barite or baryta. Barium derivatives are used as fillers for rubber, linoleum, etc., as pigments in paint, in glass manufacture, in the ceramic industry and in various other industrial applications. These applications concerned the main source of risk, primarily to workers in the appropriate industries. Barium salts are highly toxic. They cause vomiting and diarrhoeas, which may be associated with stomach, intestinal and kidney haemorrhage. They also affect the central nervous system, causing convulsions and have been implicated as a cause of pneumoconiosis. Despite the known toxicity of barium salts, barium sulphate is used to coat the alimentary tract for X-ray photographs since barium absorbs X-rays strongly and thus increases the contrast. The insolubility of barium sulphate minimizes its toxicity, and makes its use acceptable.

BERYLLIUM

A light, strong metallic element. When alloyed with copper, it is valuable in a variety of types of electrical apparatus and non-sparking tools. Highly toxic, especially by inhalation of dust. A known carcinogen. Tolerance, 0.002 mg per cubic metre of air.

Beryllium is released during the burning of coal, but the main environmental hazards is to those working in industries where beryllium is produced or used, *e.g.*, the maiufacture of nuclear reactor, aircraft and rockets. In humans, beryllium has been shown to damage skin and mucous membrane. It accumulates in the lungs where it causes beryllium disease (**berylliosis**). It may cause cancer in lungs and bone marrow. It is not exerted from mammalian tissue and, therefore, its effects are cumulative. At the biochemical level, beryllium competes with magnesium for enzyme sites and has been shown to inhibit DNA polymerase, thymidine kinase and alkaline phosphatase.

CAESIUM

A metallic element in the alkali metal group, namely sodium, potassium etc.

The isotope (Cs 137 ($^{137}Cs_{86}$) is one of the main radioactive pollutants produced in the nuclear power plants and during testing of nuclear weapons. Cs 137 emits beta and gamma rays and has a half life of 30 years. These radioactive emissions can cause mutations and cancer.

Caesium, being chemically similar to potassium, can be mistaken for potassium by algae. Therefore it can proceed up the marine food chain and eventually can become concentrated in higher animals.

CADMIUM

Cadmium is a silvery white metal with properties similar to zinc. It occurs in various minerals in association with zinc. It is used in industry for galvanic plating to protect iron and steel from corrosion, for low-melting point allowed, and to moderate the chain reaction in nuclear power stations. Cadmium sulphide (yellow) and cadmium selenide (deep red) are basic substances for dyes, which are used in ceramic glazing, porcelain dying and plastics. In the agricultural field, cadmium occurs in artificial phosphate fertilizers and in certain plant protection agents.

Cadmium is produced and utilized almost exclusively in the industrial countries of the world. Cadmium is a toxic metal. In the environment, cadmium is dangerous because many plants and animals absorb it efficiently and concentrate it within their tissues. Normally, however, retention from food by mammals is low but absorption is increased if the mammals are on a low calcium diet. Once absorbed, cadmium associates with the low molecular weight protein, metallothionein, and accumulates in the kidneys, liver and reproductive organs. Very small dose can cause vomiting, diarrhoea and colitis. Continuous exposure to cadmium causes hypertension, heart enlargement and premature death. There is some evidence suggesting that cadmium can induce chromosome abnormalities and may exert a carcinogenic effect on the lungs.

Cadmium poisoning manifests itself for the most part in damage to the bone and joint structures. Over the years 1940-1958 in the Tojama district in Japan, 130 people died from an undiagnosed disease ("itai-itai" disease). The victims died in paroxysms of pain from neuralgia. The disease was regarded in the first years as something to be ashamed of, so that cases were concealed as far as possible from public knowledge. At postmortem examinations, cadmium was found in the kidneys, liver and skeleton in amount in the range of 4000-6000 ppm. The first symptoms of "itai Itai" disease are albumin in the urine, kidney trouble, and a decline in the phosphorus content of the blood serum. The patients suffer pains in the lumbar region and in other joints. The cause of this is disintegration demineralization of the bones in consequence of cadmium poisoning. As these symptoms were found for the most part in older, physically ailing women, the cadmium catastrophe did not attract the same attention as the Minamata catastrophe. That mainly women were afflicted with this disease is explained by the fact that during pregnancy a woman's body absorbs cadmium like a sponge, and if the diet is poor in vitamins and calcium, the body builds cadmium up instead of the calcium necessary for the bone substance of the fetus. As far as is known at present, the pregnancies took a completely normal course. But 10 to 30 years later the "itai-itai" disease developed in the aging mothers. Further manifestations of this malady are fatigue, kidney trouble, pains in the ribs, loss of teeth, and pains in the spinal column and the pelvis. Owing to the buildup of cadmium in the bones, they turn brittle and break under the slightest strain (see fig. below) The victims in Tojama had drawn all their drinking water from a river into which cadmium was fed in waste water from chemical factories.

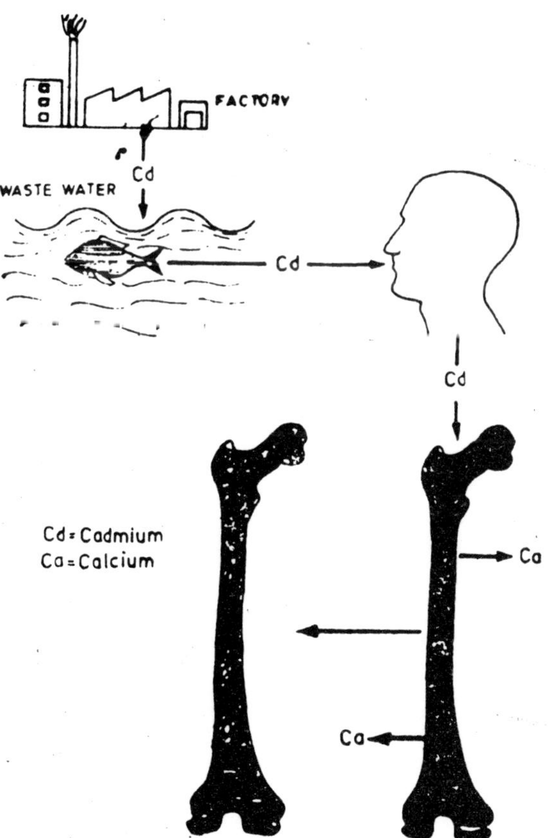

Cadmium Works its way into the Bone Structure and makes the Bones Brittle.

A similar cadmium catastrophe occurred along the Tama River in Japan. Some 20 metal alloy undertakings discharged cadmium in waste water into the river. As the river water was used in the irrigation of rice fields, several hundred hectares of land under cultivation were contaminated with cadmium. The rice field was found to contain 90 ppm of cadmium, the irrigation channels 35-220 ppm and mud in the Tama River 380 ppm. Even at the river mouth 40 km away, the water contained 0.8 ppm of cadmium.

The occurrence of cadmium in river waters in different parts of the world is reported to be in the range 1 to 100 mg Cd per kg of the sample. Srivastava and Jaiswal (1989) have reported that whereas cadmium retards growth in the water plant *Spirodela polyrrhiza* is induces the formation of vegetatively propagating organ called turion. This is a district adaptive strategy to ensure successful recurrence of the species through turions after the stress is over in the water environment. They have noted that cadmium toxicity in this species is in the concentration range 0.005 ppm to 2.00 ppm and the adverse effects are on pigmentation and protein concentration.

The concentration of Cd in sea water on an average is 0.15 mg per litre 16-17 mg Cd per kg of sample in fish sold in Bombay has been reported which is of the same order reported in other parts of the world. The toxic levels are 200 mg Cd per litre of water.

CHROMIUM

Metallic element; hard, brittle grey metal obtained from chromite by direct reduction. Hexavalent chromium compounds have an irritating and corrosive effect on tissue resulting in ulcers and dermatitis on prolonged contact. It is a known carcinogen. Tolerance for chromium dust and fume is 0.5 mg per cubic metre of air.

For most organisms, it is essential as a micronutrients in trace quantities for fat and carbohydrate metabolism. In industry, it is used in making steel alloys, in chromium plating and in leather tanning. Chromates are water soluble and can poison sewage treatment processes. The chromium ion can exist in four valency states; Cr^{+2}, Cr^{+3}, Cr^{+5} and Cr^{+6}. Of these, the heaxavalent ion is the most toxic and it should be reduced to the trivalent state to form insoluble products before chromium waste is released into the environment. Hexavalent chromium has been implicated in poisoning in Japan. In this case, aerosols from chromium refining plants appear to have affected a considerable number of people causing lung cancer. Besides this, it has been shown that chromates act as irritants to the eyes, nose and throat, and chronic exposure may lead to liver and kidney damage. A characteristic effect on human beings is the appearance of perforations in the nasal septum. At the cell level it appears that hexavalent chromium may cause chromosome abnormalities. Chromium is particularly dangerous because it accumulates in many organisms. Some aquatic algae have been shown to concentrate it 4000 times above the level of their immediate environment. The maximum permissible limit of Cr in drinking water as recommended by W.H.O. is 0.05 mg per litre.

COPPER

Reddish coloured metallic element, toxic and flammable in finely divided form. Tolerance (fume) 0.2 mg per cubic metre; (dusts and mists) 1 mg per cubic metre.

Copper as a Trace Element

Copper is one of the most abundant trace metals. It is widely used in its metallic state, either in the pure form or in alloys. For almost all organisms it is an essential micronutrient.

The major copper compound in primary rocks is chalcopyrite ($CuFeS_2$) from which natural deposits of copper sulphide have probably originated. From both of these sources copper is gradually released by weathering. Bivalent copper ions are strongly adsorbed on to clay particles in an exchangeable form and copper also forms stable complexes with organic molecules. The concentration of free copper ions in soil solutions is usually low, concentration above 1ppm are toxic to many microorganisms and cause a decrease in soil fertility.

Copper, which is set free from the minerals by erosion, is combined out of the soil solution at the sorption points of the clay minerals. The bonding is a very firm one, so that copper in the ground is mobile to only a limited extent. The copper can be mobilized by strong acids or organic molecules which form complex compounds with the copper.

The absorption of copper by plants depends essentially on the quantity of copper available to the plants in the soil solution.

Copper availability decreases with increasing pH, and with increasing levels of phosphate because of precipitation as insoluble copper phosphates. Copper is not absorbed readily as organic complexes and this sometimes accounts for copper deficiency in humus rich soils.

Copper is a component of several metallo-enzymes including ascorbic acid oxidase, tyrosinase, p-diphenyl oxidase and cytochrome oxidase. The electron-transport component, plasto cyanin, is a copper containing enzyme that functions as an integral part of photophosphorylation. In addition copper is an important component in terminal oxidases, such as polyphenol oxidase, which catalyzes the oxidation of phenolics to ketones during lignin formation and in tanning. The copper appears to act as an intermediate electron acceptor in the direct oxidation of substrates by molecular oxygen through its ability to undergo reversible oxidation and reduction between the cupric (Cu^{2+}) and cuprous (Cu^{+}) forms.

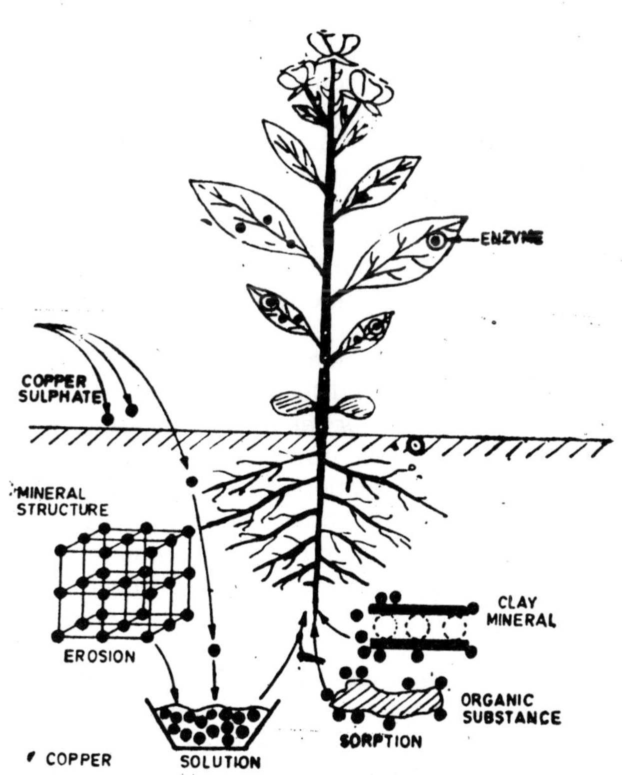

Copper Feeding of Plants.

To improve the copper supply to cultivated plants, copper sulphate and metal meal flour containing copper are often added as fertilizers. How small is the consumption of copper is shown by the absorption value of an average corn crop, which amount to no more than 20 to 30 grams of copper per hectare. To ensure adequate fertilization, however, many times this quantity must be distributed. In deficient areas, 5 to 10 kg of copper per hectare is recommended.

Copper deficiency is well known in cirtus and other deciduous fruit trees in many parts of the world. Leaves become dark green and twisted and later develop interveinal mottling and necrosis. The main shoots may die back and lateral buds grow out giving a '**witches broom**' effect. The bark becomes roughened and eventually splits, causing gum to exude, a condition knows as **exanthema**. In cereal crops like barley, wheat and oats which are most sensitive to copper deficiency the young leaves become limp and chlorotic and remain tightly rolled while the tip turns white and eventually collapses (white tip). If there is a severe shortage of copper, the ear or panicle formation is arrested, while in more favourable cases the ears remain at least partly sterile. In tomato plants the foliage is at first abnormally dark blue-green, then becomes paler and inrolled before premature withering.

Toxicity

Copper may occur in very high concentrations in water, sediments and biota in some localized areas as a result of mining activities, of intensive use of copper pellets in pig rearing, or of the application of copper fungicides. However, there is no evidence of food chain magnification. Hence, most toxic effects are due to immediate exposure to the element. All organisms are harmed by excessive concentrations, which may be as low as 0.5 ppm for algae. Most fish are killed by a few parts per million. In higher animals brain damage is a characteristic feature of copper poisoning.

The presence of copper has been reported in Periyar river water, sediments and fish from South India. The upstream region water contained 0.05 to 0.1 ppm copper and the industrial area 0.025 to 0.3 ppm while the sediment of the respective zone had 1 to 100 and 5 to 100 ppm copper. Fish contained about 0.7 ppm. Copper occurs in water in different forms. Humic acid, present in natural waters is known to form complex with copper. Copper complexes are less toxic to water plants and animals in the pH range 6-8. Copper-humic acid form is common in freshwaters while $Cu(OH)_2$ and $CuCO_2$ are characteristic forms in sea water.

IRON

Silver-white, malleable metal. Mechanical properties are altered by impurities especially carbon. Tolerance (as fume) 5 mg per cubic metre of air.

Iron is the fourth most abundant element in the earth's crust. Its greatest use is for structural iron and steel, but it is also used for making dyes and abrasives. It is an essential micronutrient required in trace quantities for the normal metabolism of plants and animals. It is a constituent of cytochromes and nonheame iron proteins involved in photosynthesis. N_2 fixation and respiratory linked dehydrogenases. Ingestion of excessive amounts may result in the inhibition of activity of many enzymes. The amounts consumed must be very large because only a small proportion of all iron ingested is absorbed from the gastro-intestinal tract. Inhalation of iron dust can cause benign penumoconiosis and can enhance harmful effects of sulphur-dioxide and various carcinogens.

Many streams are poisoned by high levels of iron in acid mine drainage. Pyrite, iron sulphide, is often found in close association with coal deposits. Upon exposure to moisture and atmospheric oxygen, the ferrous iron is oxidized to the ferric the state, a reaction which is frequently accelerated by bacteria of the **Thiobacillus—Ferrobacillus** group. The ferric iron can then react with sulphide in the presence of water to produce sulphuric acid, or react directly with water to produce a yellow, flocculent mass of ferric hydroxide. Besides being acidic, water affected in this way becomes deficient in oxygen. Such poisoning of streams is reckoned to be one of the main causes of fish kill. Although particularly associated with mining. Streams, running through iron-laden strata may become poisoned spontaneously.

LEAD

A metallic element with a bright lustre. It is more than ten times as dense as water; toxic by ingestion or inhalation of dust. Tolerance (as Pb, fumes, dusts and inorganic compounds) 0.15 mg per cubic metre of air. It is a cumulative poison. Lead dust is emitted by different industrial plants in the form of oxides and various salts. The most dangerous source for the contamination of the environment with lead, however, is motor transport. Lead is added to gasoline as an organic compound (**tetraethyl or tetramethyl lead**). Upon combustion of the gasoline the lead is converted into lead oxide, which is discharged from exhaust pipes in the form of fine dust particles capable of penetrating the lungs. Throughout the world, the total annual emission of lead by motor vehicles and industrial plants amounts to something like half a million tons.

Human beings, plants and animals absorb lead from the most diverse sources. From sewage water it gets into ground and river water; from waste gases it gets into the atmosphere and from there into foodstuffs and beverages. By far the greatest volume is inhaled with the air we breathe. While some 5-15 per cent of the lead contained in the food we consume is absorbed by the intestines, the absorption by the lungs of lead from exhaust fumes in the air we inhale amounts to almost 100 per cent.

The damage done by elemental lead released into the atmosphere is extremely difficult to eliminate. Particularly worrying, therefore, is the long-term buildup of lead in the environment. The ground on both sides of busy streets and highways is seriously contaminated with lead to a distance of 0.5 to 1 km. The oceans of the world today have a lead content something like 50 times above the natural level.

Soluble lead slats, and insoluble lead compounds which can be rendered soluble by metabolic processes, are toxic. Initially the symptoms of lead poisoning are uncertain: general upset, loss of appetite, loss of energy, etc. Later on the affected persons lose weight and suffer from anemia. In women the menstrual cycle is disturbed. Characteristic changes in the blood count set in, and the red blood cells appear granulated. In every case a blue line appears around the gums, and there is a slate-grey discolouration produced by a buildup of the metal in the oral cavity. The blue line, which is frequently the first specific sign of lead poisoning, may last a long time, well after the cause of the poisoning has been removed. At a later stage, more serious symptoms of lead poisoning will manifest themselves in such forms as lead colic, feeble pulse and high blood pressure, cirrhoses of the kidney (nephrites), acute or nagging pains in the muscle, and finally brain fever (encephalopathy) accompanied by severe headaches, epileptic-type fits and psychosis. Also established are nervous ailments displaying symptoms of paralysis, arthritis, allergies and other skin reactions, anemia and various metabolic disorders. In cases of chronic lead poisoning severe damage can also be done to the chromosomes, which transmit hereditary factors.

Typical lead poisoning with clearly diagnosable symptoms mainly affects persons working in the lead industry, who in the course of their activity have absorbed large quantities of the heavy metal. Less prominent but probably more problematic is the gradual poisoning of the environment by smaller quantities of lead and the chronic lead poisoning of people, who from childhood onward, are exposed to higher emission of lead throughout their lifetime.

MANGANESE

Brittle silvery metal. Dust or powder is flammable. Tolerance of fume is 1 mg per cubic metre of air; metal and compounds 5 mg per cubic metre of air.

Manganese occurs widely in nature and is of considerable importance in the manufacture of steel. Biologically it is an essential micronutrient for most organisms. It is required for activity of some dehydrogenases, decarboxylases, kinases, oxidases, peroxidases etc. It is required for photosynthetic evolution of oxygen. This metal is not a serious pollutant as in most waters its concentration is quite low ranging from 0.005 to 1 mg per litre. Potassium permanganate is used in very small doses to disinfect well-water particularly in Indian villages. Excessive amounts of manganese affects animals adversely, causing cramps, tremors and hallucinations, manganic pneumonia and renal degeneration.

MERCURY

Mercury is concentrated in various ores, the principal one being cinnabar (HgS). It has been mined since 700 BC and is currently used industrially in three forms : as the metal, in organic compounds and in inorganic compounds. The greatest use of mercury is in the production of electrical apparatus. The second greatest use is in the chloro-alkali industry, which produces chlorine and caustic soda by electrolysis of sodium chloride solution using mercury as the cathode of the electrolysis cell. The third greatest use worldwide is in fungicides.

Nearly all the mercury used by man eventually enters the natural environment. To this may be added amounts released during the production of mercury and other metals, from coal burning and from weathering of rocks.

The first symptoms of mercury poisoning occur in the sensory organs and manifest themselves in itching and numbness in the hands and feet and numbness in the region around the mouth. At roughly the same time, damage develops in the visual center, especially in limitation of the visual angle. The visual angle can decline within a week to a point at which vision is limited to straight ahead, as through binoculars. Damage to the hearing center leads to a reduction in hearing and eventually to deafness. Damage to the cerebellum results in physical movements not being adequately coordinated, culminating in disturbances of equilibrium and serious impediments in speech. The brain shrinks (by up to 35%), the intelligence quotient falls, growth is restricted, and the arms and legs are often deformed by muscular spasms.

All forms of mercury are potentially toxic but the toxicities vary considerably. The least toxic are the inorganic mercury compounds. They are not readily absorbed from the gastro intestinal tract. Once absorbed they may accumulate in the liver and kidney but normally they are excreted quite rapidly in the rurine. It is worth noting that mercury amalgam has long been used in dentistry to fill teeth without any toxic effects being noted. Mercury vapour is the most hazardous of the inorganic forms because it can diffuse through the lungs into the blood and then into the brain, where serious damage can occur. Arylmercurials arc as toxic as the inorganic forms since they are readily broken down to inorganic derivatives in the tissue. Alkylmercurials are the most toxic mercury compounds so far studied. They are fairly stable and have long retention times in the tissues. Therefore, they readily accumulate to high concentrations. Their lipid solubility gives them an affinity for nervous tissue which accounts for many of their harmful effects. In addition, they have been reported to cause abnormalities in cell division and to increase the frequency of chromosome breakages. Some of these abnormalities may be due to combination of mercurials with sulphydryl groups. Inhibition of enzymes in this way has been demonstrated frequently. So far, no effective treatment for mercury poisoning has been developed, though British antilewisite (BAL) or calcium ethylenediaminetetraacetate (Ca_2EDTA) may have some alleviating effect.

Unfortunately, in assessing the risk from mercury in a particular environment, it is not enough to know the form in which it entered that environment because various transformations can take place. Probably the most serious of these is the transformation of metallic mercury to methyl and dimethyl derivatives by anaerobic micro-organisms, especially *Clostridium cochlearium*, in aquatic sediments. This may also occur in decaying fish. Under aerobic conditions, this transformation can be brought about by *Peseudomonas* spp. and by the fungus, *Neurospora crassa*. In essence, it represents conversion of the least toxic form of mercury to the most toxic. Other transformation which can be brought about by bacteria are the following; phenyl, ethyl and methyl mercury can be reduced to elemental mercury and benzene, ethane and methane, respectively; phenyl mercuric acetate can be converted aerobically to elemental mercury and diphenyl mercury; mercuric ions can be reduced to elemental mercury. Not all the transformations observed are biological. Under alkaline conditions, methyl mercury converts to the more volatile dimethyl mercury. Under oxidizing conditions, in the presence of ultraviolet light, phenyl mercury, alkoxy alkyl mercury and alkyl mercury may break down to give inorganic mercury. Under anaerobic conditions, mercuric ions may combine with hydrogen sulphide to form poorly-soluble mercuric sulphide. With subsequent aeration, mercuric sulphide can be converted to the soluble sulphate which may then be methylated biologically.

Most macro-organisms are relatively insensitive to mercury and its derivatives. Nitrogen fixing bacteria in the soil require levels of about 100 ppm, before they are adversely affected. Normal soil levels are between 0.005 and 1 ppm. However, marine and freshwater phytoplankton, especially diatoms, are very sensitive to organomercurial fungicides and as little as 0.001 ppm may reduce their photosynthetic efficiency. Many algae and other plants have the ability to absorb and concentrate mercury from the surrounding environment. Droplets of elemental mercury have been found in chickweed. Such high concentrations may cause mitotic disturbances and kill the plants. Fortunately, most agricultural plants do not seem to absorb much mercury. Animals tend to accumulate mercury through their food. Pike* can accumulate a concentration of mercury 3000 times higher than that in the water in which they live. Tuna and Swordfish show the same ability. Similar observations have been made on predatory birds. Seed-eating birds accumulate mercury where alkylmercury seed dressing are being used. Much less is accumulated where alkoxylakyl compounds are used, and negligible quantities where inorganic mercury compounds are used.

The effects of mercury and its derivatives on man deserve special consideration because it was mainly these that caused concern about the effects of heavy metals released into the environment in large amounts. The first serious incident to come to light occurred at **Minamata Bay** in Japan. In this case, comparatively nontoxic inorganic mercury along with some methyl mercury was released in effluent by a chemical factory using mercuric sulphate catalysts in acetaldehyde production. The effluent entered a river running into Minamata Bay. In the sediments, the inorganic mercury was converted to methyl mercury. This accumulated in shellfish and fish which were eaten by the local inhabitants. Consequently by 1975, 115 people had died and many were left paralysed for life. Others suffered impairment of vision and hearing and other neurological symptoms. Prenatal poisoning of the foetus was observed even in the absence of symptoms in the mother. Since the Minamata Bay incident, another has occurred around the Agana River, Niigata, Japan. This led to 23 deaths. In both these cases, many domestic animals, fish, shellfish and seabirds were affected.

Mercury poisoning of human beings has also been caused by the consumption of food containing high concentrations derived from alkylmercury agricultural seed treatments used to prevent seed-borne disease. In Iraq, treated seeds, intended for planting, were used to make bread. Thousands were poisoned and hundreds died. In the USA, cattle fed on treated grain were slaughtered for human consumption. Again, severe poisoning resulted. Consequently a number of countries have now banned the use of alkylmercurical seed treatments. A committee of experts constituted by the Food and Agricultural Organization (FAO) and World Health Organization (WHO) has recommended that the use of alkylmercurials should be restricted to stocks of cereal seed used for plant breeding or seed production, and never permitted for treatment of cereal seed for export for the production of food.

* A *pike* is a large fish that lives in rivers and lakes, and that eats other fish.

Despite the major incidents referred to above, it seems fairly certain that the average person is at no great risk from exposure to mercury. The normal dietary intake is well below what is thought to be the tolerable limit of 0.3 mg per person per week, of which not more than 0.2 mg should be in the methylated form, according to WHO. For most people, the chances of appreciable exposure from air, pesticides or pharmaceuticals are very limited. Where there is a risk of occupational exposure, this should be minimized by appropriate precautions and medical screening. With regard to the general environment, there is still a need to know more about the concentration and distribution of mercury, especially with regard to those areas where localized high concentrations do exist.

NICKEL

Malleable silvery metal having excellent resistance to corrosion. Flammable and toxic as dust or fume. A known carcinogen. Tolerance. 1 mg per cubic metre of air.

Nickel is used in various forms for nickel plating, as a catalyst, as a mordant and in ceramic glazed etc. It is a micronutrient for most organisms but excessive quantities have toxic effects. In animals these include dermatitis and respiratory disorders, including lung cancer following inhalation. Amongst enzymes inhibited are cytochrome oxidase, isocitrate dehydrogenase and maleic dehydrogenase, A particularly poisonous derivative of nickel is nickel tetracarbonyl, $Ni\,(CO)_4$.

ORGANOMERCURY COMPOUNDS

Organic compounds of the element mercury *e.g.*, methyl mercury, CH_3HgCH_3 methyl mercury ion CH_3Hg^+; ethyl mercury CH_3CH_2Hg or ethyl mercury ion $CH_3CH_2Hg^+$ etc. In 1970 the Environmental Protection Agency (USA) banned all alkylmercury compounds from use as pesticides. In 1976 production of virtually all other mercury- containing pesticides was ordered stopped. Uses still permitted were to combat fungi on fabrics used only outdoors, to control Dutch elm disease and to eliminate mold on freshly sawn lumber.

Methylmercury can be made by several microorganisms from any other form of mercury, including metallic mercury itself. Most of the metallic mercury used each year ends up in the environment. The combustion of fossil fuels releases, worldwide, almost a quarter as much mercury as is produced for commercial purposes. From these sources, therefore, microorganisms will have an essentially permanent supply of raw materials for making methylmercury.

PLUTONIUM

Plutonium, chemical symbol Pu, is an artificially produced, radioactive, chemically and radiologically highly toxic metallic element in the actinide series. It also occurs in minute quantities as a result of natural disintegration processes in nature. It is formed in nuclear reactor from uranium (U) by the following process :

$$^{238}_{92}U \quad + \quad ^{1}_{0}n \quad \rightarrow \quad ^{239}_{92}U$$

$$^{232}_{92}U \quad \rightarrow \quad ^{239}_{93}Np \quad + \quad \beta$$

$$^{239}_{93}Np \quad + \quad ^{239}_{94}Pu \quad + \quad \beta$$

Plutonium has the atomic number 94; the isotopes Pu-232 to Pu-246 are known of which Pu-244, with $8.2x10^7$ years, has the longest half-life. The other isotopes also have a very long life. The most important of them, Pu-239, has a half life of 24,400 years.

As it is highly fissionable by slow neutrons, Pu-239 is used as nuclear fuel and to make atomic bombs (the atomic bomb that was dropped on the Japanese city of Nagasaki on August 9,1945, was a plutonium bomb). It is easier to extract than uranium (U-235), and its critical mass amounts to no more than 8-16 kg. Compared with a critical mass for uranium of 50 kg. Pu-239 is now obtained in large quantities from nuclear reactors, especially from fast breeder reactors. A fast breeder contains 1 tonne of plutonium in its fuel rods.

Plutonium burns in the air to form aerosols capable of penetrating into the lungs. Thus, radiologically it is one of the strongest poisons known. When they get into the human organisms, quantities of as little as one ten millionth part of a gram are sufficient to cause cancer, for plutonium emits alpha rays of limited range which are capable of very high radiation doses locally.

Plutonium is given off into the environment in limited quantities by reprocessing of nuclear fuels. In the future there is likely to be a substantial increase in the transport of plutonium to reprocessing centers by road and rail, so that the possibility of releases of plutonium by leakage or owing to terrorist activities cannot be ruled out. If a transport vehicle containing 25 kg were to overturn in an accident, the plutonium so released could cause up to 250 billion cases of lung cancer. From the same quantity of plutonium, given sufficient technical knowledge, an atomic bomb could be manufactured. The dangerous nature of plutonium is one of the main arguments against using nuclear fission as a source of energy.

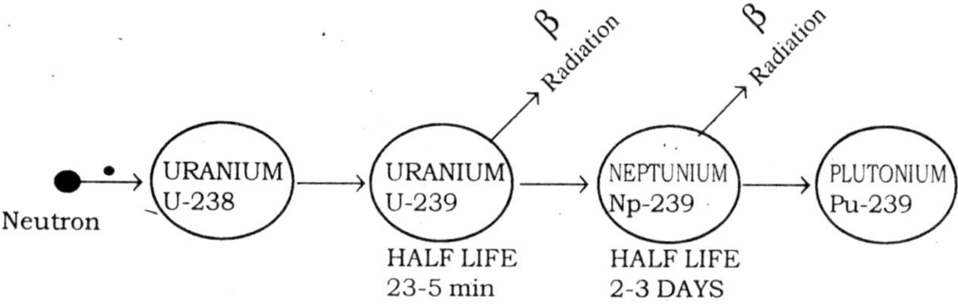

Transformation of Uranium-238 into Plutonium-239.

RADON

A tasteless, odourless, colourless gas one of the most important sources of natural background radiation. This is a 'daughter' (*i.e.*, derived) product from the decay of uranium-238 or thorium-232, and is estimated to provide up to half of the radiation dose from all natural sources. Most of the exposure results from breathing in the gas, particularly indoors.

Radon seeps out of the earth all over the world, the levels vary from place to place. Indoor readings can be much higher than outdoors, (in western countries) because well-insulated houses trap the gas and its builds up. Emissions of gas from the ground can be supplemented by the use of radioactive building materials, such as certain types of granite, alum shale, calcium silicate slag (used for concrete) or phosphogypsum (building blocks, plasterboard and cement).

SELENIUM

Strictly speaking selenium is not a metal, though it has certain metallic properties. It is a member of the sulphur-group, produced as a byproduct of the extraction of copper, nickel, gold and silver ores. It is used in electronics, and in paints and rubber compounds. For most organisms it is an essential micronutrient but it can be toxic at very low concentrations. The maximum permissible concentration in drinking water is 0.01 ppm. Poisoning of livestock has occurred where cattle have eaten plants of the Brassica family which have taken in selenium and incorporated it in cysteine and methionine in place of sulphur. Selenium itself irritates the eyes, nose, throat and respiratory tract. It can cause cancer of the liver, pneumonia, liver and kidney degeneration and gastro-intestinal disturbances.

TIN

Silver—white metal. All organic tin compounds are toxic. Tolerance (organic compounds, as Sn) 0.1 mg per cubic metre of air.

Tin is widely used mainly for making tinplate and various alloys and compounds. It is an essential micronutrient and the main cause for concern regarding its toxicity has been the development of trialkyl-tin and triaryl-tin compounds having powerful bocidal properties. These are used on growing crops as fungicides and insecticides. They are also used as antimicrobial agents and in marine anti-fouling paints. They can damage crops and, in animals, accumulate in the central nervous system with harmful effects.

TRITIUM

Tritium is a radioactive isotope of hydrogen with a half-life of 12 years. It arises in nuclear power stations in large quantities. Its distribution in the environment is very difficult to check simply because its weak beta rays cannot be measured with normal actionometers. When it is present uncombined as a form of elementary hydrogen it can penetrate steel, concrete and other protective screens. Present-day nuclear power stations (1000 megawatts) discharge between 100 and 1000 Ci of tritium into the environment in waste water every year. Still larger quantities of tritium are discharged from reprocessing plants.

Because of the weakness of its beta rays, tritium produces only a low radiation charge, and until recently it has consequently been regarded as a relatively harmless radioisotope. It is now recognized that tritium can also affect living tissue by processes known as transmutations. **Transmutation** is the conversion of one chemical element into another during the radioactive decay of an isotope. In the radioactive decay process the tritium atom emits an electron and is converted into stable helium. Through radioactive decay the chemical element hydrogen is converted into a rare gas.

Such transformation of the element character of an atom changes the chemical properties and consequently the biochemical and physiological importance of the relative molecule in which the transmuted atom is incorporated. There is thus a sort of "suicidal" effect on a molecule when such an atom is contained in it. If the molecules in which the transmuting atom is incorporated are present in large numbers in the organism, the biological risk arising from the destruction of a single molecule is negligible. If it is a question, however, of a controlling molecule that has biochemical or informative importance in a single number or in a few copies, as is the case with micromolecules of DNA, RNA and many proteins, then the transmutation of an atom of this molecule can lead to significant damage to the complete organisms or one of its parts.

In the case of tritium, transmutative damage of the chromosomal structure was observed, inter alia, in cell cultures of the hamster, and the damage was intensified as these cell cultures were subjected to radiation exposure corresponding to the relative nuclear disintegration. A higher rate of mutations than was to be expected from radioactive disintegration was established several time in bacteria in connection with a buildup of tritium. Similar observations were made regarding the rate of recessive sex-related lethal mutations in the fruit fly *Drosophila*. In the case of tritium embedded in thymidine (a component of DNA), damage was caused by transmutations 50.000 times greater than was to be expected by radiation exposure to the tritium itself. In laying down legal limits for tritium, this problem of transmutations has not been considered. Tritium occurs in water from natural sources in a concentration of 40 pCi per litre. The limit for tritium in water was neverthless set at 3.000.000 pCi per litre.

URANIUM

Metallic element of central importance in nuclear energy. Fourteen isotopes are known.

Three isotopes occur naturally in uranium ores; **uranium-234** (0.006% α–emitter; half life 247,000 yrs), **uranium 235** (0.7% α-emitter; half life 713 million years) and **uranium -238** (99% α–emitter; half life $45x10^5$ years). Uranium-235 is readily fissionable isotope of uranium used to enrich natural uranium in nuclear fuels. Its haif life is $7.13x10^8$ years. It was the energy source used in the original atom bomb.

Uranium Enrichment

Natural uranium contains 0.7 per cent uranium-235, with the rest being uranium-238. The majority of nuclear reactors need more uranium 235 in the fuel than this, so the uranium is 'enriched '— an expensive, time-consuming operation, in the early stages involving vast machinery in which atoms of uranium-238 are painstakingly split off from the fuel in order to increase the proportion of uranium-235.

Reactor fuel commonly needs a uranium-235 content of 3 to 3.5 per cent; to get this, some four-fifths of the mass of uranium-238 has to be removed. This is routinely done to provide fuel for most of the world's power stations. If, however, the process is continued and more uranium-238 is taken out of the fuel, eventually the new substance will be mainly uranium 235 and— suitable for making an atomic bomb.

Because of the mechanical processes involved, most of the difficult work of enrichment goes into the early stages, to achieve a uranium-235 content of 3.5 per cent.

VANADIUM

Silver-white metallic element widely distributed in nature. It is non-toxic as a metal. It is used as an alloying element for steel and iron, in making oxidation catalysis and in colouring agents used in the ceramic industry. Large amounts enter the atmosphere from the burning of some petroleums. It is an essential micronutrient and may be accumulated by some marine organisms to concentrations many times higher than those in the surrounding water. Excessive levels in animals inhibit tissue oxidation and synthesis of cholesterol, phospholipids and other lipids, and amino acids. Such levels may also cause precipitation of serum proteins.

ZINC

Shining white metal having a bluish-grey lustre.

Zinc makes up only 0.004 per cent of the earth's crust. Its most important use is as a protective coating on other metals, particularly in galvanizing iron and steel. It is an essential micronutrient and an essential constituent of alcohol dehydrogenase, glutamic dehydrogenase, lactic dehydrogenase, carbonic anhydrase, alkaline phosphatase, carboxy peptidase B, and other enzymes (Noggle and Fritz, 1986.) In most natural waters zinc is found in traces (less than 1 mg per litre i.e., well within the safe limits). Concentrations above 5 mg per litre causes disagreeable taste. In drinking water the level of zinc usually ranges from 0.005 to 1 ppm or mg per litre, but in certain regions it may exceed upto 7.0 mg per litre. Zinc is generally regarded as one of the less hazardous elements, though its toxicity may be enhanced by the presence of arsenic, lead, cadmium and antimony as impurities. Toxic effects have been observed from the inhalation of fumes from galvanizing baths. The 'zinc fever' produced is characterized by chills, fever and nausea. Removal from the fumes leads to complete recovery. Zinc chloride fumes have sometimes caused fatal oedema of the lungs. Zinc or galvanized containers are not recommended for food storage but are acceptable for storing drinking water. This is because acidic foods can dissolve enough zinc from the container to cause poisoning. A factor which serves to minimise the risk of zinc poisoning is that it appears to be lost along food chains, unlike methyl mercury or cadmium for example, which accumulate.

Part IV

Land and Animals

- Land Use
- Landslides
- Pedosphere
- Soil
- Soil Profile
- Soil Structure
- Soil Texture
- Soil Organic Matter
- Soil Chemistry
- Soil Organisms
- Soil Erosion
- Agricultural Wastes
- Soil Pollution
- Soil Development
- Soil Conservation
- Wind Erosion
- Organic Farming
- Clearcutting
- Water Erosion
- Weed Control
- Desertification
- Forest
- Deforestation
- Forest Haze
- Forest Conservation
- Fertilizers
- Tillage
- Animals
- Endangered Species
- Project Tiger
- Wild Life
- National Parks
- Pest Control
- Pesticides
- Insecticides
- Biological control of Plant Pathogens
- Acaricade
- Aldarin
- Benzpyrene (Benzo (a) Pyrene)
- BHC
- Bordeaux Mixture
- Choerdane
- DDD
- DDE
- DDT
- Dieldrin
- Endosulphan
- Endrin
- Fumigant
- Fungicides
- Herbicides
- Lindane
- Nematicides
- PCBs

LAND USE

We are by nature and tradition creatures of the earth, the rain, and the sky. Only a century ago, most people lived on farms, and most of those who did not were close enough to the wide open spaces to come in direct contact with that part of nature that was still beyond the touch of man. Today, only a small percentage of the people remain in rural areas. The vast majority live in the "concrete jungle" or its suburbs.

But the rush to cities is not always the most important influence on the quality of life. The way we plan and manage the resources of the landscape determines the degree to which they can be preserved for permanent use. It is not always the number of people but, instead, their disrespect for the environment and their careless and callous use of machines that have the greatest destructive effect on natural resources.

All around us the landscape changes as each new construction project is completed on the available open space—more streets, more buildings sprawling suburbs, airports, power plants, transmission lines, and factories. Trees and shrubbery disappear and pastures fade away in the murky environs of industrial encroachment as the demands increase for more land for more people. Land development swallows up the open landscape at a very fast rate. The need for more water for people and industries has made it necessary to dig canals and to build dams for temporary impoundment of the water. The need for electricity has forced the construction of new power plants.

As population grows the congestion in recreational areas becomes critical. Beaches and related recreational facilities have been the first to feel the brunt of the encroachment of more people than they can accommodate. Many national parks require reservations several months in advance and some of the mountain areas are just as congested. In short, there are too few outdoor recreational areas for too many people, many of whom are entrapped for most of the year in the pressure-cooker of urban life. People have an increasing need to get out of city apartments. Rapid expansion in the number and size of outdoor recreational areas is one of the urgent needs brought about by the combination of population growth and urban concentration But, too often, recreational and commercial developments are in conflict with good conservation practices. This means, in effect, that desirable open spaces must be selected and preserved before their full usefulness can be nullified by the destruction caused by planless commercial enterprises.

LANDSLIDES

Land slides are one of the major large scale erosional processes on mountain slopes, that have become a prominent hazard with increased laying of mountain roads. The various forms of down slope movement of material under the pull of gravity are collectively termed **mass wasting; soil creep, earthflows, mudflows, rockfalls** and **landslides** are various types of mass wasting. **Soil creep** involves slow down slope movement of soil along a very gentle slope. **Earthflows** are encountered in areas with relatively steep slopes and having heavy rainfall; the soil slides in the form of a sheet within a few hours, exposing bedrock below. **Mudflow** is similar to earthflow, but encountered in semi-arid regions where a thunderstorm may bring a heavy rain, turns barren soil surface into a mud which flows down the slope often causing considerable damage. Rock fall includes pieces of rock breaking away and falling at the base of a rocky cliff and accumulate in the form of a talus.

Landslides are the most predominant forms of land wasting wherein rock, regolith and soil with accompanied vegetation breaks loose without any advance warning to block highways often burrying the unsuspecting labour force or the hutments down slope. A landslide, unlike earthflow does not have an internal flowage. Landslides are a very common hazard in mountain regions where slopes have been cut to construct roads, thus weakening the structural base of the slope above the road. The landslides are generally triggered by a heavy rainfall, that loosens the soil of the slopes. Whereas a rockslide commonly occurs on a solid, relatively flat rock surface, the landslide generally carves in curved surface, tilting the contents of the land slide, and is also termed **slump block.**

PEDOSPHERE

The upper most surface of the earth consists of loose and solid stone and rock and humus. Changes of temperature, the freezing and thawing of water in rock fissures, and the dissolving and crystallization of salts reduces the rock to such an extent that loose weathering products result. Under the influence of weathering agents such as oxygen, carbon dioxide and water, many minerals on the upper surface so created are subject to chemical erosion, as a result of which further decomposition products are formed. The layer of weathered material so created comes to be occupied by lower-and higher-order organisms, by whose exudations and intake of mineral nutrients the decomposition is carried further. The dead remains of plants, and animal form the organic substance which is decomposed by microorganisms to form carbon dioxide and release minerals.

The outcropping geological seams are thus reconverted into soil by weathering and new clay formation, the decomposition of organic matter and humus formation, and the transposition of soil components (by wind or water).

The pedosphere is no precisely demarcated zone, but it passes gradually into the other zones and is in turn infiltrated by them. The pedosphere is a meeting ground for air from the atmosphere, water from the hydrosphere, minerals from the lithosphere, and, with the settling of the products of weathering, the organisms of the biosphere.

In contrast to chemically defined substances, the soil components constitute a three-phase system consisting of a solid element (minerals), a fluid element (water) and a gaseous

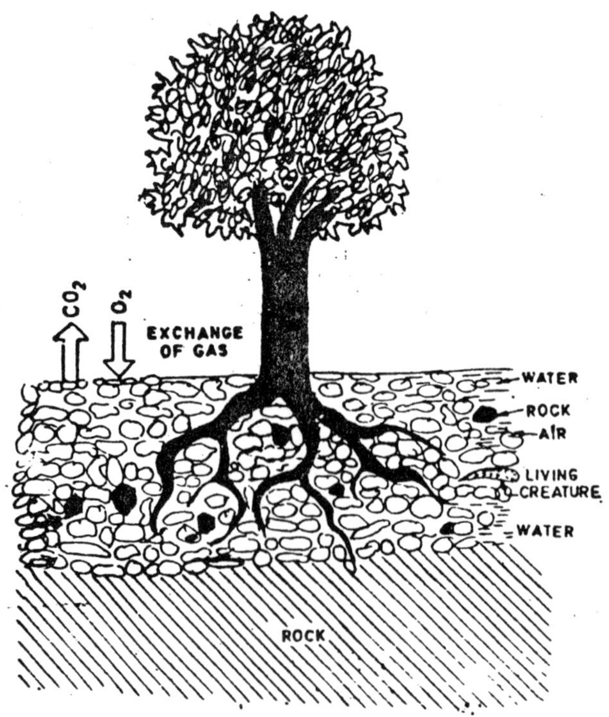

The Pedosphere.

(ground air), which are in a state of constant flux. The parent material from rocks in the form of sand silt and clay, newly formed clay minerals, and salts belongs to the inorganic components of the ground. Living organisms such as bacteria, fungi, mites and worms belong, like the roots of plants and dead organic matter, to the organic components of the ground. A distinction is drawn between mineral soils, which are formed from rocks and stone, and organic soils (peat) which come exclusively from dead organic substances.

The soil is the habitat for plants and offers living space to a number of ground animals. The plants draw from it their needed nutrients, water and oxygen. In return, from the roots of the plants the earth draws carbon dioxide and absorbs their waste substances. These properties are combined in the term ***soil fertility.***

SOIL

The nature and quantity of plant life that can be grown on a given piece of land depend strongly on the characteristics of the soil. Study of the properties and distribution of the soil types is therefore essential both for an understanding of existing patterns of vegetation and for determining strategies and prospects for success in increasing the yield of plant materials used by human society.

Soil is the weathered top surface of earth's crust constituted by mineral matter (sand, silt and clay), organic matter in different stages of decomposition (humus), microorganisms (bacteria, fungi, etc.) mixed together in such a way as to have capillary and non-capillary pore spaces filled with moisture and air. These components are not simply mixed up to constitute soil but, organise themselves, through a series of processes to give proper structure to the soil, commonly in the form of layers or horizons. The distribution of these horizons makes up its soil profile. A soil has a district origin, undergoes development to form a distinct structure. It is thus more appropriate to define soil as a natural body of mineral and organic constituents, differentiated into horizons, which differ among themselves and from underlying material in morphology, physical make up, chemical composition and biological characteristics. Soil is the medium of anchorage and supply of nutrients and water for the plants and in turn the plants are the ultimate source of food for all animals and man. Thus soil is the basic and most important life support component of biosphere.

Normally, we do not feel concerned about soil as we tend to think that it is most abundant and a never exhausting resource found all over the land surface. A city dweller may even regard it only as a nuisance in the form of dust. But with the rise in human population requiring more and more production of food, the importance of soil and its fertility has now been well recognized. Even in ancient times the importance of soil as a resource for crop cultivation was recognized and human settlements established on plain cultivable lands where good soil and abundant water were available round the year. Man has also realised that all living beings are ultimately the product of soil, and after death, the body materials are returned to the soil.

More than 3000 years ago Mosses had said... "dust thou art and unto dust shall thou return" emphasizing the fact that out of dust the food is produced which goes to build our body and after death, the body slowly decomposes and becomes part of soil. The cycle may be long for the nutrient elements. From the absorbed surface of colloidal clay and humus it enters the roots, gets transported to leaf, enters biosynthetic pathways to become a complex organic compound that goes to build the body of the plant or some organelle. Then some of it is grazed by a herbivore, and some others remain stored in stem for a long period and still some others are returned to the soil in the form of leaf and twig litter. Some of it goes into the formation of seeds and propagules. Herbivores, carnivores and decomposers ultimately derive their food from one of these sources. Thus soil is, the foundation of biosphere and the fountain head of life on the earth. Soil is a product of climatic, physiographic and biotic forces acting upon the rock surface in geological and recent past. The process of soil formation is called **pedogenesis**. One very important fact that must be borne in mind is that soil is a living system maintained at its position by the binding and protective effects provided by plants. Its nutrient or fertility status is delicately balanced by the input of dead organic matter and the decomposer microorganisms that harbour soils. A soil devoid of its microfauna and microflora or of organic matter is no good. A soil, if kept devoid of plant growth for a few years is bound to get eroded and lost. While it takes several decades or hundreds of years to build a think column of soil, it would take only a few years of misuse to get it lost for ever. Eroded soils with exposed rocks cannot support life. Deforestation leads to erosion and erosion leads to desertification.

SOIL PROFILE

The various processes of soil development result in the differentiation of a soil into a series of horizontal zones, which differ in morphology, chemical composition and biological activity. These zone are called soil horizons. The arrangement of horizons in a soil is termed soil profile and can be studied by taking a vertical section of a soil. The upper-most layer or zone is called the **A horizon**. It is the home of most of the soil organism and the location of the greatest abundance of roots. The uppermost layers of A horizon consisting of freshly fallen **litter** and partially decomposed duff are usually designated A_{00} and A_0 respectively (also O_1 and O_2 respectively). The mineralization of humus occurs in A horizon and the released minerals are leached down by percolating water. The zone is accordingly called **zone of leaching** or **zone of eluviation**. The next zone proceeding downward is the **B horizon**. It received downward-moving minerals leached from above and, often, upward-migrating substances from weathered parent materials below; therefore it is called the zone of accumulation or **zone of illuviation**. The **C horizon** consists of the weathered rock material. The bedrock, below usually but not always the true parent material, is the **D horizon**. (see fig. below)

In soil studies more detailed than our treatment here, the horizons are often subdivided by the addition of subscripts running 1 to 3, top to bottom, (that is , A_1, A_2, A_3).

Soils in which the layered structure is well developed and distinct are called **zonal**; those without this well-developed vertical profile are called **azonal**. Alluvial and colluvial soil are often azonal.

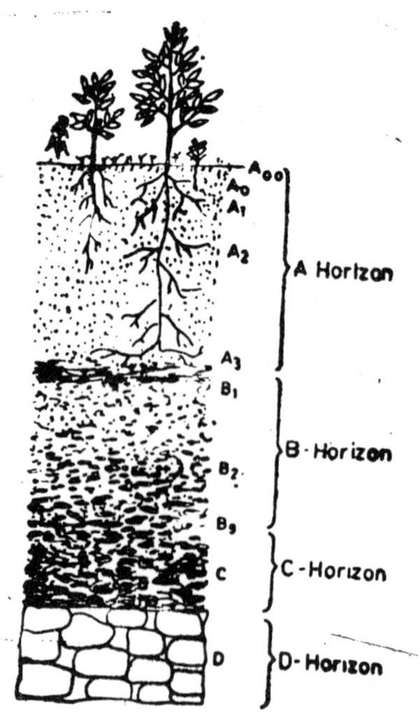

Soil Profile Showing Different Zonations.

A and B horizons together constitute **solum** or **true soil.** The terms **top soil** and **subsoil** do not have a uniform usage. **Top soil** commonly refers to the upper fertile layer of the soil including A horizon (in that case B horizon is sub soil) or both A and B horizon (C horizon being the subsoil).

SOIL STRUCTURE

Soil structure is an important property which enables two soils with essentially similar texture to behave differently. The coarse textured soils though allow better percolation of water and aeration have low binding capacity and nutrient level. The fine textured soils, though impeding water movement and aeration can provide better foothold for the plant roots due to better binding capacity and can hold water and nutrients more effectively. The development of a soil structure enables soils of different types to have their particles aggregated into structural units called **peds** which can hold water and nutrients efficiently. The spaces developed between the peds afford better aeration and water movement.

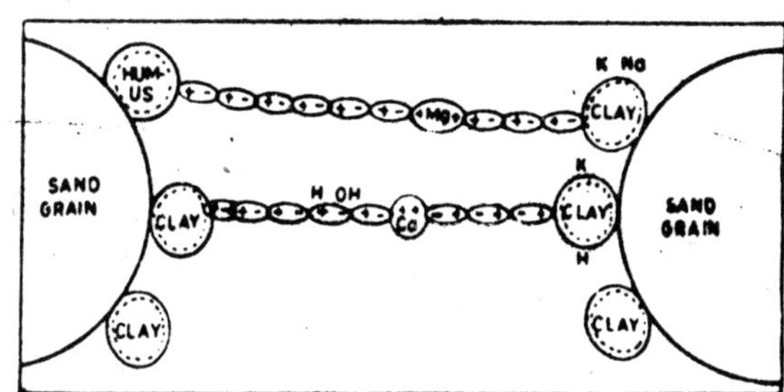

Development of soil structure is a slow natural process. The soils are accordingly graded as structureless (single grained) with no peds, weak structure, moderate structure and strong structure with peds which do not breakup on handling. The structured soils are differentiated on the basis of the shape of peds. These peds may be horizontally flattened (platy), vertically aligned with rounded (columnar) or pointed (prismatic) top, like a cube (blocky) or shape like a sphere (spheroidal). When the structure of a wet soil is destroyed by trampling, the soil becomes ***puddled,*** creating waterlogged situation.

The formation of soil structure if facilitated by the presence of water, divalent cations of calcium and magnesium and Micelle—the clay and humus particles in colloidal range. Numerous clay and humus particles stick to a larger sand grain. These micelle have negative charge on their surface enabling adsorption of cations. The water molecules, being dipolar form long chains which bind on either end to micellar surface, the divalent cations of calcium and magnesium reversing the polarity of the water chain so that single chain is linked to two opposite micelle. The dehydration results in shortening of chains of water molecules. As several such chains are linked to the sand grains, the shortening of water chain pulls the micelle and the sand grains nearer, ultimately leading to their aggregation into peds.

SOIL TEXTURE

Soil is a mixture of small grains of mineral matter formed from decomposing rocks combined with organic matter resulting from the decaying bodies of dead organism. Most of the minerals that plants use come ultimately from rock particles in the soil. Soils are classified by the relative proportion (by weight) of the particles of different sizes they contain (*Soil texture*). The finest particles (diameter less than 0.002 millimetre) give rise to clay soils, lager particles (between 0.002 and 0.02 millimetre in diametre) to silt, and still larger particles (between 0.02 and 2 millimetre in diametre) to sand soils. The type of the soil, however, is not dependent on numerical proportions of the particles (see fig. below). At least 85 per cent of the sand fraction qualifies a soil to be called a sandy soil. On the other hand often 40 per cent clay fraction is sufficient to call it a clayey soil. This is primarily based on the fact that one gm of sand would have fewer particles in comparison to 1 gm of clay. The large number of clay particles thus exert a much stronger influence on the properties of a particular soil. Further more gravel and larger fraction are generally excluded as they have minor role in the characteristics of a soil.

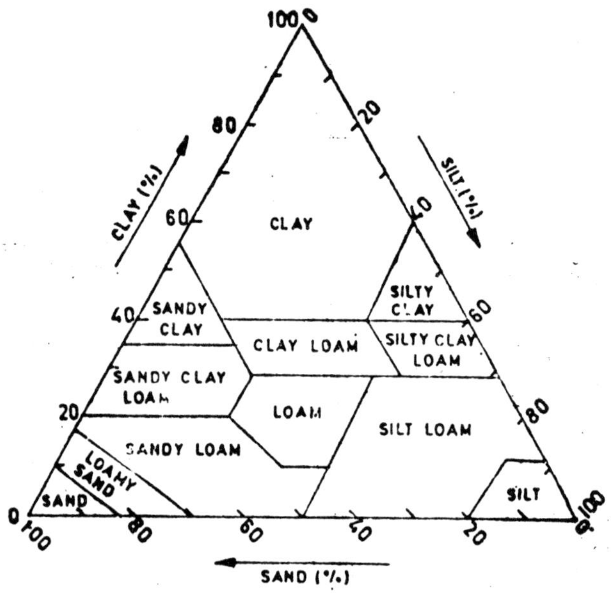

Soil Classification by Texture.

Soil texture influences the rate at which water percolates through soil and the amount of water a soil can hold. Coarse soils are characterized by rapid infiltration, and hence low surface runoff, but they cannot retain much water. The fine textured clays are penetrated by water only slowly but have a high storage capacity. The pore space in soil, which is filled in varying proportions by water and air, is typically around 50 per cent in many kinds of soils; what is more important than the total volume of the pore space is the characteristic size of individual pore spaces. Few large pores make a much less satisfactory soil than many small ones. Soil organism and organic matter are crucial in preventing excessive coagulation of soil particles into large clods.

SOIL ORGANIC MATTER

Organic matter constitutes dead remains of plants and animals, which fall upon the ground.

Organic matter in soil not only influence pore size, but itself serves as a sponge that soaks up and retains moisture. Through its decomposition by bacteria, it acts also as a source of carbon dioxide, water, and mineral elements. Certain constituents of dead organic matter, such as waxes, fats, lignins, and some proteinaceous materials, resist decomposition and are converted instead into the dark colloidal substance called ***humus***. (A colloidal substance consists of particles larger than molecules but small enough to remain suspended in solution). The physical and chemical properties of humus affect the character of soil out of proportion to its fraction by weight, in part because of the large surface-to-volume ratio associated with particles of such small size. The fraction by weight of dead organic matter in soil is often in the range of 0.4 to 1.1 per cent.

The organic matter, deposited upon the soil surface often forms three distinct layers, freshly fallen litter on the surface, partially decomposed duff with intact fibres and other hard tissues and finally humus, a dark amorphous mass. Under conditions favourable for microbial activity organic matter content of the soil is generally less than 50 per cent. Such soils are termed mineral soils. In cold climates, especially under conditions of poor drainage, the organic matter decomposition is very slow, resulting in piling up of partially decomposed organic matter. The soils with more than 50 per cent organic matter are termed **peat**. Peats are generally named after the source-**moss peat, sedge peat, woody peat** and so on. A submerged peat is termed **low moor peat**, while one above the water level as **high moor peat. Bog peat** is low in nutrients (**oligotropic**) acidic and develops in areas of high rainfall having waterlogging Fen peat is rich in bases (**mesotropic**) neutral in reaction, and develops in areas where waterlogging results from high water table.

SOIL CHEMISTRY

The chemistry of a soil is governed largely by the properties of its clay particles and the similarly tiny particles of humus. Clay particles are platelike, with a layered internal structure that leaves negative electric charges arrayed on the surfaces of the plates. Humus particles are also negatively charged on their exteriors. These electrical properties account for the characteristic of soil exchange capacity (the ability to retain and exchange cations such as H^+, Ca^{2+}, Mg^{2+}, K^+, and Na^+).

This function is crucial in governing soil fertility. Without the negative charges on particles of clay and humus, the positively charged nutrients cations released to the soil by the decay of organic matter or added to the soil in fertilizer would travel with water down below the root zones, where it becomes unavailable. This movement of the free ions into the sub soil is called leaching. Bound to the negative charges on clay and humus, however, those ions are made available to plants gradually when they are replaced at the negatively charged sites by hydrogen ions from the soil. There is actually a replacement hierarchy—hydrogen, calcium, magnesium, potassium, sodium—in which each ion listed tends to displace any that appears to its right in the list, if the two are present in equal amount; hydrogen tends to replace all of the metallic ions. If a particular cation is present in very large quantities, however, it can even replace ions to its left in the hierarchy by sheer force of numbers (that is, by mass action).

The magnitude of exchange capacity depends on the number of negatively charged sites that are available in a soil. The unit of measure is milliequivalents per 100 grams, which means the number of milligrams of hydrogen ions (H^+) that will combine with 100 grams of dry soil. Different types of clay have widely varying exchange capacities (in the range of from 10 to 100 milliequivalents/100 g), whereas humus has the greatest of all (from 150 to 300 milliequivalents/ 100 g). This chemical role of humus is perhaps even more crucial than the roles humus plays in maintaining soil texture and retaining water. The importance of all three functions provided ample reason for viewing with alarm the depletion of soil humus by certain agricultural practices.

A major source of hydrogen ions in the soil is the production of carbonic acid from the solution of carbon dioxide in water. Another source is nitric and sulphuric acid added by polluted rainfall or formed from nitrogen and sulphur compounds produced by decomposition of organic matter or added in fertilizer. Yet another is organic acids exuded by plant roots or produced by decomposition. Excessive concentrations of hydrogen ions (acidic soil) replace nutrients cations on clay and humus colloids faster than the rate at which the nutrients can be taken up by plants; the result is that the nutrients are leached away. Acidic soils are common in humid climates. A shortage of hydrogen ions (strongly alkaline soil), on the other hand, can leave some nutrients ions too tightly bound to clay and humus to be absorbed by plants. Other nutrients, such as iron and manganese, become trapped in compounds that are extremely insoluble in basic solutions.

Crops differ widely in the pH ranges they tolerate. Most of the grains flourish best in slightly acidic soil, potatoes and berries in quite acidic soil, alfalfa and asparagus in neutral soil. When soil is too acidic for the crops desired, it is common practice to neutralize the soil by adding lime.

SOIL ORGANISMS

Soils are not just collections of crushed rock; they are extraordinarily complex ecosystems. The animals of the soil are extremely numerous and varied. In forest communities upto 300 million small invertebrates live in each hectare of soil. Some 70 per cent of them are mites, a group of arthropods that may eventually be found to be as diverse as insects. In a study of pasture soils in Denmark, as many as 45,000 small oligochaete worms, 10 million nematodes (roundworms), and 48,000 small arthropods (insects and mites) were found in each square metre. Even more abundant are the microflora of the soil. More than a million bacteria of one type may be found in a gram of forest soil, as well as almost 100,000 yeast cells and about 50,000 bits of fungus mycelium. A gram of fertile agricultural soil can yield over 2.5 billion bacteria, 400,000 fungi, 50,000 algae, and 30,000 protozoa.

The chief activity of the soil organisms, mainly bacteria is the decomposition of organic matter, resulting in conversion of freshly fallen litter into duff, humus and finally releasing the minerals which are added to the soil complex through two processes: humification (formation of humus) and mineralistaion (release of minerals from complex molecules). Different sets of bacteria are often involved in the final stages. Thus ammonifying bacteria convert the complex nitrogenous compounds (*e.g.,* proteins) into ammonia compounds. The nitrifying bacteria are involved in the conversion of ammonia compounds first into nitrites (*Nitrosomonas*) and finally nitrates (*Nitrobacter*). Soil bacteria may also fix atmospheric nitrogen symbiotically (*Rhizobium* in root nodules) or nonsymbiotically in aerated soils (*Azotobacter*) or in aerated soils (*Closteridium*).

The soil organisms help in improving soil aeration and mixing of soil. The burrowing animals such as rodents, worms and insects play an important role in aeration and soil mixing. As much as 25 tons of soil may pass through alimentary canals of earthworms and deposited on surface of 1 hectare of soil in one year. Ants and termites are also active, often locally. In Amazon valley, and locally called Saubas, can build galleries 15-30 m long, 3-6 m across and 0.5 to 1.5 m high. Termites in Turkestan build hills 5 m in diameter and upto 2 m high.

In most natural situations, the plants, animals and microorganisms of the soil are absolutely essential for its fertility. The beneficial effects of earthworms are well known, but most people are unware of the complex ecological relationships within the soil that make it suitable for the growth of oak trees, sagebrush, corn, or any other plants. The soil contains microorganisms that are responsible for the conversion of nitrogen, phosphorus and sulphur to forms available to the plants. Many trees have been found to depend on associations with fungi. The fungi get carbohydrates and other essential substances from the roots, and the root-fungus complex is able to extract from the soil minerals that could not be extracted by the root alone. Such mycorrhizal associations are just beginning to be understood, but it is clear that in many areas the visible plant community would be drastically altered if the mycorrhizal fungi were absent from the soil.

Recognizing that most of the complex physical and chemical processes responsible for soil fertility are dependent upon soil organisms, environmental biologists are opposed to the continuing treatment of soils with heavy dosages of deadly and persistent poisons.

However, not all activities of soil organisms are beneficial. These cause injury to the roots of plants, are the causes of several diseases, remove water and important nutrients from the soil as also some denitrifying bacteria convert nitrogenous compounds into free nitrogen, which escapes into the atmosphere. These are nevertheless important activities to maintain the delicate balance of our biosphere.

SOIL EROSION

In the recent past world food output has more than doubled. This was an impressive achievement. But this remarkable feat has a high price. Some of the agricultural practices that boosted food production have also lead to excessive soil erosion.

Because of both population growth and rising affluence, world demand for food climbs higher each year. In the face of this continuously expanding demand and the associated increase in pressures on land, soil erosion is accelerating. Soil erosion is a natural process, one that is as old as the earth itself. But today soil erosion has increased to the point where it far exceeds the natural formation of new soil. As the demand for food rises, the world is beginning to mine its soils, converting a renewable resource into a nonrenewable one. Even in an agriculturally sophisticated country like the United States, the loss of soil through erosion exceeds tolerable levels on about 44 per cent of the cropland.

Over most of the earth's surface, the thin mantle of top soil on which agriculture depends is 15 to 20 cm thick. The ever-increasing demand for agricultural products contributes to soil erosion in many ways. Throughout the Third World farmers are pushed onto steeply sloping, erosive land that is rapidly losing its topsoil. Elsewhere such, as the American Midwest, many farmers have abandoned ecological stable, longterm rotations, including hay and grass, as well as row crops in favour of the continuous row cropping of corn or other crops. In other areas farming has extended into semiarid regions where land is vulnerable to wind erosion when plowed.

The loss of topsoil affects the ability to grow food in two ways. It reduces the inherent productivity of land, both through the loss of nutrients and degradation of the physical structure. It also increases the costs of food production. When farmers loss topsoil they may increase land productivity by the use chemical fertilizer, on through irrigation to offset the soil's declining water absorptive capacity. Farmers losing topsoil may experience either a loss of land productivity or a rise in costs. But if productivity drops too low or costs rise too high, farmers are forced to abandon their land.

Depletion of soil nutrients and humus need not follow from intensive cropping, as demonstrated by the continued fertility of rice paddies in Asia that have been cultivated continuously for thousands of years. That success has been due to several factors : the high clay content of the soil, the annual augmentation of the alluvial topsoil in the paddies by silt eroded from hilltops, the nitrogen flxation by blue-green algae in the flooded paddies, and the practice of returning human and animal waste to the soils. Intensive cultivation without either natural or artificial augmentation of nutrients can only exhaust soil fertility, however, and even the use of inorganic fertilizers alone does not prevent the depletion of the humus. The practical application of the detailed study of soils is to discover both the potential and the limitations of different soil types for the production of vegetation, and the procedures appropriate to using each type as a permanent or renewable resource rather than as a "mine" to be exhausted and abandoned.

Kinds of Erosion

Soil erosion in nature is a slow geological process, which under stable conditions is slower than soil forming processes or at least in equilibrium with these. This is termed *normal or geological erosion.* When, however, the process of soil erosion is speeded up by rare natural causes or the activities of man, it is known as *accelerated soil erosion.* It may be in the form of water erosion, wind erosion, Land slides and stream bank erosion. The physical erosion, chemical erosion and biological erosion which concern breakup of rocks are more appropriately covered under physical weathering, chemical weathering and biological weathering.

The Causes of Soil Erosion

The apparent increase in soil erosion over the past generation is not the result of a decline in the skills of farmers but rather of the pressures on farmers to produce more. In an integrated world food economy, the pressures on land resources are not confined to some particulars countries; they permeate the entire world. Many traditional agricultural systems that were ecologically stable recently (around 1950) when there were only 2.5 billion people in the world, are breaking down as world population moves toward 5 billion. The exploits of animal husbandry resulting in increased cattle have put a lot of pressure on postiure lands, destroying much of natural ground cover, making large areas barren and susceptible to erosion loss. In an effort to find more space for growing crops, forests have been cleared. The uncontrolled deforestation has led to reduction in catchment area. The resultant increase in floods has been one of the major causes of erosion.

Over the past thousand years as the demand for food increased, farmers devised ingenious techniques for extending agriculture onto land that was otherwise unproductive while skill keeping erosion in check and maintaining land productivity. These techniques include terracing, crop rotations, and fallowing. Today, land farmed through these specialized techniques still feeds much of humanity. Although these practices have withstood the test of time, they are breaking down in some situations under the pressure of continuously rising demand.

In mountainous regions such as those in Japan, China, Nepal, Indonesia, construction of terraces historically permitted farmers to cultivate steeply sloping land that would otherwise quickly lose its topsoil. Centuries of laborious effort are embodied in the elaborate systems of terraces in older settled countries. Now the growing competition for cropland in many of these regions is forcing farmers up the slopes. There has been no disciplined construction of terraces of the sort their ancestors built, when population growth was negligible by comparison. Hastily constructed terraces on the upper slopes often begin to give way. These in turn contribute to landslides that sometimes destroy entire villages. For many residents of mountainous areas in the Himalayas fear of these landslides has become an integral part of daily life.

Research in Nigeria has shown how much more serious erosion can be on sloping land that is unprotected by terraces. Cassava planted on land of a 1 per cent slope lost an average of 3 tonnes per hectare each year, comfortably below the rate of soil loss tolerance.

On a 5 per cent slope, however, land planted to cassava eroded at a rate of 87 tonnes per hectare annually — a rate at which a topsoil layer of six inches would disappear entirely within a generation. Cassava planted on a 15 per cent slope led to an annual erosion rate of 221 tonnes per hectare, which would remove all topsoil within a decade. Intercropping cassava and corn reduced soil losses somewhat, but the relationship of soil loss and slope remained the same.

Throughout the Third World increasing population pressure and the accelerating loss of topsoil seem to go hand in hand. In some parts of the world farmers have been able to cultivate rolling land without losing excessive amounts of topsoil by using crop rotations. Typical of these regions is the mid-western United States, where farmers traditionally used longterm rotations of hay, pasture, and corn. Fields planted in row crops, such as corn, are most susceptible to erosion. By alternating row crops with cover crops such as hay, the average annual rate of soil erosion was kept to a tolerable level. Not only do crop rotations provide more soil cover, but the amount of organic matter that binds soil particles together remains much higher than it would under continuous row cropping.

As the demand for foodstuffs increased and as cheap nitrogen fertilizer reduced the need for legumes, American farmers throughout the Midwest, the lower Missipi Valley, and the Southeast abandoned crop rotations to grow corn or soybeans continuously. The resultant higher rate of erosion was responsible for the loss of considerable top soil in most regions.

Fallowing has permitted farmers to work the land both in semiarid regions and in the topics, where nutrients are scarce. In vast semiarid areas—such as Australia, the western Great Plains of North America, and the drylands of the Soviet Union—where there is not enough moisture to support continuous cultivation, alternate-year cropping has evolved. Under this system land is left fallow without a cover crop every other year a accumulate moisture. The crop produced in the next season draws on two years of collected moisture.

In some situations this practice would lead to serious wind erosion if strip cropping were not practiced simultaneously. Alternate strips planted to crops each year serve as windbreaks fro the fallow strips. This combination of fallowing and strip-cropping permitted wheat production to continue in the western U.S.

In the tropics--such as parts of Africa south of the Sahara, Venezuela, the Amazon Basin, and the outer islands of Indonesia—fallowing is used to restore the fertility of the soil. In these areas more nutrients are stored in vegetation than in the soil: When cultivated and stripped of their dense vegetative cover, soils of the humid tropics quickly lose their fertility. In response to these conditions, farmers have evolved a system of shifting cultivation (also known as jhum cultivation in tribal areas of India). They clear and crop land for two, three, or possibly four years and then systematically abandon it as crop yields decline. Natural vegetation soon takes over the abandoned field. Moving on to fresh terrain, farmers repeat the process. When these cultivators return to their starting point after 20-25 years, soil has regained enough fertility to support crop production for a few year. Under increased population and economic pressures it is, however, often practiced too frequently in same plots and becomes not only unproductive out even destructive.

Another source of accelerated soil erosion in recent years has been the shift to larger farm equipment, particularly in the Soviet Union and United States. In the United States, for example, the shift to large-scale equipment has often led to the abandonment of field terraces constructed to reduce runoff on sloping lands. In dryland farming regions, tree shelter belts that interfere with the use of large-scale equipment have also been removed. The enlargement of fields to accommodate huge tractors and grain combines also reduces border areas that have traditionally served as checks on erosion.

This transformation of agricultural practices has been fueled by the growing worldwide demand for U.S. feed crops, particularly corn and soyabeans, and by the availability of cheap chemical fertilizer. The rise in demand, in turn, has been amplified by population growth that has hastened the deterioration of traditional agriculture in many countries. As a result, agricultural systems throughout the world are now experiencing unsustainable levels of soil loss.

Assessment of the Problem

The first assessment of world soil erosion was carried out by geologist Sheldon Judson who estimated in 1968 that the amount of river-borne soil sediment carried into the oceans had increased from 9 billion tonnes per year before the introduction of agriculture, grazing, and other activities to 24 billion tonnes per year.

Data compiled in 1980 by three Chinese scientists working for the Yellow River Conservancy Commission in Beijing indicated that river was carrying 1.6 billion tonnes of soil to the ocean each year (Table 1). Hydrologists estimate that an average one-fourth of the soil lost through erosion in a river's watershed actually makes it to the ocean as sediment. The other three-fourths is deposited on foot-slopes, in reservoirs, on river floodplains or other low-lying areas, are in the riverbed itself, which often causes channel shifts.

The Ganga of India deposits 1.5 billion tonnes of soil into the Bay of Bengal each year. The Mississippi, the largest U.S. river, carries 300 million tonnes of soil into the Gulf of Mexico each year, for less than the Yellow or the Ganges. Yet it represents topsoil from the agricultural heartland and is thus a source of major concern for U.S. agronomists.

Scientists have recently documented that vast amounts of wind-borne soil are also being deposited in the ocean as sediment. Island-based air sampling stations in the Atlantic, along with recent satellite photographs, indicates clearly that large quantities of soil dust are being carried out of North Africa over the Atlantic. Visible from satellite, these huge plumes of fine soil particles from the arid and desert expanses of North Africa at times create a dense haze over the eastern Atlantic. Estimates of the amount of African soil being carried west in this way reported in four studies between 1972 and 1981 (Prospero, 1981), range from 100-400 million tonnes annually, with the latest report being at the upper end of the range.

Table 1 : Sediment Load of Selected Major Rivers

River	*Countries*	*Annual Sediment Load (million tonnes)*
Yellow	China	1,600
Ganges	India	1,455
Amazon	several	363
Mississippi	United States	300
Irrawaddy	Burma	299
Kosi	India	172
Mekong	several	170
Nile	several	111

Sources : S.A. El-Swaify and E.W. Dangler, "Rainfall Erosion in the Tropics : A State of the Art," in American Society of Agronomy, *Soil Erosion and Conservation in the Topics* (Madison, Wisc : 1982).

Air samples taken at the Mauna Loa Observatory in Hawaii from 1974 through 1982 indicated a continuous movement of soil particles from the Asian mainland, with a peak annual flow consistently occurring in March, April and May, a time that coincides with a period of strong winds, low rainfall, and plowing in the semiarid regions of North Asia. Scientists at Mauna Loa can now tell when spring plowing starts in North China (Parrington, 1983).

India is one of the few other countries to compile a national estimate of soil loss. In 1975, Indian agricultural scientists collected data on local soil erosion from each of the research stations in the national network maintained by the Indian Council for Agricultural Research. Using these figures, they estimated that 6 billion tonnes of soil are eroded from India's croplands each year (Tejwani, 1982). From this and from an estimate that 60 per cent of the cropland is eroding excessively, the excessive topsoil loss can be calculated by subtracting from the total a tolerance level of five tonnes per acre. This yields an excessive topsoil loss from Indian cropland of 4.7 billion tonnes per year, more than twice the U.S. level. This estimate rests on far less data than does the figure for the United States but it is based on information from agricultural scientists familiar with local soils and it is corroborated by data on siltation of hydroelectric reservoirs, river sediments loads, and other indirect indicators.

As in the United States, soil erosion in the Soviet Union has been spurred by the shift to large, heavy equipment and the enlargement of fields, which eliminated many natural boundary constraints on erosion of soil by both wind and water. Each year an estimated half-million hectares of cropland are abandoned because they are so severely eroded by wind that they are no longer worth farming.

Like the United States, the Soviet Union has an extensive dryland farming area and a substantial irrigated area. The European Soviet Union, which accounts for a large share of total farm output, has moisture levels similar to the U.S. Midwest. In terms of rainfall intensity, topography, and erodibility of prevailing soil types, nothing indicates that soil erosion in the Soviet Union would be markedly less than in he United States (Table 2). Where cropping patterns are concerned, the Soviet Union relies much more heavily on small grains, whereas the United States relies relatively more on row crops, such as corn and soyabeans.

Table 2 : Estimated Excessive Erosion of Topsoil from World Cropland

Country	*Total Cropland (million acres)*	*Excessive Soil Loss (million tonnes)*
United States	421	1,700
Soviet Union	620	2,500
India	346	4,700
China	245	4,300
Total	1,632	13,200
Rest of World	1,506	12,200
Total	3,138	25,400

Source : Worldwatch Institute estimates.

When most of the topsoil is lost on land where the underlying formation consists of rock on where the productivity of the subsoil is too low to make cultivation economical it is abandoned. More commonly, however, land continues to be plowed even though most of the topsoil has been lost and even though the plow layer contains a mixture of topsoil and subsoil, with the latter dominating. Other things being equal, the real cost of food production such on land is far higher than on land where the topsoil layer remains intact.

Indirect Effects of Erosion

When farmers lose topsoil they pay for it in reduced soil fertility, but unfortunately the costs of erosion are not confined to the farm alone. As soil is carried from the farm of runoff, it may end up in local streams, rivers, canals, or irrigation and hydroelectric reservoirs. The loss of topsoil that reduces land productivity may also reduce irrigation, electrical generation, and the navigability of waterways.

In India, scores of hydroelectric and irrigation reservoirs have been constructed, many of them with assistance from international development agencies. Two combination of watershed deforestation and steep slopes being cleared for cultivation is yielding record siltation rates. The siltation rates are now several time as high as the rate that was assumed when the projects were designed (Table 3). Not only is the life expentancy of these projects being reduced, but in most cases there will be no alternative sites for dams once the existing ones are rendered useless. A dam site is often unique. Once lost, if cannot be replaced.

Table 3 : Siltation Rates in Selected Reservoirs in India

	(in acre-feet)	
Reservoir	*Assumed Rate*	*Observed Rate*
Bhakara	23,000	33,475
Maithon	684	5,980
Mavurakshi	538	2,000
Nizam Sagar	530	8,725
Panchet	1,982	9,533
Ramganga	1,089	4,3661
Tungabhadra	9,796	41,058
Ukai	7,448	21,758

Sources : S.A. El-Swaify and E.W. Dangler, "Rainfall Erosion in the Tropics: A State of the Art," in American Society of Agronomy, *Soil Erosion and Conservation in the Tropics* (Madison, Wisc,: 1982), and Center for Science and Environment, *The State of India's Environment 1982* (New Delhi : 1982).

The third major indirect effect of soil erosion in the loss of navigability. Perhaps the most dramatic case occurs in the Panama Canal. The combination of deforestation and the plowing the steeply sloping land in the watershed area by landless people is leading to an unprecedented siltation of the lakes that make up part of the Canal. If the trends of the late seventies and early eighties continue, the capacity of the Panama Canal to handle shipping will be greatly reduced by the end of the century, forcing many ocean—going freighters that have relied on its 10,000 mile shortcut to make the trip via Cape Horn.

AGRICULTURAL WASTES

Very broadly, any refuse of any form from agricultural operations of any kind. In common usage, the term generally includes manure and other wastes from farms and the operation of feedlots or poultry houses: slaughterhouse wastes; fertilizer runoff from cropland; harvest wastes; pesticides that escape in the atmosphere or into the water bodies; and salt and silt drained from irrigated land or eroded land.

Before synthetic fertilizers were available, the wastes or manure from small feedlots were not in fact wastes. This manure was spread onto fields for fertilizing next year's crop. Recycling the wastes from large feedlots in uneconomical. It costs energy and materials to load the wastes and distribute them onto fields considerable distances away. Commercial fertilizers have been more economical than manure.

Feedlot wastes have high biological oxygen demands. Their nutrients upset natural ecological communities and nourish plant growth. Water purification plants in communities down-stream have increased burdens of sludges and have more difficulties in eliminating objectionable tastes and odours from the water. Feedlots runoff may also introduce disease-causing organism to the water.

Farmers distribute fertilizers extensively on croplands and pastures. These fertilizers are important to high-yield agriculture. The extensive use of fertilizers, however, creates some problems. Some of the phosphates and the nitrates from fertilizer runoff enter the water bodies causing eutrophication. Nitrates are especially susceptible to runoff with each rainfall and they enter groundwater. Where they are extensively used, they appear in well water. Pesticides in soil are carried into lakes and streams in the dissolved state, as well as on soil particles in runoff.

SOIL POLLUTION

The effect of pollutants on soil are difficult to evaluate. Air borne pollutants emitted by factory stacks travel long distances and slowly deposit on soil. Sulphur oxides present in the fumes are responsible for acid rains and consequent lowering of soil pH. Chlorine and nitrogen oxides are other common gaseous pollutants which combine with water and pollute the soil. Acid rain has been described earlier in the air pollution section. Particulate matter particularly near cement factories, coal transhipment, mining belts, etc., reach soil surface or neighbouring regions. Magnesite dust affect very adversely the soil property such as rise of pH, decrease in exchangeable K, Ca, Mg and available P and K to almost a critical level. Major sources of soil pollution are the seepage of sewage, industrial waste water and sanitary landfills. The materials get leached to also pollute ground water, wells, lakes and ponds.

A number chemicals have now become essential parts of agriculture. Pesticides are now regarded as the number one pollutant on a global scale. In many countries heavy dose of pesticides are used very often. These, indeed prevent crop and vegetable infection and help in increasing production, but their pollution effect to soil is too high. We have to be very selective and careful in deciding the minimal dose and about the mode of application and to strike a balance between the advantages and disadvantages. DDT reaches the human system via soil and water. In India, the illiterate and uninformed farmer takes a handful of pesticide and throws over the crop unmindful of inhaling the chemical and ingesting it through his contaminated hands. Some of the pesticides are very persistent and non-biodegradable and therefore remain in food chain over long periods of time. Under normal agricultural conditions, chlorinated hydrocarbons seem to persist for three to five years, whereas organophosphates and carbamates are gone in one to three months or less.

Considerable evidence already exists that the use of insecticides may reduce soil fertility, especially in woodland soils, which are subject to spraying but not artificial cultivation. Populations of earthworms, soil mites, and insects are dramatically changed, and they in turn affect the soil fungi, which are their principal food. Even if bacteria were not affected directly, there is no question that the general effects on the soil ecosystem would carry over to them and to other microorganisms. But it would be unwarranted to assume that the bacteria were not directly affected. It is known that a few microorganisms can degrade DDT to DDD under the proper conditions; a few can degrade dieldrin to aldrin and several other breakdown products of unknown toxicity.

SOIL DEVELOPMENT (PEDOGENESIS)

Differences in soils around the world arise from the mineral compositions of the parent materials and from differing climatic conditions, which together influence the organic and inorganic processes of soil development. Many important processes are involved in soil development. These include disintegration of rocks into parent material (weathering), breakdown of organic matter (decomposition) by soil organisms, mobility of nutrients by percolating water (hydrogic process), oxidation and reduction. Several major types of soil-forming regimes have been identified: the main ones are **podzolization, laterization, calcification, gleization** and **salinization.**

Podzolization is the set of processes associated in its extreme form with cool climates, abundant precipitation, and acid upper soil layers strongly leached of mineral nutrients and the oxides of iron and aluminium. The nutrients, oxides, and humus accumulate in the deeper layers. Fungi are the main soil-forming organisms. Those soils are characteristic of northern forests, although they exist in some circumstances well into temperate and subtropical regions. The soils developed are known **podzols** and are greyish in colour due to silica in the A horizon.

Laterization also known as **Ferrallitization** is a set of processes associated with the humid tropics and subtropics. High mean temperatures in these regions permit sustained and rapid bacterial action, which minimizes the accumulation of plant litter and humus. In the absence of the organic acids associated with humus, the soil is neutral, rendering the oxides of iron and aluminium relatively insoluble; those oxides accumulate in the upper soil horizons as hard clays and rocklike material called laterite (from the Latin for brick), a mixture of $Fe_2O_3.nH_2O$ and $Al_2O_3.H_2O$.

Calcification occurs in climates where evapotranspiration exceeds precipitation. There is little leaching of the metalliccations, and microbial activity is slow, so the soils tend to be alkaline and rich in humus. Calcium carbonate in solution is carried upward from the water table by capillary action in the season of little surface moisture and is left in the upper soil horizons in solid form when the water evaporates.

Gleization is the set of processes characteristic of poorly drained environments in cool or cold climates. The low temperatures permit heavy accumulation of organic matter, and the excessive wetting produces a sticky clay underneath.

Salinization is the regime characterized by accumulation of highly soluble salts in the soil. The situation arises naturally from poor drainage in regions of low precipitation and high temperatures (deserts), and it can be brought about by faulty agricultural practices—irrigating too parsimoniously in dry climates, or using salt-laden irrigation water. Soils in the salinization regime are weakly to strongly alkaline.

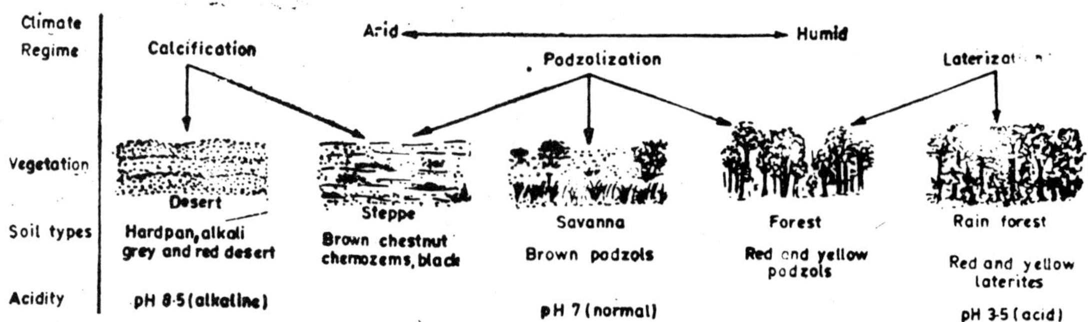

Relation of Soil-forming Regimes, Vegetation, and Soil Types. (Adapted from Janicket. al., 1974).

The geographical boundaries separating the spheres of influence of the different soil-forming regimes are often not distinct. An enormous variety of soils results from variations within regimes and from overlaps between regimes, and a quite complicated classification scheme and toxonomy for soils has been devised to help soil scientists cope with this diversity. Fig. is a schematic illustration of how a few of the principal **soil types** emerge from the soil-forming regimes.

SOIL CONSERVATION

Eager to maximize food output today, we are borrowing from tomorrow. The loss of over 25 billion tons of topsoil from our cropland each year is the price we pay for shortsighted agricultural policies designed to boost food output at the expense of soils, and of failed or non-existent population policies.

Soil erosion is a physical process, but its consequences are economic. As soils are depleted through erosion, the productivity of labourers working the eroding land becomes more difficult to raise. In agrarian societies, deterioration of this resource base makes it more difficult to raise income per person. Further, as growth in food output slows, so does overall economic output. In largely rural, low-income societies with rapid population growth, this can translate into declining per capita income, as it already has for a dozen countries in Africa.

Over the long term, world agricultural trade patterns and the international debt structure will be altered. As soils are depleted, countries, are forced to import food to satisfy even minimal food needs. Scores of countries in the Third World and Eastern Europe find their international indebteness further aggravated by their chronic dependence on imported food. And the loss of topsoil will force an energy for topsoil substitution as it increases the need for fertilizer and fuel for tillage. Other things being equal, land with less topsoil requires more energy to produce our food.

Soil erosion will eventually lead to higher food prices, hunger, and quite possibly, persistent pockets of famine. Although the world economy has weathered a several-fold increase in the price of oil over the past decade, it is not well equipped to cop with even modest rises in the price of food. Although the immediate effects of soil erosion are economic, the ultimate effects are social. When soils are depleted and crops are poorly nourished, people are often under-nourished as well. There is thus an imperative need to save the top fertile layer where it is not fully lost and develop one in soils that have already lost it. The soil conservation, thus becomes an area of major human activity.

The first major step in the soil conservation involves to check uncontrolled felling of trees, to bring back the original forest areas—which were cleared—under forest cover (**reforestation**) and extend forest cover to more new areas not previously under forest cover (**afforestation**). The chipko movement of Himalayas and the Van Mahotsov are important steps in this direction. Over the recent years the expansion of social forestry where in the economic trees are planted along village sides and roads, had gone a long way in helping conservation programmes.

The eastward spreading of Rajasthan desert due to wind erosion has been of major concern in India. This has been partly controlled by growing rows of trees perpendicular to the prevalent wind direction. These **wind breaks** or **shelterbelts** check wind velocity and prevent fertile top soil from being blown away.

Several agronomical practices are useful in soil conservation. **Contour farming** with ridges and furrows along the natural contour lines of a slope helps in collecting water in furrows and reduces erosion loss. **Contour terracing**, wherein terraces are cut along the contour is especially adapted for rice cultivation on slopes. **Stubble mulching** is a practice of leaving the base of the plant with roots intact and depositing stems and leaves on the field, reduces erosion as well as the moisture loss from the soil. **Crop rotation** involves growing different crops, alternated in between by legumes, and helps retain fertility of the soil. **Strip cropping** involves planting different crops side by side in parallel belts to check erosion loss.

In areas with moderate rainfall where normal farming is not advisable **dry farming** is practiced. Rainfed crops such as corn and grazing fields combined with animal husbandry provide suitable alternative conservational method.

The utilisation of erosion-resistant grasses such as *cynodon dactylo* in an important agrostological method of soil conservation and involves either using grasses in rotation with field crops (lay farming) or using lands for grassy pastures.

The methods of soil conservation also include construction of dms along gullies, basins along contours to retain water or ridges along sides of a channel.

These methods of soil conservation not only help in control of soil erosion, they also help in water conservation, flood control and help in attaining maximum productivity from the land.

WIND EROSION

Wind plays a major role in removal of the top fertile layer of the soil in arid and semi-arid climates where vegetational cover is too spares to provide any protective cover, and hardly significant tree canopy to act as wind breaks. The winds, consequently blow strong enough to cause considerable damage. The two common forms of wind erosion include **abrasion** and **deflation**.

Abration or sandblast action results from sand grains and dust particles being lifted and thrown against exposed rock or land surfaces producing pits, grooves and hollowed surfaces. Since sand grains, being heavier do not rise more than few metres, the erosion is limited to generally less than 3 m height from the ground. Strong abrasion effect may result in wearing away the base of telegraph poles, automobile paint or the surface of buildings.

Deflation refers to the lifting of sand and silt particles into the air and their deposition elsewhere. The deposited masses of sand are called **sand dunes**, and those of silt, **loess**. When materials are blown away from a place, it leaves hollowed-out space called blowout or **deflation basin**. When deflation is active in alluvial fans or terraces, the blowing away of silt and sand leaves back pebbles and stones which are closely packed by rolling action of wind, and form desert pavement. A black, iridescent coating of oxides often appears on these pebbles, and stones and is termed **desert varnish.**

The sand particles being heavier in weight generally move close to the ground. The grains may move by bouncing repeatedly along the ground or against other sand grains (**saltation**,Fig.) or especially when wind is slower they creep along the ground surface (**surface creep**).

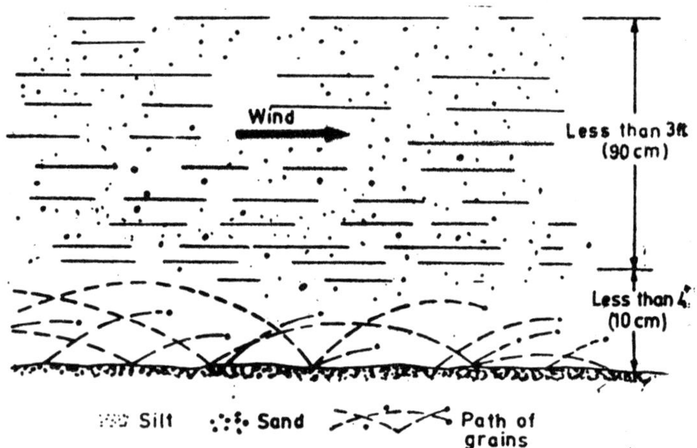

Saltation: The Movement of Sand Grains by Wind.

ORGANIC FARMING

Most farmers in developed countries over the last few decades have used ever-increasing amounts of inorganic fertilizers, coupled with herbicides and pesticides to get maximum yield from their land. Others have felt that these practices not only are harmful to the environment but don't even make sound economic sense. This argument states that the use of herbicides and pesticides is an expensive and inefficient way to control pests and that the loss of humus associated with a programme of chemical fertilizers ultimately reduces soil productivity. It has also been noticed that chemical fertilizers are quickly leached down with percolating waters and readily reach ground water, lakes and rivers. In lakes these cause an early eutrophication bringing down fish farming. *An organic farmer* is one who uses only organic fertilizers such as manure, bone meal, and waste plant products and applies no synthetic pesticides or herbicides.

In recent years, some large-scale commercial farmers around the world have switched to organic techniques even though organic fertilizers are heavy and therefore expensive to spread, and the threat of pest invasions is always a concern. In some cases, yields have improved with the change, but in many other regions total yields decreased somewhat. However, in many cases in which yields were diminished, the farmer's profit went up because operating costs decreased. By reverting to organic techniques, great quantities of energy can be saved and the people would still be adequately fed, but overall yields would decrease. This can prove a viable solution in areas with less land and population pressures. If only short-term factors are considered, agriculture based on fossil fuel energy is the most productive system available. On the other hand, the problems raised by this system are potentially so severe that many believe that the gain of extraordinarily high yields at present is more than offset by the threat of a serious decline in productivity in the future. Organic fertilizers help to maintain top fertile layer of the soil and soil retains its valuable characteristics to reduce losses by erosion and help preserve the soil microorganisms, so vital for the functions of the soil.

CLEARCUTTING

One of the methods of harvesting a forest in which all the trees on a given area of land are cut down in a short period of time. The term "**patch cutting**" is used if the given area is less than an acres.

Clearcutting is a valid means of harvesting a wide variety of timber, and the method has been used extensively in the Scandinavian countries, Germany, japan, and North America. However, it was never properly managed.

In many forests of the east the land was cleared for crops. Clearcutting was the first step in securing land. It was not done for wood. That practice was actually a form of deforestation, and it should not be confused with clearcutting. Deforestation of this type was practiced to provide more land for crops and grazing animals, particularly sheep and goats. Then overgrazing robbed the land of its potential for crops. Clearcutting today is a method for obtaining wood, not land; and when properly managed, it is always done with a view to ensuring that further timber harvests will be possible.

When the trees are mature they may be harvested through either, a **partial cut** or a clearcut. In a partial cut selected trees are harvested, other trees are left. The objective is to provide seed trees to reforest the area. With shadetolerant trees some trees are left as a canopy, as shelterwood. On slopes particularly vulnerable to rapid erosion, a partial cut is desirable to leave something to protect the soil. The partial cut is more costly than the clearcut because generally more expensive, more sophisticated equipment must be used.

Clearcutting requires fewer roads than partial cutting. A clearcut area offers certain advantages. The openings in a forest created by a clearcut provide food for a number of forest animals and birds. If clearcutting is managed poorly, streams nearby become choked with sediments and uninhabitable by desirable game fish. If the clearcut is not taken to the stream edge and if good logging roads are built that minimize sediment-laden runoffs, the deterioration is stream quality does not last going. In the most modern best managed timber harvesting, the roads are built well away from streams. The timber is felled to lie uphill and is stacked uphill instead of rolled downhill, which causes the greater soil surface damage. Tree pulling equipments are also available to save-timber losses caused by splintering and also inflicting less damage to the ground. These practices help reduce the damage both to the ground structure and to nearby streams.

More water runs off from a clearcut area partly because a greater percentage of the rainfall reaches the ground. In a forest, much of a rainfall evaporates from foliage surfaces. If total deforestation and sustained defoliation of other species is used to keep plant life away from a watershed, not only is the runoff considerably higher, but the level of soil nutrients in the runoff also increases.

WATER EROSION

Water erosion is significant in the areas with high rainfall with low vegetational cover and on mountain slopes with fast running water along the down slope. The freshly exposed ground generally experiences the removal of soil in uniform thin layers, when runoff of surface water is considerable. The process is termed **sheet erosion** or **sheet wash**. In areas with steep slope the sheet erosion progresses with more intensity resulting in the development of closely spaced channels or grooves. This constitutes **rill erosion**. The rill erosion may in course of time lead to **gully erosion** in which the rills become larger and integrate into large gullies, ultimately resulting in rugged, barren mountain slopes.

A flat non-slopy ground where, gravitational force is not moving the water faster, the raindrops strike unprotected soil particles, lift and drop these into new positions. When these fallback upon the ground these are carried away by the runoff water. The Phenomenon termed **splash erosion** removes the important silt, clay and humus fraction of the top soil, often lodging them in spaces to further impede the infiltration of the water into the soil.

Stream erosion, the removal of the material from banks and floor of a channel may result from hydraulic action, abrasion (corrasion) and corrosion. Hydraulic action involves the force of moving water dragging sand and gravel which erode poorly consolidated alluvial material from stream bank or its bed. The moving sand and gravel may also strike against rocks and boulders, breaking the pieces away and smoothening the surfaces. The process involved is **abrasion**. The rock material may also be subjected to chemical changes-**corrosion**. Corrosion is very effective on limestone.

WEED CONTROL

Control methods for restricting the growth of weeds. There are various methods to control the growth of weeds.

Chemical control

Chemical weed-killers (herbicides) are used for the most part, especially in agriculture. They are sprayed on the weeds, which wither away under their effects. The herbicides can also have unfortunate incidental effects on other plants, however, and so on the quality of food crops. (See Herbicides).

Mechanical Control

By working the soil (clearing it of weeds, plowing etc.), the weeds can be exterminated or destroyed by mechanical means. The traditional method is very expensive and laborious, however.

Use of Animals

Great success has been achieved by the employment of animals that consume weeds as fodder. Some imported weeds have been treated in this way. One example among many is the reduction of the prickly-pear cactus (*Opuntia*) is parts of Australia. About the year 1840 several *Opuntia* species were imported from Central America into the Australian states of Queensland and New South Wales, where they grew unexpectedly rapidly and by 1925 covered an area of some 24 million hectares. From 1913 attempts were made to check the spread of these cacti. Of some 160 animal species from North and South America known to prey on the prickly pear, 48 species were examined in a quarantine station for possibilities of damage to the ecological equilibrium of their new habitat. Of the 23 species ultimately released, 13 species (12 insect species and 1 mite) were able to propagate. Most successful was an Argentine small butterfly, the caterpillars of which ate seams out of the cacti, while putrefactive bacteria resulted in the death of the cacti. By 1936 a stretch of land in Queensland that had previously been overgrown by a dense cactus thicket was practically cleared of the weeds, and in New South Wales up to 10 per cent more land was available for cultivation.

Use of Other Plants

Eradication of weeds by other plants is based on the assumption that the plants that supersede are not in the nature of weeds themselves but have the effect, as far as possible, of improving the soil conditions. As the term *weed* is purely relative, many plants can have a harmful, neutral or beneficial effect on the cultivated plants and the soil according to their nature and conditions. For the most part a plant has a weed effect if it has a thick root structure which restricts the root growth of useful plants and deprives them of nutrients and trace elements. If a thick-rooted plant is replaced by one with light roots, which allow developing plants to germinate, and if by symbiosis with root bacteria combining with atmospheric nitrogen (as in the case of papilionaceous, or leguminous, flowers) it contributes to improving the soil by enriching it with nitrogen compounds, in most cases the weed problem will be solved. Covering the ground with an intermediate fruit growth of this order will also help to prevent drying out of the soil. The growth of microorganisms and the living conditions of earth-worms that help to break up the soil are also promoted in this way. For these reasons the suppression of weeds by introducing other plants to the area is certainly the most presentable and ecologically most satisfactory method of weed control.

Use of Rusts as Mycoherbicides

Rust diseases are damaging economically but they have been utilized as microbial agents for weed control. In two of the cases it application has been successful. Skeleton weed (*Chondrilla juncea*) has invaded millions of hectares of wheat and rangeland in SE Australia and Western United States. It can be a limiting factor in wheat production under certain conditions. *Puccinia chondrillia* (Chondrilla rust) indigenous to Mediterranean, has been introduced and proven successful in control of skeleton weed in Australia, Development of rust in skeleton weed is accompanied by defoliation and subsequent death of the plant.

Rubus Spp (Blackberry was introduced by European immigrants into Chile in 19th century. It covered some 5 million hectares (12.4 million acres) of farming and grazing land in the country by 1973. *Phragmidium violaceum* a rust of blackberry in Germany was introduced into Chile in 1973. By 1976 it had spread over much of the country. The infected plants show reduced vigour and the loss in competitiveness so is being replaced by more desirable species for grazing.

DESERTIFICATION

Desertification refers to a loss of productivity of the land leading to the creation of a desert where before there had been thriving agriculture. The phenomenon is caused by poor land management and environmental pressures, by human rather than natural factors. It currently affects almost 250 million people. Climate change (the Greenhouse Effect) may aggravate the problem.

The world has five main desert zones : the desert basins that run from north-western Mexico to the south-western United States : the Atacama Desert in south-western South America; the great desert belt running from the Sahara in Africa through the deserts of Iran, the USSR, Pakistan, India, Mongolia and China; the Kalahari desert in southern Africa; and most of the continent of Australia.

Desertification does not occur in these desert areas, but in the arid and semi-arid lands nearby. (Arid land has 200-250 mm of rain per year. Semi-arid has 250-600 mm of rain per year.) Desertification can start almost anywhere, but fertile land at the edge of existing deserts is at particularly high risk. In the Sudan for example, the edge of the Sahara has moved southwards about 100 km between 1958 and 1975.

There are a number of reasons for desertification, the main being overcultivation, deforestation, overgrazing, and unskilled irrigation.

Countries need to grow cash crops: this displaces traditional agriculture, which moves to marginal land. The marginal areas, previously unused, either decline in productivity or remain at a basic productive level until a climatic shock such as drought finishes off the process of degradation.

Low livestock prices force producers into rearing more cattle, which overgraze the rangelands. The soil is stripped of vegetation and nutrients, becoming a dry dust which will blow away in the wind.

Forests are removed either as a cash crop or to make room for increasing cattle or grain farms. Desertification commonly follows **Deforestation**, because water runs off the bared hills too quickly, taking much of the topsoil with it. The eroded remains are unable to maintain a viable agriculture.

Irrigation schemes are started to service the new crops, but are inefficiently planned and result in waterlogging of crops, or salinization (*i.e.* salting-up) of the soil. Continual evaporation from the soil can bring salt up from the subsoil, leaving it on the surface of the ground which becomes unusable.

Any of these factors can be compounded by, or caused by, a large increase in local population such as is happening over large parts of the globe.

The phenomenon of desertification is not new : it played a role in the downfall of the Sumerian, Babylonian and Roman civilizations. Much of the Sahara Desert, for example, was at one time well vegetated and highly populated, with flourishing forests and plains supporting a good variety of wild game. What is new, however, is the scale on which it is occurring, and the threat that is posed for the future.

In human terms, desertification is a tragedy mainly affecting those who can least afford it: people in low-income developing countries, in areas already climatically, geographically and economically disadvantaged. Agriculture in these countries is often the main source of jobs, income and Gross National Product, so the effects of desertification can lead to famine and political turmoil. After the drought of the early 1970s, every government fell in the affected Sahel countries. For the rural poor in the affected areas, survival depends on the success of a few crops or the sale of a few animals. As productivity of the land falls, their living conditions worsen, Crops fail, water sources dry up, animals die and fuelwood becomes more difficult to obtain. Grass becomes exhausted, farmers move on to areas previously considered infertile, water sources become polluted with silt and salt. The area becomes too degraded for continued use, and the people move elsewhere.

Migration is the only solution for the affected populations. During the Sahel drought of the early 1970s, nearly a million 'environmental refugees' fled from Upper Volta (now Burkina Faso). Between 50,000 and 250,000 people died in Sahel desert.

The intensity of the drought, according to some observers, may have been aggravated by the expansion of the groundnut (*i.e.* peanut) cash crop in Niger and other Sahelian countries in the 1960s. The amount of fallow land in the agricultural zone shrank rapidly; nomadic pastoralists had to increase their grazing in the marginal lands of the north. This caused further desertification, and increased the nomads, exposure to the effects of the drought.

The Sahel is an example of the circular nature of desertification. Pressure for land leads to over-intensive use, which leads to land loss: as a result, the pressure increases on the remaining agricultural areas.

The Sahel disaster of the early 1970s prompted the United Nations Conference on Desertification in Nairobi, Kenya, in 1977. The UN conference brought together technical experts and senior diplomats who made a number of recommendations. Central of these was the need to spend $18 billion in developing countries over the next 20 years, to rehabilitate all damaged irrigated land, half the affected rangeland, and three-quarters of rainfed croplands. Two million hectares of sand dunes were to be stabilized too.

The year 2000 was set as the goal for halting desertification, but the target is now, more than a decade after the UN conference, seen as impossible to achieve.

FOREST

Forest is a community of plants dominated by trees, with crowns almost touching each other. A more open community dominated by trees at least having a crown coverage of 40 per cent is more appropriately termed woodland, though often loosely included under forests and often called open forest. Trees in a forest are generally taller than 5 m. A community dominated by shorter trees and shrubs is called scrub. Two main types of forests are generally recognised. Evergreen forests which retain leaves throughout the year, the mature old leaves continuing to drop the year round. Deciduous forests lose their leaves towards the end of a growing season and are generally without leaves in winter, fresh leaves appearing in spring.

Distribution of forests, which commonly from climax formation in many areas, is determined by the climate conditions of the area. Wet tropical climate with more than 200 cm of annual rainfall supports tropical rain forests which show large diversity of species dominated by broad leaved evergreen trees which are very tall and have close canopy. Such forests are distributed in Malbar region of India, Colorado in Panama, Burma, Indonesia, Amazon and Cango basins. Many important timber trees Teak, African Mahogany, Indian rosewood and sandalwood grow in these forests. The tropical regions with progressively less annual rainfall support tropical semievergreen forests, tropical deciduous forests and tropical thorn forests. Coniferous forests from the dominant forest vegetation of temperate zone between 500 and 600 north latitudes. These also occur at lower latitudes at higher attitudes generally between 2000 m and 3000 m altitude in Himalayas. The dominant species of needle leaf conifers such as *Pinus, Abies, Picea* in Himalayas. *Pseudotsuga* and *Sequoia* are characteristic of forests in United States.

There has been considerable reduction in the forest area of the world having come down from nearly 7000 Mha in 1900 to about 2900 Mha in 1975, the maximum reduction, to the tune of 40 per cent having occurred in tropical forests, the maximum reduction having been witnessed in India, Sri Lanka and Bangladesh. This has mainly been the result of population increase and rapid urbanization.

DEFORESTATION

'Deforestation' refers to the disappearance of forests from large parts of the world's surface. An estimated 290 million hectares of forests and woodland were lost between the early 1960s and the early 1980s. Deforestation has been occurring steadily over the course of this century, with a rapid acceleration since the Second World War.

Forest areas in the more temperate parts of the world appear to be stable or even slightly increasing—with the exception of Central Europe, where large tracts of forest are dying, presumably as a result of air pollution. The brunt of deforestation has been borne by woodland and tropical forests in the Third World. In some regions, the rate of forest loss has accelerated so quickly that the result is complete deforestation. The disappearance of a single tropical forest is an environmental tragedy because of the loss of an enormous number of species, some possibly unique. The global trend of destroying tropical forests, however, has extremely serious, potentially catastrophic, consequences.

Tropical rainforests cover some 7 per cent of the world's land area, about 900 to 1,200 million hectares. 58 per cent of them are in South America, with Africa (19 per cent) and Asia and Oceania (23 per cent) having the rest.

The main countries with tropical forests are Brazil, which contains almost 33 per cent of the total; Zaire and Indonesia (each with 10 per cent;) Papua New Guinea, Columbia, Venezuela, Gabon, Peru and Burma, each with over 200 square kilometres of tropical forest. Three countries—Brazil, Zaire and Indonesia—together possess more than half of all the world's tropical forests.

The tropical rainforests are of crucial environmental importance because of the diversity of plants and animals that inhabit them. The Ice Ages destroyed vast areas of forest in the more temperate zones but not nearer the equator; many rainforests maintained their existing populations and provided shelter for those species that managed to escape from the freezing northern wastes.

As the temperature never sinks below zero in the tropical forests, plant species do not need to concentrate their energy on survival during a cold winter, but can go for maximum growth. The intense competition for survival means that species do not tend to dominate, as in parts of the north but proliferate in a shoulder-to-shoulder existence.

The world is commonly assumed to contain some 5 million species, of which 1.75 million are properly recorded by science. About 155,000 of the 250,000 known species of plants are to be found in the rainforests, 80 per cent of all insects and nine-tenths of the world's primates (monkeys and related animals).

There are two general reasons behind global deforestation: survival, and economics.

Survival: It refers to the population growth, pressure to clear land for farming and government-inspired settlement schemes that have led to much deforestation. ***'Slash and burn'*** technique has destroyed large areas of forest in different parts of the world. It is an agricultural system practiced by primitive tribes. New forest areas are cleared, burned and used for cultivation. These are abandoned after a few years and new forest area selected. As the pioneers move on, the forest can begin to re-establish itself and it is possible that primary forest species may return after 50 years or so. The forest is thus ready for fresh slash and burn cycle. The increasing pressures of agriculture have, however, led to shortening of cycle so that forests do not have sufficient time to rejuvenate.

In some areas of the world, the trees are destroyed because the inhabitants need to burn them for fuelwood. The FAO estimates that 16 of the 45 countries in sub-Saharan Africa face fuelwood deficits on all or part of their territory, which could only be met by cutting trees faster than they are growing: another 18 could not even meet their demands by overcutting, and had to go short.

Economics : In terms of forest area, more than 50 per cent of deforestation is caused by fuelwood gathering and settler expansion. According to estimates of the World Wildlife Fund, between 11 and 15 million hectares of tropical forest are lost every year, around the world. Six million hectares of this is rainforests. It is reported that nearly all primary tropical forests of India, Sri Lanka, Bangladesh and Haiti have been destroyed, what exist now are secondary forests. Between 40-50 per cent primary forests are destroyed in Thailand and Phillippines, complete deforestation is expected by the end of 2000 A.D. in Malaysia and Nigeria unless corrective measures are taken.

Deforestation has significant consequences. The forests fix soil to the earth: when the trees are removed, the soil blows away or is removed by rains. This erosion of the soil makes the area useless for farming; clogs up local water-courses, to the extent of silting up major canals and dams; and leads to floods in the rainy season, as the soil is no longer able to retain the rains in sufficiently large quantities (this is one of the main reasons why the **Chipko movement** was founded in India—erosion, soil loss, and floods after over-intensive tree-felling). In extreme cases, the deforested area will become a desert.

Deforestation will also have, an effect on local climate. Forests absorb more of the sun's energy than open land: deforestation can disrupt local weather patterns by warming air that was previously kept cool. Deforested zones experience greater fluctuations in air and soil temperature than the forest that they have replaced. Humidity is another factor that is altered.

In the 1980s, the environmental movement and the international community have realized the threat to the tropical rainforests. In 1982-83, the World Wildlife Fund and the IUCN (International Union for the Conservation of Nature and Natural Resources) launched a Tropical Forest Campaign. By 1986, 40 per cent of WWF's programme of international project expenditure involved tropical forest conservation. In 1985, an International Task Force composed of the World

Bank, the World Resources Institute, the IUCN and the United Nations Development Association launched Tropical Forests : a Call for Action, proposing a global plan to prevent deforestation, save the tropical forests and avert the fuelwood crisis by increasing the number of plantations and better managing of exploited forests.

A number of measures have been undertaken in India to save the existing forests and bring large areas under tree cover. Chipko movement in Garhwal, Himalayas and **Appiko movement** in Karnataka are voluntary efforts to save the natural forests. Social forestry and Agroforestry aim at growing trees on private and common land around villages, pathways etc. National Forests Policy, 1952 had target of bringing one-third of the geographic area of the country under forests. As the deforestation could not be checked a **Forest (conservation) Act** came into force in 1980, amended in 1988 to include stricter penal provisions against the violaters.

FOREST HAZE

Ozone is a contributor to the hazy atmosphere frequently seen in forested mountain areas. Terpene hydrocarbons (a class of unsaturated organic compounds) are emitted by vegetation. The volatility and concentration of these compounds is such that if calm weather there may be significant concentrations in the atmosphere is and above wooded areas. F.W. Went and R. Rasmussen made a detailed study of the emanations from vegetation. They believe that the terpenes undergo photo-chemical polymerization to an atmospheric aerosol (finally divided droplets of particulate matter), resulting in the "blue haze" seen in many forested areas. Went estimated that the total release of terpene-type hydrocarbons from vegetation is about 170 million tons per year. Ozone in the atmosphere is probably responsible for most of the terpene reactions that cause the bluish smoke. Thus, a natural haze or "smog" is formed that may be seen especially on warm, calm days in forested or other heavily vegetated areas.

FOREST CONSERVATION

Forests are a renewable source and contribute substantially to economic development. They have a major role in enhancing the quality of environment. India has an area of 752.9 lakh hectare notified as forests. Of this, 406.1 lakh hectare are is classified as reserve and 215.6 lakh hectare as protected. Unclassified forest area is spread over 131.8 lakh hectare. About 19 per cent of the total geographical area is under forest cover.

Increasing destruction and degradation of forests and tree-lands especially in the Himalayas and other hill areas, is leading to heavy erosion of top soil, erratic rainfall and recurring floods. It is also causing acute shortage of firewood and, what is more important, loss of productivity due to eroded and degraded land.

India is one of the few countries which has a forest policy since 1894 which was revised in 1952. As the forest destruction continued a Forest (Conservation) Act was enacted in 1980. It was amended in 1988 to make it more stringent. Main plank of the revised Forest Policy 1988 is protection, conservation and development of forests. Its aims are : (i) maintenance of environmental stability through preservation and restoration of ecological balance; (ii) conservation of natural heritage; (iii) check on soil erosion and denudation in catchment area of rivers, lakes and reservoirs; (iv) check on extension of sand dunes in desert areas of Rajasthan and along coastal tracts; (v) substantial increase in forest/tree cover through massive afforestation and social forestry programmes; (vi) steps to meet requirements of fuel wood, fodder, minor forest produce and soil timber of rural and tribal populations; (vii) increase in productivity of forests to meet national needs; (viii) encouragement of efficient utilisation of forest produce and optimum substitution of wood and (ix) steps to create massive people's movement with involvement of women to achieve objectives and minimise pressure on existing forests.

Recognizing the urgent need to halt dangerous trends where forests have reached a stage of regression very near the irreversible threshold, the Department is giving a new orientation to entire gamut of forests-related activities. Some of the important areas for immediate effective protection of forests include afforestation and development of wastelands, reforestation and replantation in existing forests, forest settlement, restriction on grazing, encouragement for wood substitutes and supply of other kinds of fuel, elimination of forest contractor, discouragement of monoculture practices, etc.

FERTILIZERS

For centuries farmers fertilized their crops with dung, decaying leaves or straw, long before they understood the chemistry of fertilization. Today, fertilizers are used so extensively that they have become a major component of most agricultural systems.

Of the mineral elements needed by plants, nitrogen is most often deficient. Plants need nitrogen in large quantities because it is a major constituent of proteins and nucleic acids. Although nitrogen makes up about 78 per cent of the atmosphere by volume, and the supply is therefore virtually unlimited, but gaseous nitrogen cannot be used by most plants. Some bacteria in soil (*Azotobacter, Closteridium*) and others in root-nodules of legumes (*Rhizobium*) can fix atmospheric nitrogen and convert it into organic molecules. Thus soil fertility can be maintained by rotating with legume crops. One typical rotation scheme would be to grow corn, a plant that depletes soil nitrogen, one year, and then cultivate a nitrogen-enriching crop such as alfalfa the following season. Unfortunately, there is a greater market for corn than for alfalfa, and in many regions the farmer would realize more profit if corn were planted every year. To offset this problem, chemists have learned to convert atmospheric nitrogen to synthetic plant fertilizers by producing ammonia which can be used in soils directly or converted to other usable compounds such as nitrates (NO_3-).

Large quantities of phosphorus and potassium are also added to commercial farms. Both are mined commercially from mineral deposits. Of these, the world's supply of phosphorus is more limited and the price of this fertilizer is linked to the availability of the deposits. In addition, soil is often treated to adjust its pH. Lime is applied to acid soils to raise the pH, whereas sulphur or ammonium sulphate is applied to alkaline soils to lover the pH. These treatment have other effects besides changing the pH. For example, lime provides calcium, and ammonium salts contribute nitrogen.

No doubt, commercial fertilizers have increased agricultural yields all over the world and thus have helped feed the expanding human population, but several problems have arisen as well : fertilizers are expensive and are therefore unavailable to farmers in many less developed countries; they also cause pollution of ground water and premature eutrophication of lakes and ponds and even ground water. High nitrate concentration is especially harmful to infants, who may die from drinking formula milk prepared with water polluted in this manner. In a few agricultural areas in the United States, domestic wells and municipal reservoirs have been contaminated by nitrates. However, it is difficult to blame agricultural runoff unequivocally because the same materials may leach from sewage disposal systems or open dumps.

Perhaps the most serious concern about the use of chemical fertilizers is that the soil humus is continuously depleted and is not replaced. If humus is destroyed, water losses will increase, leaching problems will become more severe, trace minerals will be removed, and the agricultural system will become totally dependent on applications of large quantities of fertilizer. One solution to this problem is to combine use of synthetic fertilizers with organic fertilizers.

TILLAGE

Soils are treated in various ways to improve the conditions for plant growth. One of the oldest agricultural practices is tilling, the turning over of soil before planting seeds and turning it over between plants as they grow. Tilling mixes up the nutrients and loosens soil particles, making it easier for roots to penetrate the soil. It also gives crop plants, which are left alone, a competitive advantage over weeds, which are deliberately disturbed to damage their root systems. Tilling also improves aeration of the soil, allowing irrigation water to penetrate more easily.

On the negative side, tillage exposes deep layers of the soil to the air and therefore speeds up the decomposition of humus. Exposure of soil also leads to more rapid evaporation of water, although the destruction of the weed plant cover reduces water loss from leaf surfaces. Therefore, sometimes tillage leads to net loss of water and sometimes it increases soil moisture. In times of heavy rainfall or dry winds, bare soil is vulnerable to erosion because the root structure that normally anchors the soil particles has been disrupted.

Many primitive societies have traditionally practiced no-till agriculture, primarily because plowing is difficult or expensive. A benefit of this approach is that soil quality is maintained. In recent years, no-till agriculture has become popular in developed counties as well. The farmers who practice it often find that they need less irrigation water and fewer fertilizers. The machinery required is expensive, however, and the shift to no-till agriculture is slow.

ANIMALS

The various activities of human beings have resulted in increasing number of threats to the survival, of animals. They have died as their habitats disappear due to ***deforestation, desertification***; they have been poisoned unintentionally by the widespread distribution of harmful substances; or they have been hunted, often to extinction. Animals in the wild are dying, or dying out, in large numbers. We use animals for food, for scientific research, or in zoos. A number of environmental organizations have become involved in the issue of animal rights; some are opposed to the keeping of any animals for food, experimentation or in zoos, while others have no objection to the eating of animals but strongly protest the living conditions under which millions of beasts and birds are now kept.

Animals as Food

Large scale rearing of animals for food (factory farming) is a relatively recent development. If animals are to be kept indoors in large numbers without widespread outbreaks of disease, artificial heat and light are necessary; so the arrival of electricity was essential before factory farms were conceived. Once arrived, however, the industry has grown very rapidly.

Chickens used for eggs are reared in a number of ways. Battery hens are economic because they are packed into a small space. In the UK, they are usually allocated 10 cm of cage width per bird, four or five birds per cage. Each bird spends its life with a living area roughly the size of a book page. As the birds' wing span is about 80 cm, being confined four to a cage 45 by 50 cm in dimension leaves them unable adequately to stretch or exercise their wings.

Cages are usually stacked three to four tiers high in windowless buildings, with lighting up to 17 hours per day, temperature kept high, and birds fed large quantities of high-protein food. The conditions are designed to maximize egg — laying; a battery hen will lay approximately 250 eggs a year, compared to 199 from a free-range bird.

A day after birth, chicks are separated into sexes, and the females sent off to be bred for egg-laying, They can expect to live 15 months, or in some cases two years: though they have many more years of potential egg-laying left, they cannot fill the large quotas expected of them, and are killed.

Males are genetically ill-equipped for fast fattening into chicken flesh are destroyed by crushing in a special mill capable of mincing 500 chicks a minute or gassed with carbon dioxide or chloroform or suffocated by tying them into sacks or boxes.

In free—range method hens must have continuous daytime access to open-air runs, in which the ground is mainly covered with vegetation. The amount of space in their living conditions is also controlled.

Intensive rearing and feeding practices, however, can have side-effects on humans. The salmonella outbreak in the UK in recent years is an example of this.

A number of organizations publish information relating to battery hens and free-range eggs: FREGG, **Free Range Eggs Association** is probably the best-known.

Chickens used for meat is another big business. Profit margins for these birds are extremely tight, so space is at a premium. Broiler breeding units usually house 8—10,000 birds (although some flocks are 25,000 strong), which are easily accommodated in the early stages in a space 15x15 cm per bird.

Broiler chickens are bred for docility and quick growth. Disease and injury is a constant problem, with a high rate of heart attacks, leg and back injuries caused by bone weakness, bone fractures and dislocations, and infectious illness. More than 6 per cent of all birds die before their full seven weeks' growth. In the USA, where some 3 billion birds are produced a year, this means 180 million premature deaths.

Turkey production is the largest growth industry among the meat products, with the number of birds in Britain increasing from 3 million a year to more than 23 million between the 1960s and the 1980s.

Production is similar to that for broiler chickens, though not as overcrowded. Female turkeys are killed at 13, 15 and 18 weeks, for the over-ready market, while males are killed at 24 weeks for processed food like turkey ham or sausage.

A number of practices concerning dairy cattle have aroused concern. The first is that, for a cow to produce milk, she must calve once a year, but it is regarded as uneconomic for the calf to be naturally raised. The calves are taken from their mothers between three and seven days after birth, causing apparent distress to both, and then transported to market in conditions that are often unsatisfactory.

To increase milk production a new hormone (**bovine growth harmone**, also known as bGH or bovine somatotrophin, BST) is introduced. This will increase milk yield by 25 to 40 per cent but the practice reduces the productive life span of the cow considerably.

Artificial insemination is now common (at least 60 per cent of conceptions), and 'embryotomy' (where a foetus of a 'high quality' cow is transferred to the womb of a lesser animal) is increasing. The latter practice, and that of mating young heifers at increasingly early ages (between 18 months and two years) can cause painful births, as the calf may be too big for the mother to deliver comfortably.

Of concern to humans is the common practice of dosing cattle with antibiotics. This over-use will encourage certain strains of bacteria to become immune to the drugs.

Growth hormones used in cattle have also had unfortunate effects on humans. The most notorious is reported DES—more commonly known as stilbenes—a synthetic hormone originally injected into pregnant women in the UK and USA in the belief that it would avert miscarriage. As with Thalidomide, most mothers and babies escaped without damage—but at least 429 girls born to these women developed vaginal cancer in their teens or early twenties. 79 died, mostly in the USA where the drug was used more commonly, and the drug was banned from use in humans.

It was till permitted in some countries for use as a growth hormone in cows, however. Stilbenes offer unusually quick weight gain, are cheap, and can be injected easily into muscle tissue. In 1980, a number of children in Milan were reported with abnormalities; infant boys were growing breasts, and girls were developing mature sexual organs. Italian health inspectors discovered that all the children had bene eating veal baby food contaminated with the female growth hormone, stilbenes. The drug was illegal in Italy and France, but was probably obtained and used illicitly.

The animal movement not only protests the living conditions of 'farm' animals : it also objects to the way in which they are transported to the slaughterhouses and then killed.

Animals Used in Scientific Research

Some 10,000 experiments a day are carried in registered laboratories of the world. The 'human health' argument, to some extent, divides the public response to vivisection. Many animal rights organizations argue for a complete ban on all animal experiments : others argue for research essential to the protection of human health to be carried out, with a drastic reduction in total numbers of experiments. The problem is whether the assurances of scientists as to the validity of their experiments can be accepted.

The anti-vivisection movement has been successful in exposing some examples of vivisection that appear to be unnecessary and cruel. The most well-known of these are : the **draize test**, where irritant substances are applied to the eyeballs of living animals, usually rabbits, to determine their damaging effects,. A substance (such as liquid bleach, a shampoo or mascara) is placed in the rabbit's eye by holding out the lower lid and placing the liquid into the 'cup' that is formed. The eyelid is then closed, the animal restrained so it cannot scratch the eye and then the rabbit is observed for up to three weeks to determine the extent of ulceration, discomfort, or pain. The animal is then killed.

Another well known example is **LD 50 test** : This stands for Lethal Dose 50 per cent, the amount of a substance which will kill half of the animals being tested. Defenders of this test say that it is essential to find the toxicity of the substance before it is released to human beings; opponents criticize it on the grounds that different species react differently to many substances; and that often the quantity of substance administered is so large as to make the results meaningless. Campaigns by the animal movement against these tests have rendered them less popular with industry although they are still in use. Scientific research on animals will undoubtedly continue, although considerable pressure is being exerted for reform. The Animals (Scientific Procedures) Act was passed in the UK in 1986, to become law the following year. This extends control to large areas of experimentation requires stricter licensing for competence of experimenters, and, for the first time, licensing of individual experiments, and demands that one person be named in each laboratory to supervise day-to-day care of animals, in co-operation with a named vet who will be on the staff or on call. Both will be answerable for abuses.

Whereas the concern for human health may not allow any reduction in experimentation with animals, new methods are being developed. In IOS (**Isolated Organ System**) method developed by a French firm Dei Lierre isolated organ fragments are used for conducting pharmaceutical research with the help of robotised computerized system. The method offers several advantages: a single animal can provide several organs that can be cut into numerous small fragments. As a result, a single animal can be used for a large number of experiments. The isolated organ is kept alive to develop a specific reaction, without the entire animal having to be involved, thus making it possible to approach the organ in pure state.

ENDANGERED SPECIES

There are over 2 million species of living organisms today, of which 1.5 million are animals and 0.5 million plants. Extinction is a natural feature of the evolution of life on earth, the best-known examples being the disappearance of the dinosaur, seed ferns and cycadeiods.

In the last 400 years, however, human activities have been responsible for the loss of most of the animals and plants that have disappeared. Gone for ever, for example, are seventeen species or subspecies of bears, five of wolves and foxes, four of cats, ten of cattle, sheep, goats and antelopes, five of horses, zebras and asses, and three of deer. These are large mammals: there are thousands of insects, plants and small mammals which have suffered the same fate.

Although clearcut lines of demarcation are difficult to draw, extinction of species can be rated into three categories: those already **extinct** at least from their natural habitats, endangered species and threatened species. The distinction between endangered and threatened species is rarely made, and often difficult to make. Both stand at the edge of extinction. An ***endangered species*** is any species which is in danger of extinction throughout or a significant part of its range. A ***threatened species*** includes any which is likely to become an endangered species within a foreseeable future throughout or a significant part of its range. Dodo once very common is already extinct from Mauritius. The passenger pigeon which used to congregate in enormous flocks over North America barely a hundred years ago, was wiped out for food early this century.

The brown pelican has disappeared from Louisiana (USA) the "pelican state", and hatching of young birds has declined at its principal rookery on Anacapa Island off the coast of California, probably due to contamination with DDT and the plasticizer PCB (Polychlorinated biphenyl) which cause thin eggshells and poor reproduction. The Guadalupe fur seal was twice thought to be extinct, the first time after heavy exploitation for its fur during the last century when several scientific expeditions found and collected what they thought were the last remaining specimens; and, again, when a disgruntled fisherman and zoo collector in a frenzy swore to kill the whole herd and reportedly did so. The grizzly bear—the official symbol of California—can no longer be found in California except on the state flag.

And the bald eagle—the official symbol of the United States—is in danger of extinction in many parts of the country, though still abundant in Alasaka. Plants are no exceptions. The weather-carved bristlecone pines cling tenaciously to their last stronghold in the White Mountains where some of them have held out for 5,000 years, but now need protection.

The United States is not the only part of the world where wildlife is suffering from human encroachment. Many of the species of big game in Africa are now being protected in game preserves. The Indian Tiger has been hunted so aggressively that its numbers have dwindled from 40,000 at the turn of the century to fewer than 2,000 which are now under protection by the Indian government. Even monkeys, used extensively in medical research, are in critically short supply. Nearly 60,000 monkeys are used each year in the United States alone to keep medical research going.

The great mammals of the sea, the whales, also face extinction. Eight kinds of whales are in danger. Those in most danger of extinction are the blue whale and the humpback, but the fineback, sei, and sperm whales are also threatened. The blue whale is the largest animal that ever existed. Another sea mammal, the porpoise, suffers a terrible toll by being inadvertently caught in tuna nets. About 200,000 porpoises were destroyed every year during 1969-72, but improved techniques could greatly reduce the loss.

At least 400 species of animals the world over are facing the threat of extinction. These are officially classified as *endangered species*, those whose prospects for reproduction and survival are in immediate jeopardy. They include nearly 300 species of fish and wildlife and about 100 species of fishes, reptiles, amphibians, birds, and mammals native to the United States.

The actual number of animals is only one criterion for deciding that a species is threatened with extinction. In some cases a declining population is all that is needed to indicate that a species is in danger. But some species, that still exist in large numbers, may be in peril because they are under some form of stress such as a polluted environment, disease, predation, excessive hunting or fishing, over-exploitation for commercial purposes, or destruction of their habitats by human intrusion.

Many of the plant species extinct from their natural habitats are now maintained under cultivation. A famous example in *Ginkgo biloba*, the maiden hair tree, that survived total extinction as the tree was grown in Chinese temple gardens, from where the propagation was done to other parts of the world. *Franklina alatamaha* was last found in wild in 1803 and now exists only in gardens. Other species like *Betula uber* of South-West Virginia and *Shortia galacifolia* of Georgia, once thought to be extinct, have been rediscovered. The number of plants and animals which are listed endangered or threatened is very large. 2000 species of plants are listed from continental America, the common examples being showy lady slipper *Cyperipedium reginae, Orchis spectabilis*, sundews, pitcher plants, venus fly trap and others. In India the list includes species such *Saussurea lappa, Balanophora involucrata, Picrorhia kurroa* and *Atropa acuminata*.

The concern for protection of endangered species of the world resulted in ***CITES*** (Convention on International Trade in Endangered Species of wild Fauna and Flora) agreement in 1975, and now has 95 member countries. The commercial trade of endangered species listed in appendix I such as apes, lemurs, cheetah, Asian elephants, canes, sea turtles, giant panda, some orchids and cacti is totally banned. Those listed in appendix II facing serious risk can be traded after obtaining proper permits.

PROJECT TIGER

In India the Centrally Sponsored Plan Scheme 'Project Tiger' was launched on 1st April, 1973 to achieve the following objectives ;

(i) To ensure the maintenance of a viable population of the tigers in India for scientific economic, aesthetic, cultural and ecological values and

(ii) To preserve for all times, areas of such biological importance as a national heritage for the benefit, education and enjoyment of the people.

To achieve these objectives, 18 Tiger Reserves have so far been established in 13 states covering over 28,017 sq. km. forest area. The 18th Tiger Reserve at Valmikinagar was created in the West Champaran district of Bihar in January, 1990. The area is continguous to Royal Chitwan National Park of Nepal and would provide protection to the population of tigers in Northern Bihar. Table 4 gives at a glance the distribution of areas of Tiger Reserves in the country.

A Steering Committee functioning under the Chairmanship of the Prime Minister provides guidelines for the management of the Tiger Reserves. The non-official members of the Steering Committee (Project Tiger) and the four scientific institutions nominated by the Steering Committee (Project Tiger) review the project Tiger biannually.

The Department of Environment (DOE) provides 100 per cent financial assistance for establishing field research centres as well as veterinary centres in each Tiger Reserve to boost up research activities in these reserves. Besides this, 100 per cent financial assistance shall also be provided for establishing Nature Interpretation Centres in each Tiger Reserve for providing educational and scientific information.

The total expenditure during the year 1989-90 for the maintenance and development of the existing 17 Tiger Reserves was about Rs. 7.00 crores, out of which the Central assistance was about Rs. 4.7 crores. Valmiki Tiger Reserve the eighteenth one has been notified only recently.

Table 4. Tiger Reserves in India

Sl. No.	Name of the Tiger Reserve	Area (in sq. km.)			Year
		Core	Buffer	Total	
1.	Bandipur (Karnataka)	523	343	866	1973
2.	Corbett (Uttar Pradesh)	338	183	521	1973
3.	Kanha (Madhya Pradesh)	940	1,005	1,945	1973
4.	Manas (Assam)	470	2,370	2,840	1973
5.	Melghat (Maharashtra)	448	1,170	1,618	1973
6.	Palamau (Bihar)	213	715	928	1973
7.	Ranthambhore (Rajasthan)	392	433	825	1973
8.	Simlipal (Orissa)	845	1,905	2,750	1973
9.	Sunderbans (West Bengal)	1,330	1,255	2,585	1973
10.	Periyar (Kerala)	501	276	777	1982
11.	Sariska (Rajasthan)	498	302	800	1982
12.	Buxa (West Bengal)	315	444	759	1983
13.	Indravati (Madhya Pradesh)	1,258	1,541	2,799	1983
14.	Namgarjunasagar (Andhra Pradesh)	1,200	2,368	3,568	1983
15.	Nadapha (Arunachal Pradesh)	1,808	177	1,985	1983
16.	Dudhwa (Uttar Pradesh)	648	163	811	1987
17.	Kalakad-Mundanthuari (Tamil Nadu)	571	229	800	1988
18.	Valmiki Tiger-Reserve (Bihar)	336	504	840	1990
	Total	12,634	15,383	28,017	

Source : DOE Annual Report 1989-90.

An eco-development programme was launched during the year 1989-90, in the buffer area of the Ranthambhore Tiger Reserve to solve the problem of the people living in and around the Tiger Reserve. It is proposed to implement similar eco-development programmes in the buffer areas of all the 18 Tiger Reserves. The activities proposed to be undertaken as a part of the eco-development programme aim at increasing the biomass productivity of the buffer areas to ensure the availability of adequate firewood and fodder.

WILDLIFE

Wildlife, in its broadest sense, includes all living things except man and domesticated plants and animals. In this sense, wildlife includes organisms such as pathogenic bacteria, parasites, pests, and a few predators that are inimical to human welfare and, therefore, are not wanted. Considerable human effort is directed toward stamping them out. But in the more common use of the term, wildlife refers to those wild organisms—totaling nearly 2 million species—that are beneficial and desirable for various reasons or, at most, only insignificantly harmful.

A widely recognized principle of biology is the concept of the balance of nature—the checks and balance of the ecosystem that maintain a dynamic equilibrium in the populations of plants and animals in their natural environment. Throughout the history of life, various natural conditions have upset the balance many times, at least partially being responsible for the extinction of millions, perhaps billion, of species during geological time. Climatic changes, disappearance of the food supply, increase in effective predation, epidemics, or the inability to compete, reproduce, or adapt to new conditions brought about the natural death of entire populations and their replacement by species that had more suitable qualities for adaptation. But of all the forces that have upset the balance in the recent past, man has been by far the most disruptive element. Human manipulation of the environment and mindless aggression against once-abundant species of wildlife has had effects that have been appropriately called catastrophic.

During the last 200 years, about 600 species of animals have declined nearly to the point of extinction; and in the last 2,000 years, 110 species of mammals have disappeared from the earth—never to be seen alive again. The commencement of modern extinction is taken to be the year 1600 by the International Union for Conservation of Nature and Natural Resources. Since then, 75 per cent of all species extinctions have been those in which human being played a major role. Any major disturbance by man whether cutting down of trees, destruction of unwanted organisms, large scale road constructions on hilly areas or industrial activities creates a chain reaction that can have catastrophic consequences.

India, thanks to its location in the tropical belt and a wide variety of climatic conditions, is among the few countries richest in biological heritage. As a simple comparison and vast continent of Europe has no more than 7,000 species of flowering plants, India with one ninth land surface has atleast twice as many flowering plants. An estimated 45,000 species of plants and 65,000 animals grow in India. Among animals species there are 20,000 insects, 4,000 molluscs, 2,000 fish, 140 amphibians, 420 reptiles, 1200 birds and 340 mammals, with a view to protect this rich national heritage a wildlife (protection) Act was enacted in India in 1972. More and more areas are now being brought under the umbrella of protection by being declared as sanctuaries and National parks. Special projects for the protection of tiger, Gir lion, crocodile, rhino, snow leopard have been launched. 12 sites in India have been selected as potential biosphere reserves, where the whole ecosystem will be monitored, legally protected and intensively researched, to help better protection of wildlife.

NATIONAL PARKS AND SANCTURIES OF INDIA

National wildlife action plan adopted in 1983 provides the framework of strategy as well as programme for wildlife conservation. Different types of protected areas are recognised by the IUCN Commission on National Parks and Protected Areas (CNPPA). These include (a) Areas of particular interest to CNPPA (Scientific Reserves/Strict Nature Reserves, National Parks/Provincial Perks, National Monuments/Natural Landmarks, Nature Conservation Reserves/Managed Nature Reserves/Wildlife Sanctuaries and Protected Landscapes); (b) Areas of interest to IUCN in general (Resource Reserves, Anthropological Reserves/Natural Biotic Areas, Multiple use Management Areas/Managed Resource Areas and (c) Internationally Recognised/Affiliated designations (Biosphere reserves and world Heritage sites).

Of these Sanctuaries National Parks and Biosphere Reserves deserve special mention. A sanctuary is generally species oriented, boundaries not sacrosant with limited biotic interference. A National Park is hitched to the habitat for a particular wild animal like lion, tiger, etc., boundaries circumscribed by legislation with no biotic intereference. Biosphere Reserve is oriented towards whole ecosystem, large area, boundaries circumscribed by legislation, no biotic interference, tourism normally not permissible, Research and Scientific Management and attention given to gene pool conservation. A National Park may be located within a sanctuary or not.

At present, protected area network in India comprises 69 national parks and 398 sanctuaries covering four per cent of the total geographical area of the country. It is proposed to be increased to 4.6 per cent by setting up more parks and sanctuaries.

The **Wild Life (Protection) Act, 1972,** adopted by all : except Jammu and Kashmir (which has its own Act), govern wild life conservation and protection of endangered species. The Act prohibits trade in rare and endangered species. India is also a signatory to the Convention on International Trade in Endangered Species of Wild Flora and Fauna. Under this, export or import of endangered species and their products is subject to strict control. Commercial exploitation of such species is prohibited. The Centre provides financial assistance to states for : (i) strengthening management and protection of infrastructure of national parks and sanctuaries; (ii) protection of wildlife and control of poaching and illegal trade in wildlife products; (iii) captive breeding programmes for endangered species of wildlife; (iv) wildlife education and interpretation; (v) development of selected zoos; and (vii) conservation of rhinoceros in Assam.

Some of the important National Parks and Sancturies of the country are given in Table 5.

Table 5 : National Parks and Sanctuaries of India

State	National Park and Sanctuaries
Andhra Pradesh	Kawal, Pocharram Pakhal, Neelpatattu
Arunachal Pradesh	Namidapha
Assam	Kaziranga, Manas
Bihar	Hazaribagh, Betla
Goa	Mollen
Gujarat	Gir, Velavadar, Wild Ass, Nal Sarovar
Haryana	Sultanpur Lake
Himachal Pradesh	Gobind Sagar
Jammu and Kashmir	Dachigam
Karnataka	Bandipur, Nagarhole, Ranganthitoo
Kerala	Periyar, Waynard, Neyyar
Madhya Pradesh	Kanha, Shivpuri, Bandhavgarh
Maharashtra	Tadoba, Pench, Bori, Karnala Dhakna—Kolkaz, Yawal
Manipur	Keibal Lamjao
Meghalaya	Balpakram
Mizoram	Dampa
Nagaland	Intangki
Orissa	Simlipal, Satkasia, Chilka lake
Punjab	Abohar
Rajasthan	Ranthambore, Ganga, Sariska
Sikkim	Kanchenjunga
Tamil Nadu	Guindy, Mundumalai, Annamalai, Vedanthangal, Vettangudi
Uttar Pradesh	Corbett, Dudhawa
West Bengal	Sajanakhali Mahanadi, Jaldapara, Sunderban, Deer Park

PEST CONTROL

The control of any plant or animal that, in its location, is an economic, aesthetic, physical or biological threat or annoyance to humans or their possessions. Generally people do not include control of germs and viruses in this category, although they certainly are threats.

Insects and weeds are the two largest groups of pests. There are around 8,00,000 species of insects in the world. Even though less than 1 per cent of all insect species are pests, yet they are responsible for roughly half of all human disabilities and death caused by disease. They eat or ruin crops and fibre. Weeds compete with crops for nutrients and water. Nematodes worms and larvae feed on plant roots. Playing and hopping insects eat and leaves and fruits of plants, and very often lay their eggs up on them.

Methods

The chief methods of pest control are Chemical, Biological and Cultural. **Chemical Pest Control** involves the use of chemicals (Pesticides) and is the most familiar.

Biological Control of Pests

This method involves the control of pests by exploiting their living biological enemies, by they predators, parasites for pathogens.

Biological controls are generally very specific. Non-target organisms, including people, are not harmed, and the amounts of synthetic pesticides that otherwise would be used are reduced. Unlike the use of pesticides, biological controls are far less likely to lead to the development of resistant species (at least not as quickly as have many pesticides) With rate exceptions in isolated areas, biological controls may not expected to wipe out a pest population. Instead—as the meaning of "control" implies—they reduce the pest population and a new equilibrium population is reached in an integrated approach. The various techniques are combined with the use of pesticides.

Predators

Typical predators are lizards, falcons, bats, wild and domestic cats, various insects such as ground beetles, ladybugs, assassin bugs and their larvae, and the larvae of syrphus flies and of certain gall gnats.

The tendency of insects to prey on one another works to the advantage of people whenever the prey is a pest and the predator is not. For instance, the ladybug (ladybird battle) thrives on soft-bodied insects such as aphids but does not harm to people, livestock, or crops. The *Rodolian* battle attacks the cottony cushion scale insects in citrus crops and has been used successfully in South India. The chrysoline beetle imported from Australia helps control the Klamath weed in the western part of the United States.

Moreover, success has been achieved in suppressing apple woolly aphid by a parasitic wasp in Kashmir and Himachal Pradesh and prickly pear by cochineal insects in South India. Limited control has been possible in case of san Jose scale of apple by a parasitic wasp in Himachal Pradesh and Kashmir, South African Giant Snail by a predatory snail in Andaman islands and Lantana weed by a tingid bug in Uttar Pradesh. Mass release of indigenous parasites of coconut caterpillar have been able to partially suppress the pest in South India.

Parasites and Parasitoids

A parasite is an organism that lives on or in another organism from which it obtains its food and to which it gives on benefit in return. The true parasite is generally considerably smaller than the host and does not kill it. Like lice, it may live out its whole life on the host; it may be host- independent for part of its life cycle (*e.g.*, fleas and mosquitoes), or it may require several hosts at various stages of developments (*e.g.*, tapeworms). The parasitoid (often lumped with parasites) usually has about the same size as its host, kills its host, then moves on to be a free-living adult. In one sense a parasitoid is a predator, with one difference that the predator's use of a particular prey is apparently optional. A true predator can take a variety of prey, selecting what happens to be available. One highly useful parasitoid is a tiny, pinhead sized wasp (*Ooencyrtus clisiocampae*), which does not bother people but lays its eggs inside eggs of spanworms. When the wasp's eggs hatch, the larvae eat the yolks of the spanworm eggs. The spanworm defoliates several kinds of —trees—elm, oak, maple, hickory, linden, apple, and ash.

Insect Pathogens

Diseases of insects may be caused by bacteria or viruses and sometimes by other forms such as fungi and protozoa. *Bacillus popilliae* is a species of bacteria which attacks the grubs of Japanese beetles. The beetles' blood turns into a milky liquid, giving the grubs an opaque look. This disease of the Japanese beetle is called the milky spore disease. Preparations of these bacteria are registered for the control of the Japanese beetle. The spores are ingested by the grub; they germinate, then penetrate the alimentary canal. Grubs that die from the disease release new spores.

Bacillus thuringiensis causes disease in a number of insect species of the order Lepidoptera. This spore-forming bacterium (available as spray) produces a number of substances that are toxic to insects, and these substances are used against the tobacco budworm, the Eastern spruce budworm, the gypsy moth, the cabbage looper, the cabbageworm, the alfalfa caterpillar, and the tomato hornworm. In South India a variation is effective against certain mosquitoes.

Bacterial insecticides are advantageous in that they do not harm beneficial insects, or other living things—people, wildlife, livestock, fish, on crops. Moreover, target insects apparently do not become resistant. However, the disadvantages are that the compounds are expensive; they do not act rapidly; and under unfavourable weather conditions they may not be effective.

In early 1976 the EPA (USA) registered the first insect virus for use as an insecticide. After years of testing, the nuclear polyhedrosis virus (NPV) was found to be both safe and effective, and it was approved for use against two cotton pests, the cotton bollworm and the tobacco budworm, as well as against the tussock moth. The indigenous NPV has given effective control of several caterpillar pests in South India. In its many slightly different strains, NPV is effective against other insect pests—the gypsy moth and a number of insects that attack food crops. It may eventually be cleared for wider use. Viruses of other kinds are also being tested. These viruses are specific and safe, and insects largely remain in forms susceptible to them. They are expensive, and unlike most major insecticides are not contact poisons. The viruses must be ingested to work, and do not work as rapidly as insecticides. They are not as effective as chemicals in knocking down a severe infestation. NPV is easily inactivated by ultraviolet rays or by heat, and it must be carefully packaged and stored. Knowledge of the habits of the target insects is needed because the virus must be applied during one critical stage in the development of insect larvae.

Elicitors

Just as we have natural defenses against invading bacteria or viruses, other living things are similarly equipped. Plants, for instance can produce toxins called phytoalexins that stop the growth of attacking pathogens. The pathogens carry something called an elicitor, which stimulates the plant to produce the phytoalexin. The first isolation of an elicitor substance, was reported in 1976. The elicitor did not appear to be very specific. Elicitor isolated from a pathogen responsible for stem and root rot in soybeans stimulated and production of phytoalexin on soybeans and red kidney beans.

Work on establishing several exotic parasites/predators in India for controlling several crops is underway. A coordinated project under the aegis of the Indian Council of Agricultural Research and also by the Directorate of Plant Protection Quarantine and storage is underway. This is an enduring community service best undertaken by technically trained professionals. Sporadic efforts at controlling certain pest species by mass release of its sterile (irridated) males have not made any impact in this country.

Use of Animal and Plant Odour Compounds in Biological Pest Control

Many animal species secrete particular stimulating substances, known as ***pheromones,*** which influence relations between individual members. They are most strongly developed in gregarious species such as ants, bees, mice, rats. But they also occur in non-gregarious groups in the form of sexual lures which affect only the mating partner.

Thorough investigations have been made recently into pheromones for insects. These substances are effective for the most part in extremely low concentrations: the male of the American cockroach, for instance, reacts to a mere 30 molecules of the sexual lure emitted by the mating partner. Such substances can be perceived over areas many kilometres in extent. It has now been possible to isolate the sexual pheromones of the following pests, among others : coddling moths, Asiatic cotton worms, sugarcane bugs, night moths, various bark beetles and various kinds of moth that endanger the corn crop.

In Principle there are two possibilities for controlling damage by pheromones :

1. The lure substance, extracted from the animal or produced synthetically, is inserted in traps or snares, into which the male of the species concerned is thus enticed. One example of this is the bark beetle, the pheromones of which can be reproduced synthetically,. When this lure substance, which affects both sexes, is inserted in selected tree traps, the bark beetles in the vicinity are attracted. By felling these lure trees and removing the bark, the pupa of the beetle can be readily destroyed. The use of sexual pheromones for pest control by this method is likely to succeed, however, only when the population density of the pest in the locality concerned is low, so that the lure substance in the traps will not be masked by the lure substance of the particular insect.

2. By large-scale distribution of a lure substance during the nuptial flight of butterflies and moths, the males, of the species can be so confused that they can no longer locate the female with its similar emission pheromone. This process was put to the test successfully for the first time in the United States in connection with control of the vegetable owl (*Mamestra oleracea*) caterpillars. In this case a concentration of about 10-10 grams of pheromone per gram of air was used. In order to maintain this concentration, about 1.5 grams of substance were needed per hectare per night.

In contrast to the lure substance, certain odiferous agents given off by many plants have a repellent effect on various insects. Such repellents can be employed for plant protection in one of two ways : (1) The repellents from particular resistant plant forms can be isolated, or produced synthetically, and then sprayed over the field of cultivated plants in need of protection. From resistant species of potatoes, for instance, glucosides such as solanine and tomatine can be isolated, which act as deterrents to potato beetles. (2) Very much simpler is the method of siting plants that emit repellents in a mixed culture with selected useful plants, so that attacks by pests on the useful plants are avoided. If tomatoes and cabbages are grown in the way in mixed cultures, the repellents emitted by the tomatoes will prevent attacks on the cabbage plants by caterpillars of the cabbage butterfly.

Cultural Methods

Cultural methods can be adopted or revised to exercise an influence, in the landscape and environment, on the population density of parasites in agriculture and forestry. The principle underlying such measures is to render the living conditions of the harmful organisms less favourable, with a view to restricting their propagation and so keeping their population density below the danger threshold. In the long run these preventive measures, by means of which a solitary intervention in the landscape can produce long-term results, may provide the most elegant and at the same time a considerable—labour and energy—saving solution. Their application, however, requires a precise knowledge of the ecology of the useful plants concerned and of the harmful organisms affecting them.

Many insect, pests, especially smaller ones, avoid the wind, as they are easily carried away by it. By planting carrots and cabbages in fields exposed to the wind, it has been possible to reduce the infestation of these cultivated plants bait carrot flies and cabbage flies considerably.

Especially vulnerable to insect pests are monocultures of a particular species of plant, in which the insect can propagate an multiply unhindered. On the other hand, mixed cultures, in which two or more kinds of plants are grown together, offer reasonable possibilities for preventing the spread of insect pests from a fallen plant to a neighbouring one, as the distance between the individual plants of the same kind is increased. An additional safeguard is to place two plants in a mixed culture which protect one another against insect pests. The planting of carrots and onions together in a mixed culture would thus provide effective protection against the carrot fly and the onion fly, since the onion fly avoids the essential oils of the carrot and the carry fly avoids the odour of the onion.

Another very effective methods of protecting plants is the cultivation and use of resistant kinds of culture plants. It was thus possible to arrest the severe damage to vineyards in Europe caused by the incursion of the vine louse (*Phylloxera*) in the second half of the nineteenth century by grafting the European vine onto American stock, which was not vulnerable to the vine louse. Today there are numerous plants that are not susceptible to damage from insect pests, such as potatoes that are resistant to viral diseases, and cereal crops that are invulnerable to rust fungi. Nevertheless, efforts in this area in future must be intensified considerably.

Another measure for the culture destruction of insect pests is the planting of shrubs and hedges, which provide birds that prey on insects with cover and a breeding ground. The general procedure for cultural defense against insect pests consists, therefore, in increasing the diversity of the landscape and thereby improving its regulation capacity.

The great advantage of the biological destruction of insect pests, as opposed to their destruction by chemical means, lies in the fact that it does not impair the quality of foodstuffs. Viruses and parasites tend to attach themselves to a particular host, and so remain completely neutral with regard to other animals and to human beings. Another advantage of biological destruction of pests consists in the cheapness and simplicity of its methods. No complicated machinery is required by applying it and no great labour costs are incurred in frequent spraying. The biological destruction of pests leaves nature to do its work and makes full use of natural principles.

PESTICIDES

Pesticides are a variety of chemical substances used to control organisms which may adversely affect public health, or organisms which attack food and other material essential to mankind. Such organisms include rodents, insects, nematodes and fungi. Chemicals used against infectious bacteria causing human animal or plant diseases as well as those employed against viruses, protozoa and internal parasites of animals plants and humans are generally not classified as pesticides.

Pesticide Types

There are many types of pesticides : **insecticides** which kill all or specific species of insects; **herbicides** used against weeds; **fungicides** against fungi; **acaricides** against mites; **nematicides** against nematodes; **rodenticides** against rodents; **molluscides** against smails and slugs.

The Growth of the Use of Pesticides and Toxicity

Modern pesticide research started in the last century, with the development of an insecticide which checked the spread of Colorado beetle in the USA. The rapid growth of pesticide research occurred before and during the Second World War, and the current generation of pesticides have been developed within the last 40 years.

Pesticides have made possible great increases in food production and improvements in human health. Where harm has occurred, this has largely been due to ignorance of the properties of these chemicals and misuse, *i.e.*, wrong quantity, application or timing. Since it is impossible to know every property of every pesticide it is inevitable that some environmental damage will occasionally result from their use. However, this is no argument against their use, so long as the benefits can be clearly shown to outweight the damage. The importance thing is to be aware that there is a finite risk of damage, to minimize it by appropriate toxicity testing of chemicals before they are applied on a large scale and to ensure that the application is carried out correctly. When pesticides are applied on a large scale, a constant watch must be kept for environmental damage so that corrective steps may be taken as soon as possible.

Misuse of pesticides can often lead to major problems. Wearing protective clothing—in some cases respirators in others aprons, boots and face masks—are a must as pesticides enter the skin through feet and hands if left unprotected. This could lead to beefy dermatitis. Containers used to store pesticides should never be used for storing grain.

All pesticides should be treated as what they generally are—dangerous substances. This includes pesticides sold for garden or home use. These international hazard symbols will often give useful information about a particular product :

Pesticides and the Third World

When a pesticide is banned in the home country there is great commercial pressure to export it. The developed countries are becoming more aware of the dangers of pesticide misuse, so banned pesticides are most frequently directed at the Third World.

Everyday cases of poisoning by pesticides in developing countries are reported. In 1975, hundreds of people in Karnataka's Shimoga district in India were struck by a mysterious attack of arthritis which wastes away limbs and brings about dwarfs. Studies indicate that pesticides that were sprayed in fields were ingested by crabs which, in turn, were eaten by farmers, poisoning them as well. According to an WHO estimate about 750,000 people are poisoned by pesticides every year, 14000 of which are fatal. Three-fourths of these occur in the Third World. The use of pesticides has been increasing at snowballing rates. In India, pesticide consumption rose from 2000 tonnes in 1955 to 80,000 tonnes in 1984—a 40-fold increase. By the end of the current year, the estimated consumption is projected at 100,000 million tonnes.

Current calculations are that up to 1 million pesticides accidents occur every year, causing 10,000 to 20,000 deaths. Most Third World accidents are caused by lack of knowledge as to the effects of the pesticide concerned.

Cosmetic pesticides are sprayed indiscriminately on fruits and vegetables in major cities in India to improve the 'looks', Methyl parathion on cauliflower gives an extra white look; ladyfinger ('bhindi') is dipped in copper sulphate to look greener. Washing vegetable with a 5 per cent solution of edible soda or vinegar, followed by a rinse of water will remove residues.

The United Nations has resisted attempts to enforce a system of 'prior informed consent', in which exporting countries have to notify the importing nation of any domestic restrictions on pesticide use. The onus is still on the purchaser to find out the harmful effects of the substance that is being bought. In March 1987, thallium sulphate killed several hundred people in Guyana after it was used as a poison to keep rats away from the sugar crop. Thallium sulphate has been outlawed in Britain and other developed countries for more than 20 years.

INSECTICIDES

One of a large class of substances used to kill insects that are harmful to man, either directly as disease vectors or indirectly as destroyers of crops, food products or textile fabrics.

General Families of Insecticides

1. *Inorganic* : arsenic, lead and copper (inorganic compounds and mixtures).

2. *Synthetic Organic Compounds* :

(a) *Chlorinated Hydrocarbons* such as Aldrin, DDD, DDE, DDT, chloride, Lindane etc.

(b) *Organophosphates* (Organic esters of phosphorus) such as malathion, parathion and related substances.

(c) *Carbamates* such as carbaryl.

3. *Natural Organic Compounds* : Rotenone, pyrethrins etc. These are less toxic and quickly decompose to nontoxic substances.

Inorganic Insecticides

These are generally aresenates, sulphates, or chlorides of copper, lead and mercury, and certain sulphur preparations. Some examples are given in Table 6.

The use of inorganic insecticides has diminished in the recent years because of the development of more effective types that are less toxic to man.

Table 6. Inorganic Insecticides

Compound	*Formula*	*Toxicity*
Calcium arsenate	$Ca_3(AsO_4)_2$	Estimated lethal dose 100 mg/kg
Lead arsenate	$PbHAsO_4$	Lethal dose 10 mg/kg
Paris Green	$Cu(O_2C_2H_3)$ $3Cu(AsO_2)_2$	Ld_{50} 22 mg/kg in rats
Sodium arsenite	$NaAsO_2$	Highly poisonous
Thallium sulphate	Tl_2SO_4	LD_{50} 25 mg/kg in rats

Chlorinated Hydrocarbons

A family of insecticides also known as Organochlorines.

They include all persistent insecticides—those that do not break down quickly in the environment and are passed along the food chain of various species including man. These are largely combinations of carbon, hydrogen and chlorine. **DDT** is probably the most famous member of this family. Other chlorinated hydrocarbons (organochlorines) include **lindane, aldrin, dieldrin** and **chrlodane**. The principal chlorinated hydrocarbons used as insecticides are listed in fig. on p. 173

Toxic Effects : Chlorinated hydrocarbons are the most prevalent pesticides in the environment because of their wide use and persistence. The most persistent of these compounds are DDT are derivatives such as DDD and DDE. DDE has an environmental half life of ten years or more, DDE can persist for decades. Almost as persistent are lindane (BHC) and heptachlor. Less persistent are aldrin and dildrin, but even these require 2 ½ years in the soil for 95 per cent degradation. All of the compounds mentioned are currently in use throughout the world, though DDT, aldrin and dieldrin have been banned for most purposes in the United States, and their use has been discouraged in the United Kingdom and a number of other countries. Reaction against the use of those compounds has followed awareness of their toxicity to non-target organisms and the discovery that some target insect populations were becoming resistant. However, substantial quantities of chlorinated hydrocarbons are still used in wood preservation and where there is, at present, no satisfactory substitute. The most extensive use of DDT now occurs in tropical counties because it is cheap, persistent, generally effective and with minimum harm to human beings. Until a suitable alternative is available, its use will continue. At present, possible substitutes are must more expensive, less persistent and more toxic to human beings.

Diphenylethane Group

Structure	*Trivial Name*	*Specific name*
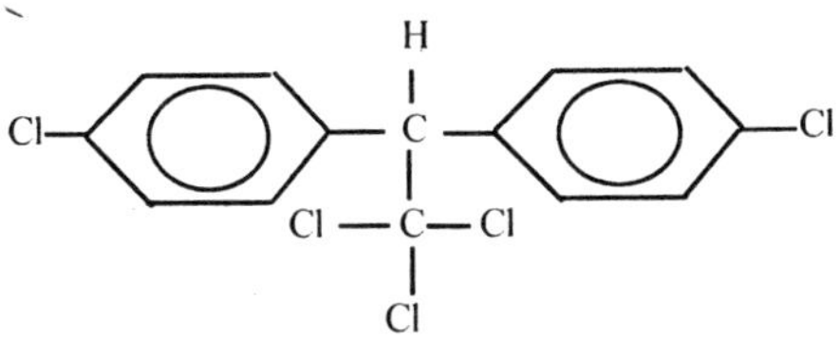	DDT	2, 2-bis (p-chlorophenyl) 1, 1, 1-Trichloroethane
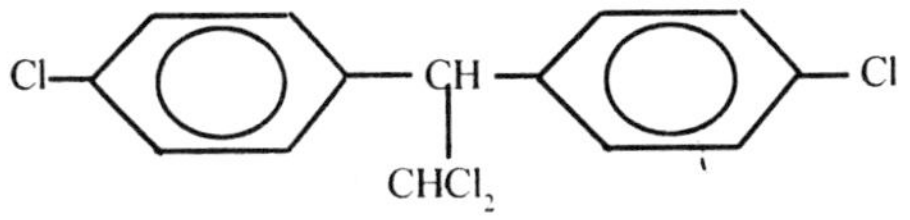	DDD	2, 2-bis (p-chlorophenyl) 1, 1-Dichloroethane
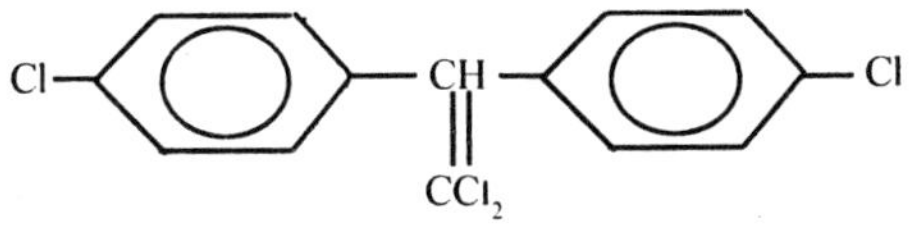	DDE (a break-down product of DDT)	2, 2-bis (p-chlorophenyl) 1, 1-Dichloroethane
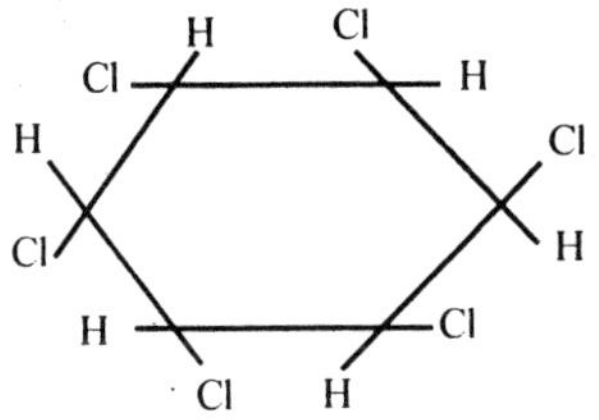	Lindane (BHC)	1, 2, 3, 4, 5, 6-hexa-chlorocyclohexane (γ-isomer)
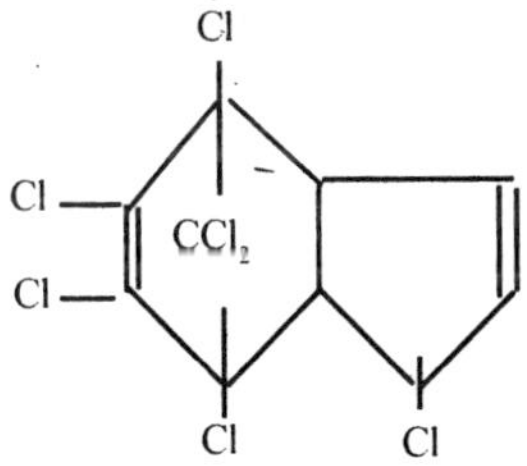	Heptachlor	1, 4, 5, 6, 7, 8, 8-hepta-chloro-3α,4,7,7α-tetra-hydro-4,7-endometha-noindene
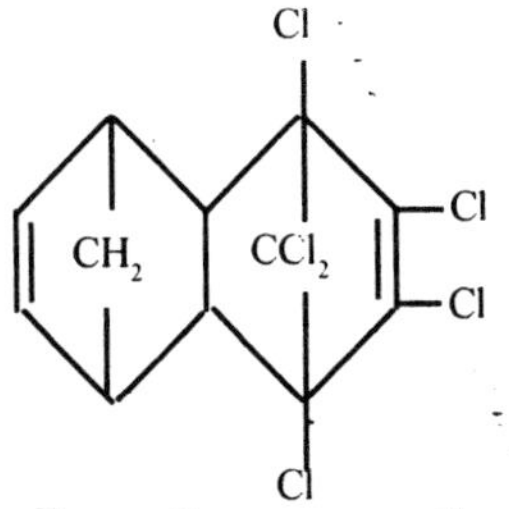	Aldrin	1, 2, 3, 4, 10, 10-hexa-chloro-1, 4, 4-α, 5, 8, 8a-hexahydro-1,4-endo-exo-5,8-dimethanon-apthalene
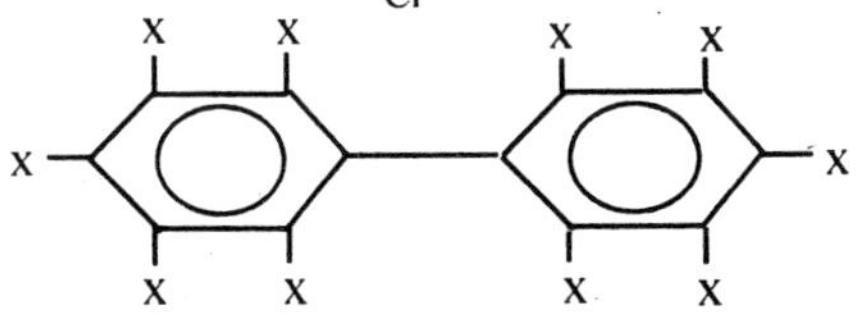	PCB	Polychlorinated biphenyl

x = Possible site of chlorine atom

Structures of Selected Chlorinated Hydrocarbons.

Most of the problems caused by effects of DDT on nontarget organisms are the result of its being used in excessive amounts. The excess finds its way into ponds, lakes, rivers and ultimately, the sea. In some cases, DDT is applied directly to fresh water insect habitats; in others, accidental spraying to such habitats may occur. Otherwise, DDT may enter waterways in surface run-off, or be washed out of the atmosphere in rain or snow. Despite this, the concentration in natural waters is low. However, many plants and animals tend to accumulate DDT and its more persistent derivative DDE. This process of accumulation can continue through food chains until harmful levels are reached in the ultimate predators. The effect is, of course, greater in the case of DDE owing to its greater persistence and prevalence. Accumulation occurs largely because of the affinity of DDT and DDE for fats. This leads to their localization in animals in adipose tissue where the turnover rate is low. Stress may cause mobilization of adipose tissue. The subsequent release of the accumulated residues may lead to toxic levels in the blood of animals even after exposure to the pesticides has ceased.

The harmful effects of chlorinated hydrocarbons was first noted in birds of prey where reproductive failure led to a marked decline in many populations. This was partly due to impaired calcium metabolism which resulted in fragile eggs with abnormally thin shells, and partly due to behavioural changes which favoured egg breakage. This has proved to be a problem in battery chicken farming too, where sawdust from wood treated with chlorinated hydrocarbons has been used as litter. Harmful effects on fish have also been noted, including behavioural changes leading to reproductive difficulties, increased mortality among the young and, in some cases, acute toxicity to adults. Further small concentrations of DDT (0.01 ppm) have been shown to reduce photosynthesis in marine plankton, while even smaller concentrations (1 ppb) in seawater can kill many brine shrimps (*Artemia salina*) within a few weeks.

Agarwal (1983) has given an account of pesticide pollution of waters and has compiled data or mean levels of DDT residue in some worked-out river and lake waters of India. During 1976-78. The DDT residue in River Yamuna at Delhi was 0.249 ppm in upstream and 0.558 ppm in the downstream off Wazirabad.

Another insecticide is this group which has caused environmental damage in the USA is kepone. This compound was used domestically as an ant or cockroach poison, but its production was stopped when it was shown to have caused brain and liver damage, sterility, slurred speech, loss of memory and eye-twitching in workers who had been exposed to it. Matters were made worse by leakage of kepone from the production plant into the main sewer system of Hopewell, Virginia USA and, therefore, into the James River. As a result, fish and shellfish stocks were contaminated and the Governor of Virginia closed more than 100 miles of the river to commercial fishing until such time as residue concentrations should fall to a safe level. Another insecticide, mirex, which is chemically identical to kepone except for one oxygen atom, is used to control ants, but there is concern about its possible carcinogenicity.

Organophosphates

A class of chemicals in which an organic group is chemically bound to a phosphate unit. Organophosphates are a family of broad-spectrum, non persistent insecticides, developed and promoted as more and more insects became resistant to DDT and other chlorinated hydrocarbons. A number of pests have, however, developed resistance to the organophosphates. The chemical name and structure are given below :

Name	*Structure*
Malathion	$C_2H_5O_2\,C\,\overset{}{C}HS\overset{S}{\overset{\Vert}{P}}\,(O\,CH_3)_2$ $C_2H_5O_2\,CCH_2$

(Contd.)

Name	*Structure*
Parathion	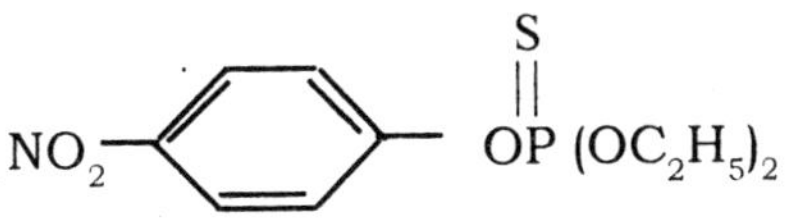 $NO_2-C_6H_4-OP(S)(OC_2H_5)_2$
Mevinphos	$(CH_3O)_2\,PO(O)C(CH_3)=CHCOOCH_3$
Dimethoate	$(CH_3O)_2\,P(S)S\,CH_2\,CONH\,CH_3$

Toxic Effects : The organophosphate insecticides are readily metabolized by mammalian enzymes and rapidly degraded in the environment. Thus, any environmental damage caused by these compounds will tend to be localized in the area of application. However, these compounds are related to the nerve gases developed for use in war and can be lethal to many different organisms, including humans. Hence, anyone applying these compounds should war protective clothing including goggles, rubber gloves and even a respirator. One should avoid treating areas which are important habitats for wildlife.

The principal action of the organophosphate compounds is to inhibit acetylcholine esterase. Thus, the action of acetylcholine released at the nerve synapses ceased to be finite with each nerve impulse. Amongst the consequences are tremors in involuntary muscles, convulsions and death.

Carbamates

A family of insecticides developed to meet the growing resistance of insects to the chlorinated hydrocarbons such as DDT.

The carbamates are nonpersistent and are quickly detoxified and eliminated from animals. Unlike the DDT or dieldrin the carbamates are not stored in fat or milk, and they do not seriously enter man's food chain.

Other carbamate insecticides are Baygon, Temik, and Zectran.

Sulphur derivatives of carbamates are herbicides (the thiocarbamates) and fungicides (the dithiocarbamates).

Common Carbamates :

Genera, Structure $G-O-C(=O)-NH-CH_3$

If G =	*Name*
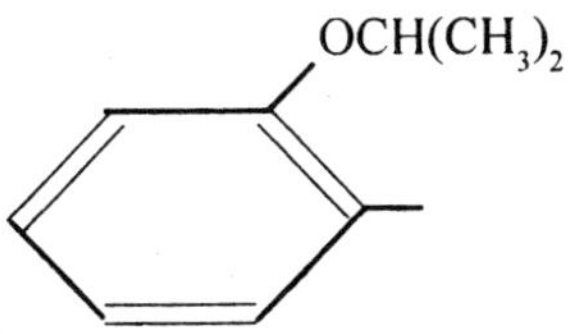 $C_6H_4-OCH(CH_3)_2$ (benzene ring with $OCH(CH_3)_2$ substituent)	Baygon

(Contd.)

If G =	*Name*
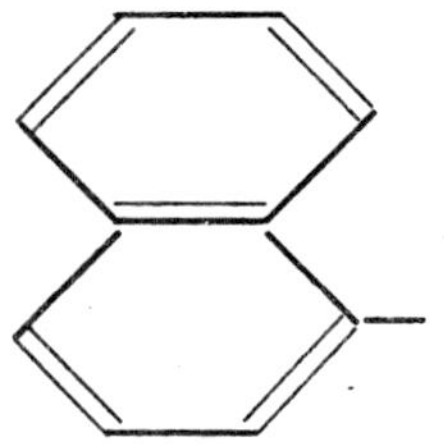	Carbaryl
$CH_3-S-\overset{CH_3}{\underset{CH_3}{C}}-CH=N-$	Temik
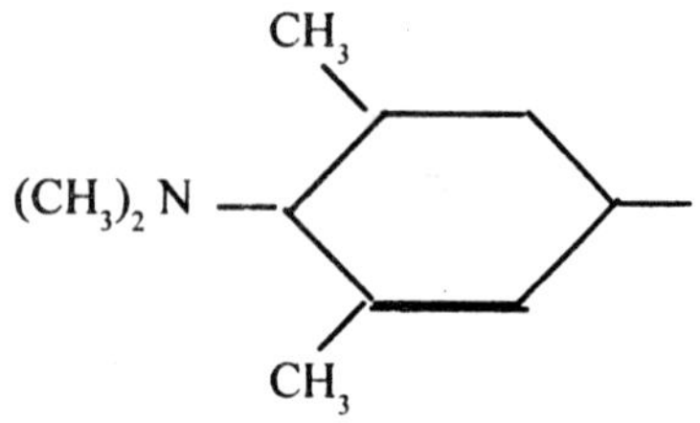 CH_3 $(CH_3)_2N$ — CH_3	Zectran

Toxic Effects of Carbamates : Among the carbamate pesticides, **methyl isocyanate or MIC (CH_3NCO)** as an environmental pollutant has drawn attention of public all over the world due to its tragic leakage from the Union Carbide Factory in Bhopal on Dec. 3, 1984 in which over ten thousand persons died and many more were seriously affected. Much of the leaked gas ultimately settled on soil and was washed out to water bodies. Although pesticide pollution of this magnitude is very rare yet smaller incidents are very frequent and together affect the health of about a million persons every year all over the world. The synthesis of MIC requires the use of phosgene ($COCl_2$) which is an extremely poisonous chemical.

Carbamate insecticides, also used as molluscicides, fungicides are herbicides, include carbaryl (sevin), baygon, temik and zectran. They are even less persistent than the organophosphates and less harmful to man. They may cause local environmental problems if used carelessly. Carbaryl, for example, is very toxic to bees. Like the organophosphates, the carbamates act by inhibiting acetylcholine esterase.

Pyrethroids

Pyrethroids are the insecticidal constituents of pyrethrum flowers (*Chrysanthemum cinerariifolium and Chrysanthemum coccineum*). Five insecticides compounds pyrethrins I and II, cinerins I and II, and jasmolin II, are present in the achenes of the flowers from which they may be extracted with organic solvents. These naturally occurring pyrethroids are unstable to the action of sunlight, air moisture and alkalis, and are rapidly degraded after application. They act as contact poison and rapidly paralyse insects such as mosquitoes, fleas, body lice and houseflies. They may be applied in the form of an emulsion with suitable emulsifiers or as dust.

Pyrethroids have a low toxicity to mammals because they are rapidly metabolized to harmless substances. However, they can cause severe allergic dermatitis and systemic allergic reactions. Large amounts may cause nausea, vomiting, headache and other disturbances of the central nervous system. Because it is nontoxic, non inflammable and leaves no oily residue, it is used to protect foodstuffs such as grains stored in commercial elevators. It has been used in mosquito repellent creams and in ointment for scabies.

The high cost of extraction of natural pyrethroids has led to the production of similar synthetic compounds, of which the best known are the allethrins. The synthetic pyrethroids are more toxic and more stable. Poisonous to fish, they are relatively indiscriminate in their effects and will kill bees where they are applied.

BIOLOGICAL CONTROL OF PLANT PATHOGENS

Biological control is the deliberate use of one organism to control another. This definition which captures the essence of the term was given by Templeton and Smith (1977). However, others give a much broader meaning to the term and intend it to cover breeding for disease resistance and use of cultural practices (including crop rotation, irrigation, and flooding). Baker and Cook (1974), define biological control of plant pathogens as their control by "one or more organisms, accomplished naturally or through manipulation of the environment, host or antagonist, or by mass introduction of one or more antagonists".

Field soils that contain organisms antagonists to plant pathogens are termed ***suppressive soils***. When added to other soils (termed ***conducive soils***), suppressive soils can bring about biological control. Root rot of papaya can be controlled by growing seedlings in suppressive soil placed in small holes dug in soil infested by the pathogen *Phytophthora palmivora*.

Soil inhabitants complete with soil invaders for subtrate, and some, such as *Trichoderma lignorum*, actually parasitize other fungi. Other soil-borne organisms produce antibiotics that impede the rate of growth of, or even kill, nearby plant pathogenic fungi. Efforts to control plant diseases by introducing antagonistic organisms into the vicinity of plant pathogenic organisms have not been successful, since most introduced organisms cannot maintain themselves indefinitely in their new environments. There are some exceptions. Species of *Chaetomium* seem to be effective antagonists against root and seedling diseases of corn. *Peniophora gigantea*, when placed on stumps of felled conifers, establishes itself quickly and grows so luxuriantly that it inhibits the growth of the root rotting pathogen, *Fomes annosus* by hyphal interference (Rishbeth, 1963). The spores of the *Paniophora* can be supplied in the form of tablets, suspended in water and sprayed onto cut stumps. Recently, they have been marketed as coinidial suspensions in sachets which is incorporated with sucrose and a dye. The low osmotic potential generated by the sucrose prevents spore germination during storage, but the sugar may actually enhance growth of *Peniophora* and the dye make the forester observe whether or not he has achieved complete cover of the stump surface. *Trichoderma harzianum* has been reported as an agent for the biological control of southern blight caused by *Sclerotium rolfuell*, and peach seedling roots when dipped in suspensions of non-pathogenic strains of *Agrobacterium radio-bacter* could not be attacked later by A. *tumefaciens*, the crowngall pathogen. In certain cases, a combination of chemical and biological control seems effective as has been observed in the control of *Armillaria* induced root rot of citrus where low concentrations of methyl bromide or carbon disulphide reduce populations of the pathogen but do not adversely affect the antagonist, *Trichoderma viride* which exerts a biological control in chemically treated soil.

Besides microorganisms, higher plants may act as biological control agents against lower organisms. For example, some green plants (***Trap crops***) that are toxic to plant pathogens, have been used to control nematodes. Root knot can be controlled by rotating row crops with the toxic pasture crop, pangola grass, with *Crotalaria spectabilis* (in which the nematode cannot reproduce), or with the toxic common marigold.

Yet another group of organisms that act as biocontrols of plant diseases are the **mycorrhizae** which are fungi mutualistically symbiotic with higher plants. They are of two kinds, the endo—and the ecto-mycorrhizae. These organisms are beneficial to their hosts, because they increase the capacity of infected roots to absorb water and minerals from the soil, they also seem to avoid pathogenic fungi. Mycorrhizae most likely exemplify plant disease controls by the principle of protection, but they are also agents of biological control. Cultural practices which amend the soil can be used to control plant diseases, for example, liming the soil control club root of cabbage, and to control potato scab the pH of soil solution can be lowered by the addition of sulphur. Green-plant materials especially legume plowed under before a cropping season, have beneficial effects on plants growing in soils infested by root-invading pathogens *e.g.*, *Phymatotrichum* root-rot of cotton and potato scab (*Streptomyces scabies*). The nature of these effects is not understood, but they obviously have to do with the microbial balance in the soil.

Also before planting, the farmer should decide whether he can prevent plant diseases with tillage or by proper crop-rotation schemes. Tilling the soil before planting corn, will reduce the amount of inoculum of *Bipolaris maydis*, the cause of southern leaf blight. But if a vertically resistant variety of corn is used, the grower, for agronomic reasons, may elect to use the 'no-till' method of growing his corn crop. Crop rotation is usually effective in reducing inculum of soil-invading plant parasites, but, continuous culture of a crop sometimes affords better control of a soil borne pathogen than does crop rotation, for example, wheat (at least in some locations) suffers least damage from the take-all disease (*Gaeuman nomyces graminis*) when grown year after year in the same soil, because the equilibrium between the parasite and antagonistic microorganisms in the soil is restored. Here the build up of microorganisms takes about three years and the disease declines to low levels about five years after the start of monoculture. Many antagonistic organisms, and viruses, that attack the fungus have been found in the soil during the decline period and through the subsequent low levels of the disease.

Cultivation, the stirring of soil near growing plants—to improve tilth and obviously to control weeds—frequently has an indirect effect on plant disease control. We may harbour plant pathogens, and weed control by cultivation (as well as by preplanting and postemergence application of herbicides certainly reduces the inoculum produced by some plant pathogens, notably viruses. Therefore, virus-disease control, particularly in fruit orchards and vegetable fields is often achieved by destruction of weeds species and other wild plants outside the fields and orchards.

Finally, cultural practices at and after the time of harvest often determine subsequent decay of harvested plant products. For example, careful handling to prevent wounding, and storage at low temperature and low relative humidity, do much to control postharvest diseases of plant products.

Biological control of soil borne pathogens has been used for many years along with other practices. However, there are still diseases for which there is no chemical control and there are reasons for seeking some means other than toxic chemicals to reduce diseases. Most probably a system of integrated control based on chemical or cultural measures followed by the biological exploitation of the disturbed equilibrium will be most successful.

ACARICIDE

Acarids include ticks and mites, which belong to order Acarina. Whereas ticks are parasites on warm blooded animals; many mites are serious plant pests. Acaricide is any preparation for killing acarids, mainly mites. These plant pests reproduce rapidly, and chemical control is difficult as they live in protected situations within buds, leaf tissues and cracks in bark. Many chemicals fail to kill mites, but do destroy their insect predators, thus further increasing mite populations. Specific chemical preparations called acaricides are needed for control of these mites.

Black currant gall mite is the most serious pest of black currant (*Ribes nigrum*) living in leaf buds forming swollen big buds. These microscopic mites leave buds temporarily between

April and June in search of new buds. These mites can be controlled by spraying with 0.5 to 1 per cent solution of limesulphur when first flowers open, and repeated three weeks later. Gall mites are controlled similarly.

Bryobia mites are particularly troublesome on apples, gooseberries and ivy. They feed on leaves which at first develop a light freckling on the upper surface and later bronze and wither. The eggs overwinter on the fruit trees and can be destroyed by thorough spraying with DNOC/ Petroleum in winter before the buds begin to swell. Sprays of dimethoate, formothion or malathion during the growing season are effective.

Bulb scale mite feeds and breeds inside daffodils and other bulbs. The pest is particularly troublesome on forced bulbs, and chemical control is difficult. Immersion of the dormant bulbs in water maintained at a temperature of 43°C for 1½ hours may eradicate infestation.

Fruit tree red spider mite causes damage to the foliage of apples, plums and damsons. Infestation is similar to bryobia mites. Tar-oil winter washes and insecticides kill natural predators, further increasing mites. Control of these mites involves winter wash of DNOC/petroleum on dormant buds to kill eggs, spraying with malathion, dimethoate or formothion after blossom fall to prevent summer attacks.

Glass house red spider mites are serious pests of green house and house plants. The withering leaves are covered with fine silk webbing. Overwintering occurs in crevices of brick work and woodwork, bamboo canes and debris. Regular fumigation with azobenzene will prevent serious increase in mite populations. Through spaying with liquid derris extract, dimethoate or formothion protects plants under glass and out doors.

Sulphur dust and lime-sulphur sprays are effective against Tarsonemid mites causing damage to Aster, strawberry, begonia, etc. and against which no other chemicals are effective.

ALDRIN

A broad-spectrum chlorinated hydrocarbon insecticide related to Endrin. It is highly toxic in rats, fish, and birds and is persistent and systemic.

Aldrin is converted in air, soil, plants and animals to equally toxic and even more persistent compounds, dieldrin. In fatty tissues of animals, aldrin is stored as dieldrin. Prolonged exposure may cause damage to both the liver and kidney.

Aldrin is banned in eight countries and severely restricted in eleven others. It can cause cancer. Classified as **highly hazardous** by the WHO. Aldrin is sold under the brand name **Alderstan, Aldrex,** and various others including the word **aldrin**.

BENZPYRENE (BENZO (a) PYRENE)

An organic compound of the aromatic hydrocarbon family found in very minute quantities in the air of nearly every city as the result of the incomplete combustion of coal, gasoline, and oil. Benzpyrene is one of the most potent carcinogenic compounds known. Probably the principal carcinogen in cigarette smoke, benzpyrene also escapes from the asphalt of roads.

BHC

[BENZENE HEXACHLORIDE $C_6H_6Cl_6$; IUPAC Nomenclature: 1,2,3,4,5,6,—Hexachloro cyclohexane]

A broad-spectrum, persistent insecticide; highly toxic to fish; moderately toxic to rats and birds. **(it should not be confused with "hexachlorobenzene" C_6Cl_6, HCB)**

Commercial BHC is a complex mixture of closely related materials formed by the action of chlorine on benzene in the presence of light. Only one component in the mixture, the gamma-form called lindane, has significant insecticidal properties. Since it is expensive to separate and purify the lindane component, the mixture is used for many purposes. Its greatest use has been against cotton pests, but it is also found in some household pesticides. (Lindane instead of BHC is normally used in the latter).

The lethal dose of BHC to man is estimated at about 30g but much lower dose can be quite harmful.

BORDEAUX MIXTURE

A fungicide mixture made by adding slaked lime (calcium hydroxide) to a copper sulphate solution.

Bordeaux mixture, is the most widely used copper fungicide throughout the world. It controls many fungus and bacterial leaf spots, blights, anthracnoses, downy mildews, and cankers but causes burning of leaves or russeting of fruit such as apples when applied in cool, wet weather. The phytotoxicity of Bordeaux is reduced by increasing the ratio of lime to copper sulphate. Copper is the only ingredient in the Bordeaux mixture that is toxic to pathogens and, sometimes, to plants, while lime's role is primarily that of a "safener". For dormant sprays, concentrated Bordeaux is made by mixing 10 pounds of copper sulphate 10 pounds of lime, and 100 gallons of water : it has the formula 10:10:100. The most commonly used formula for Bordeaux is 8:8:100. For spraying young, actively growing plants, the amounts of copper sulphate and lime are reduced, and the formulas used may be 2:2:100, and so on. For plants known to be sensitive to Bordeaux, a much greater concentration of lime may be used, as in the formula 8:24:100.

In the "fixed" or "insoluble" copper compounds the copper ion is only slightly soluble and these compounds are, therefore, less phytotoxic than Bordeaux, but also less effective as fungicides. The fixed coppers are used for control of the same diseases as Bordeaux and they can also be used as dusts. The fixed coppers contain either basic copper sulphate or basic copper chlorides, or copper oxides, or miscellaneous other formulations. Most of them, are recommended as sprays at the rate of 4 lb per 100 gallons of water or as 7 per cent copper dusts.

CHLORDANE

A broad spectrum, highly toxic persistent insecticide of the chlorinated hydrocarbon class.

Chlordane only slowly breaks down in the environment and can persist in the soil for several years. It is highly effective against most species of cockroaches and ants and is effective against flees, lice and ticks.

Chlordane is highly toxic and causes cancer and birth defects. It accumulates in human fat tissues and is transferred across the placenta from mother to child. It is classified by the WHO as 'moderately hazardous'.

DDD

[Dichlorodiphenyldichloroethane; IUPAC Nomenclature: 2, 2-bis (p-chlorophenyl) 1, 1-dichloroethane].

A persistent chlorinated hydrocarbon insecticide in the diphenylethane family; highly toxic to fish. (For chemical structure and toxicity, see chlorinated hydrocarbon).

DDD was selected because it seemed to be less toxic to fish than DDT. After the first application in 1949 the water of Clear Lake Northern California (USA), had 14 parts per billion (ppb) DDD. The Environmental Protection Agency USA banned nearly all uses of DDD in 1972.

DDE

[Dichlorodiphenyldichloroethylene; IUPAC Nomenclature: 2, 2-bis (p-chlorophenyl) 1, 1-dichloroethene].

One of the breakdown products of DDT. (For chemical structure see chlorinated hydrocarbons). Although not a commercial product used as a pesticide, it has pesticidal properties.

One of the simplest chemical changes DDT undergoes in the environment is its conversion to DDE, and several million tons of DDE is now distributed in the earth's ecosystem. Like DDT, DDE is stored in fatty tissue. DDE levels in the fatty tissue of brown pelicans off the coast of California have been found as high as 2500 ppm.

Before 1970 DDE accounted for roughly 40-50 per cent of the total of DDT—derived materials in human diets. The daily intake of DDE through the diet by the average adult in the general population was estimated to be about 0.044 milligram.

DDT

[Dichlorodiphenyltrichloroethane; IUPAC Nomenclature: 2, 2-bis (p-chlorophenyl) 1, 1, 1-trichloroethane].

A broad-spectrum, persistent organochlorine pesticide of the diphenylethane family. (For chemical structure and toxicity, see chlorinated hydrocarbons).

The insecticidal properties of DDT were discovered by Paul Mueller of J.R. Geigy, AG (Switzerland), won a Nobel Prize in 1948 for the discovery. DDT killed disease—bearing pests so effectively and with so few harmful effects on man, most other mammals, and plants that it was universally hailed as a solution to some of mankind's greatest afflictions. Several million tons of DDT has been scattered throughout the world in programmes to eradicate malaria, typhus, and various other endemic fevers, and to rid the environment of common mosquitoes and flies.

DDT is very persistent: of a kg put out into the environment is one year, about half will still be present 10-15 years later. DDT is soluble in fat and is concentrated by fatty tissue. DDT moves up food chains having at their tops such creatures as man, raptor birds, and other carnivores. From Arctic polar bears to Antarctic penguins, probably no living creature is without some DDT (or DDT—derived material) in its tissues. In 1972 the Environmental Protection Agency, USA banned all uses of DDT in the United States with the exception of certain public health purposes.

DIELDRIN

A broad-spectrum insecticide of chlorinated hydrocarbon class. It is one of the most persistent and toxic chlorinated hydrocarbon.

When flies became increasingly resistant to DDT, the cyclodiene class of chlorinated hydrocarbons were prepared Dieldrin was used extensively from the early 1950s to 1975. Dieldrin was extensively used to control soil—inhabiting insects and termities. It was used against many public health pests such as tsetse fly, malarial mosquitoes and insects on all types of crops.

Because dieldrin is a broad spectrum insecticide, it kills both pests and the natural predators of pests. Dieldrin is known to affect DNA; it causes both point mutations and large chromosomal ablerrations.

ENDOSULPHAN (Trade Name Thiodan)

A persistant insecticide in the chlorinated hydrocarbon family highly toxic to rats, fish, and birds; less toxic than DDT to honeybees.

At 1 part per billion (ppb) in water, endosulphan will kill fish. Endosulphan can cause severe damage to Concord grape vines, birch trees, geraniums, and a few other flowers.

ENDRIN

A highly toxic, persistent, systemic insecticide in the chlorinated hydrocarbon family.

Endrin is the most toxic of the commercially available chlorinated hydrocarbon insecticides. It has been widely used in India on rice and cotton, and there is a claim that substitutes would raise the cost of treating these crops by 80-90 per cent. It has been extensively used in the United States on cotton.

Endrin is known to affect genes (DNA), and it induces both point mutations and large chromosomal aberrations in experimental species.

It is banned in seven countries, severely restricted in 12 more. Environmentally persistent, it wipes out beneficial insects as well a targets and can cause nerve and brain damage to humans. Classified as 'highly hazardcus' by the WHO; the World Bank recommends that it should not be used.

Sold as Endrin 19.2 EC.

FUMIGANT

Any rapidly evaporating chemical compound used as a pesticide or a disinfectant. In agriculture fumigants are usually volatile, very toxic substances put into the soil before planning to kill insects, nematodes, and sometimes weeds. Some soil fumigants attack the microorganisms in soil that convert ammonia (or ammonium ion) to nitrites and nitrates.

Some fumigants are simple, chlorinated hydrocarbons: Larvacide (highly toxic to humans)—trichloronitromethane carbon tetrachloride; D-D (Telone)—1, 3-dichloropropene and related compounds.

FUNGICIDES

In the broadest sense, a fungicide is any chemical that kills fungi or inhibits their growth. Strictly speaking, this term should be confined to those chemicals that kill. It covers a vast range of compounds, all of which are not suitable for use in a particular environment, because of such factors as toxicity to higher plants and animals, toxicity to beneficial micro-organisms or unsatisfactory odour or colour.

Inorganic Compounds

The 1800s and early 1900s are regarded as the era of inorganic fungicides, since they saw the introduction and widespread use of sulphur, then copper and, finally, mercury-based fungicides. Many of these are still used today, especially in the Third World; for example, inorganic sulphur compounds, like flowers of sulphur or polysulphides, are applied as dusts or as suspensions to control powdery mildew diseases. Sulphur compounds are essentially **contact fungicides**, (*i.e.*, they have little ability to spread from their sites of application). This explains their continued use against powdery mildew fungi, whose mycelium is essentially superficial.

For the same reason, they are ineffective against the vast majority of plant parasites, which grow deep within plant tissues. Opinions differ as to the mode of action of sulphur; one view is that sulphur acts as an electron acceptor in fungi and therefore interferes with several aspects of normal metabolism.

The importance of sulphur declined markedly with the introduction of copper fungicides in the 1880s, such as '**Bodeaux mixture**' and '**Burgundy mixture**' mainly formulations based on copper sulphate. The use of these compounds continues to the present day (for the control of many leaf and fruit diseases). However, copper-based fungicides are all potentially phytotoxic, and attempts to reduce this undesirable side-effect often result in less effective formulations. The precise modes of action of copper fungicides are still not clear.

Mercury fungicide always had a more restricted usage than sulphur or copper, because of its acute mammalian toxicity. Mercurial fungicides were developed primarily as seed treatments, and this use persists to the present day. An interesting environmental side-line is that the use of mercurial seed dressings is now restricted for spring-sown cereal crops, but not for autumn-sown ones; the reason is that birds feed on the plentiful supplies of natural seeds and berries in the autumn and tend to eat grain to a much lesser extent, whereas in the spring they eat considerably more grain and are therefore at much greater risk.

Organic Fungicides

Organic fungicides were first discovered during the 1930s and have largely replaced many inorganic ones for agricultural uses. Two classes of chemicals stand out as being particularly important, however: firstly, the **dithiocarbamates**, like *maneb nabam, thiram* and *zineb*; secondaly, *captain* and related compounds, which belong to the **trichloromethylthiodicarboximides**.

Dithiocarbamates

The organic sulfur compounds are unquestionably the most important, most versatile, and most widely used group of modern fungicides. They include thiram, ferbam, nabam, maneb, and zineb and are all derivatives of dithiocarbamic acid. It is believed that the dithiocarbamates are toxic to fungi because they are metabolized to the isothiocyanate radical (—N=C=S), which inactivates the sulphydryl groups (—SH) in amino acids and enzymes within pathogen cells and thereby inhibits the production and function of these compounds.

$$(H_3C)_2N-\overset{S}{\overset{\|}{C}}-S-S-\overset{S}{\overset{\|}{C}}-N(CH_3)_2$$

Thiram

Thiram consists of two molecules of dithiocarbamic acid joined together. It is used mostly for seed and bulb treatment for vegetables, flowers, and grasses, but also for the control of certain foliage diseases, such as rusts of lawns, fruits, and vegetables. Thiram is also good as a soil drench for control of damping off and seeding blights. Thiram, in various, formulations is sold under many trade names. Thiram and Tersan are two examples.

$$\left((H_3C)_2N-\overset{S}{\overset{\|}{C}}-S-\right)_3 Fe$$

Ferbam

Ferbam consists of three molecules of dithiocarbamic acid reacted to one atom of iron. Ferbam is used to control many foliage diseases of fruit trees and ornamentals.

Another group of dithiocarbamic acid derivates with different molecular configurations contains the fungicides nabam (Na), zineb (Zn), and maneb (Mn). Nabam and even ferbam have been largely replaced by newer fungicides.

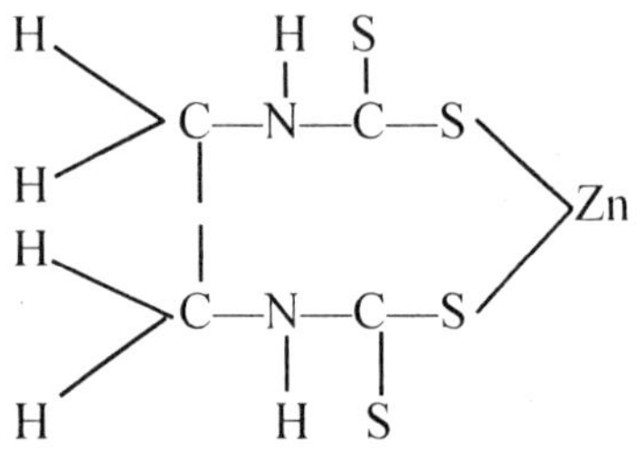

Zineb

Zineb is sold as Dithane Z-78; it is an excellent, safe, multipurpose foliar and soil fungicide for the control of leaf spots, blights, and fruit rots of vegetables, flowers, fruits trees, and shrubs.

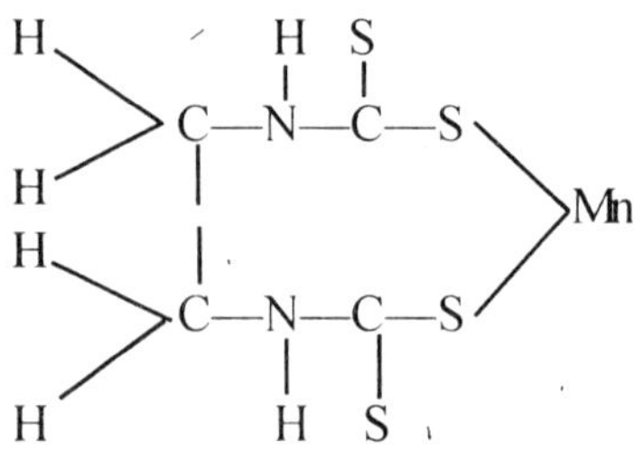

Maneb

Maneb contains manganese; it is sold as Manzate, Dithane M-22, and Tersan LSR and is an excellent, broad-spectrum fungicide for the control of foliage and fruit diseases of many vegetables, especially tomato, potato, and vine crops, and of flowers, trees, turf and some fruits. Maneb is one of the most frequently used fungicides for control of vegetable diseases. Maneb is often mixed with zinc or with zinc ion and results in the formulations known as maneb-zinc (sold as Manzate D) and as zinc ion maneb called mancozeb (sold as Manzate 200, Dithane M-45). The addition of zinc reduces the phytotoxicity of maneb and improves its fungicidal properties.

Heterocyclic Compounds. The heterocyclic compounds are a rather heterogeneous group but include some of the best fungicides, for example, the related captan, folpet, and captafol and the related iprodione and vinclozolin. Most of them also inhibit production of essential compounds containing — NH_2 and SH groups (amino compounds and enzymes).

Captan is an excellent fungicide for control of leaf spots, blights, and fruit rots on fruits, vegetables, ornamentals, and turf. It is also used as a seed protectant for vegetables, flowers, and grasses and as a postharvest dip for certain fruits and vegetables.

Folpet is similar to captan in spectrum and effectiveness. In addition, it controls many powdery mildews.

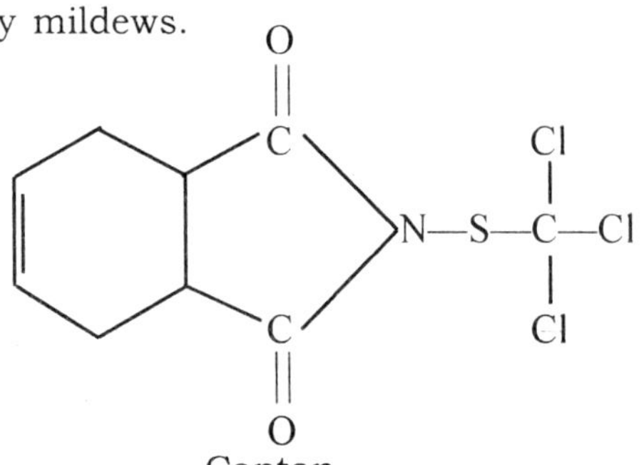

Captan

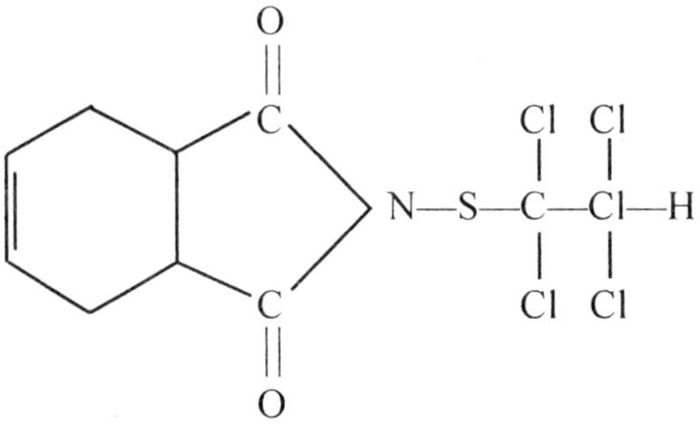

Captafol

Captafol is sold as Difolatan, and has properties similar to those of captan and folpet. Moreover, Difolatan exhibits unusual resistance to weathering, which provides extended redistribution and residual activity. These properties, combined with its low phytotoxicity, also allow the use of up of three times the regular amount of Difolatan as a single application treatment (SAT) on apples against apple scab, cherry leaf spot, citrus melanose, and scab, and against several foliage diseases of tomato. Such concentrated sprays may provide protection for longer periods and reduce the number of sprays needed.

Systemic Fungicides

The most recent and perhaps the most significant chapter in the fungicide story is the development of systemic fungicides—those that are taken up by the plant and transported to their sites of action within its tissues. Several of these compounds have appeared in the last decade, but the **benzimidazoles** have had by far the most spectacular success. These compounds are transported within the plant xylem, though less so, if at all, in the phloem. They inhibit a vast range of fungi, with some notable exceptions like the Oomycetes, some of the Basidiomycotina and some species of *Helminthosporium, Alternaria* and similar Deuteromycotina. Their modes of action are still largely unknown, although there is increasing evidence to suggest that benomyl and thiabendazole, at least, have a common mode of action, possibly based on disruption of DNA synthesis. The empirical approach to the development of fungicides is well illustrated by these compounds; (1) they are covered by different patents, despite the fact that their modes of action may be similar, if not identical; (2) it is now believed that benomyl hydrolyses to methyl benzimidazole-2-yl-carbamate (MBC), within the plant, and that this stable derivative rather than the parent molecule is the most active form; (3) thiabendazole was not developed as a fungicide at all—on the contrary, it was initially developed as a anti-helminth.

Benzimidazoles include some important systemic fungicides such as **benomyl, carbendazim, thiabendazole,** and **thiophante**. They are effective against numerous types of diseases caused by a wide variety of fungi. Most benzimidazoles are converted at the plant surface to methyl benzimidazole carbamate (MBC, carbendazim), and this compound interferes with nuclear division of sensitive fungi.

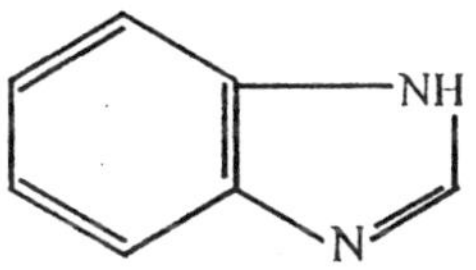

MBC

Carbendazine

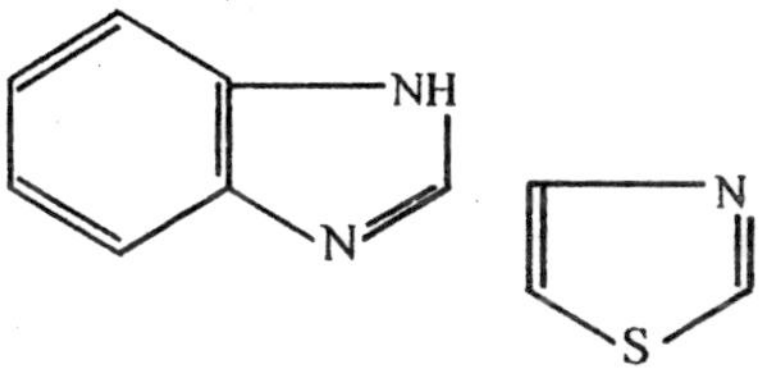

Thiabendazole

Benomyl, is sold as Benlate, Tersan 1991, etc. It is a safe, broadspectrum fungicide, effective against a large number of important fungus pathogens, and it also suppresses mites. It controls a wide range of leaf spots and blotches, blights, rots scabs, and seed-and—soil brone diseases. Benomyl is particularly effective for powdery mildew of all crops; scab of apples peaches, and pecans; brown rot of stone fruits; fruit rots in general; *Cercospora* leaf spots; cherry leaf spot; black spot of roses; blast of rice; various *Sclerotinia* and *Botrytis* diseases; and loose and covered smuts of wheat. It is highly active against and suppreses infections by *Rhizoctonia, Thielaviopsis, Ceratocystis, Fusarium* and *Verticillium.* It has no effect on oomycetes, on some dark-spored Imperfects such as *Helmintho sporium* and *Alternaria,* on some Basidiomycetes, and on bacteria. Benomyl may be applied as a seed treatment, and as a fruit dip. Benomyl seems to be mutagenic and to hasten the appearance of pathogen races resistant to it.

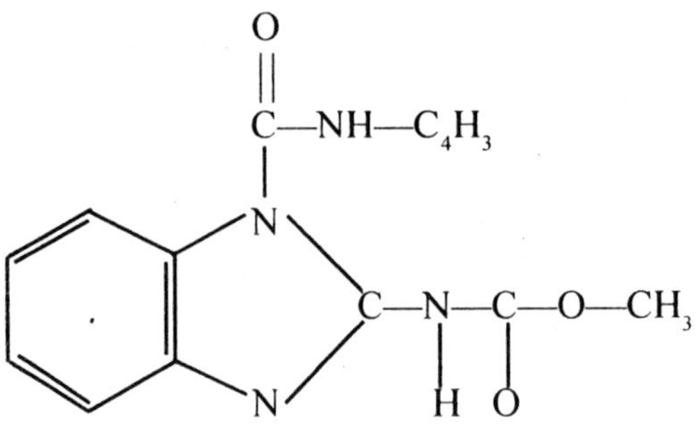

O

C—NH—C_4H_3

N

C—N—C—O—CH_3

N H O

Benomyl

One of the more notable applications of benomyl and related compounds to date is in the control of Dutch elm disease. For this purpose, the fungicide is dissolved in lactic acid or a similar acidic solvent and injected from a pressurized canister into the base of the tree trunk. This procedure leaves a ring of scars around the base of the tree-truck. The treatment is costly, however, and may need to be repeated every two years or so to confer maximum protection. Moreover, it is essentially a preventive rather than a curative measure, since the impaired water movement within a diseased tree tends to restrict access of the fungicide to the sites of infection.

Thiabendazole is sold as Mertect. It is also a broad-spectrum fnugicide and effective against many Imperfect fungi causing leaf spot diseases of turf and ornamentals and diseases of bulbs and corms. It is commonly used as a postharvest treatment for the control of storage rots of citrus, apples, pears, bananas, potatoes and squash.

Thiophanate, under the trade name Topsin, is effective against several root and foliage fungi affecting turf grasses.

Thiophanate methyl, under the trade names Fungo and Topsin M, is a broad-spectrum preventive and curative fungicide for use on turf and as a foliar spray to control powdery and downy mildews, *Botrytis* diseses, numerous leaf and fruit spots, scabs, and rots. It is also used as a soil drench or dry soil mix to control soil-borne fungi attacking bedding plants, foliage plants, and container-grown plants.

Yet again, there is inadequate knowledge of the modes of action of organic fungicides. Many of these compounds were discovered by chance and industry now conducts massive screening programmes on a wide range of chemicals solely with the aim of detecting further compounds for more rigorous investigation and development.

HERBICIDES

A class of chemicals commonly called "weed-killers" that will kill plants or otherwise interfere with their growth. Some of the more common herbicides are included in Table 1.

Herbicides can be non-selective (*i.e.* kill all type of plants commonly used to clear garden paths of weeds) or selective (*i.e.*, can be applied to a crop in order to kill unwanted weeds, without damaging the crop itself.) Snell (1986) identifies two types of herbicides: those that work by foliar application (Contact herbicides) and those that are applied to the soil (Soil Sterilants). The former are divided into two groups : seven systemic groups, which include **2,4,5,-T, paraquat** and **glyphosate** and four contact herbicides, including **ioxynil, dinoseb,** and **pentanochlor.**

There are nine groups of soil application herbicides, including **diuron, atrazine** and **nitrofen**.

A selective herbicide is one that acts only against certain types of plants and not all types (*e.g.*, 2,4-D kills broadleaf weeds in lawns with little or no damage to lawn grass). A herbicide is described as preemergent if it should be applied before weed seeds have germinated and the weed plants have emerged from the soil. Other terms that refer to the time at which a herbicide should be applied are preplant (*i.e.*, application of herbicide should be done before the land is seeded with the desired crop), postemergent, and postharvest.

Soil Sterilants

Soil sterilants are compounds that are mixed with the soil before planting, to kill plants and animal pests in the range of the substance. Methyl bromide is an example of this type. Soil sterilants soon diffuse out into the atmosphere and soil organisms eventually return to the soil.

TABLE-7— Some Common Herbicides

Name	*Structure*	*Comments*
1	2	3
2,4-I) (2-4—Dichlorophenoxy acetic acid)	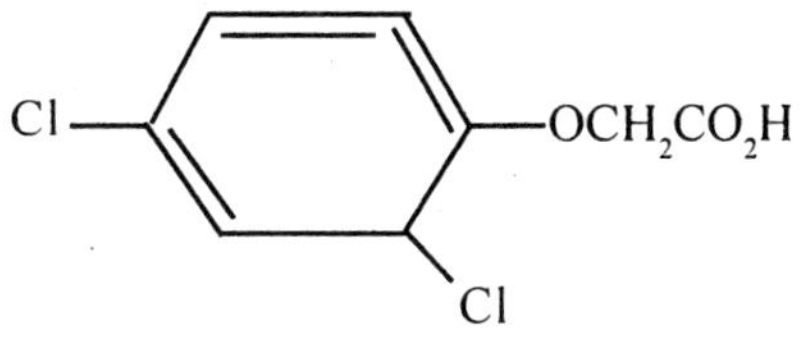 Cl–(benzene ring, Cl)–OCH_2CO_2H	Used as a salt or ester; Selective; persists in soil for 2-3 weeks. Heavily used in Vietnam War
2,4,5-T (2,4,5—Trichlorophenoxy acetic acid)	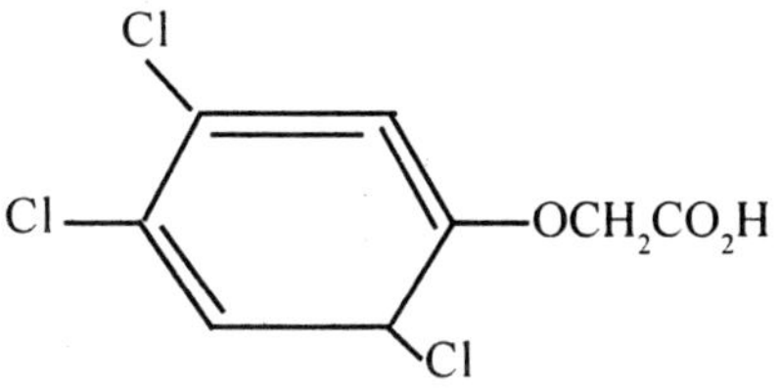 Cl, Cl–(benzene ring, Cl)–OCH_2CO_2H	Used as a salt or ester; *See* separate entry
Silvex (2,4,5—TP)	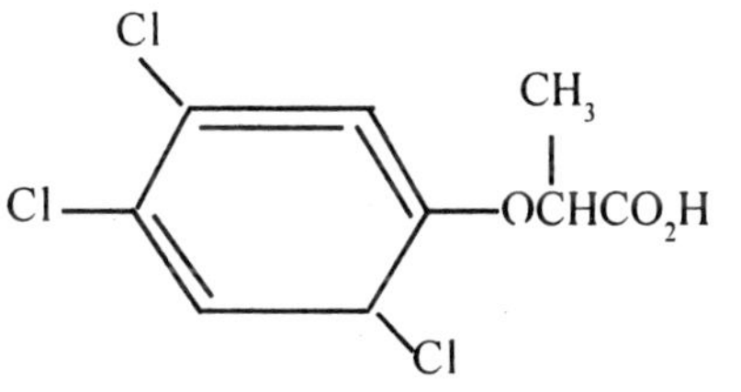 Cl, Cl–(benzene ring, Cl)–$OCH(CH_3)CO_2H$	Used as salts and esters. Mostly for control of woody plants and aquatic weeds but good (in combination with 2,4-D) against most weeds in established lawns. Activity about equal to 2,4,5-T on many woody plants, including oaks.
PCP (Pentachlorophenol)	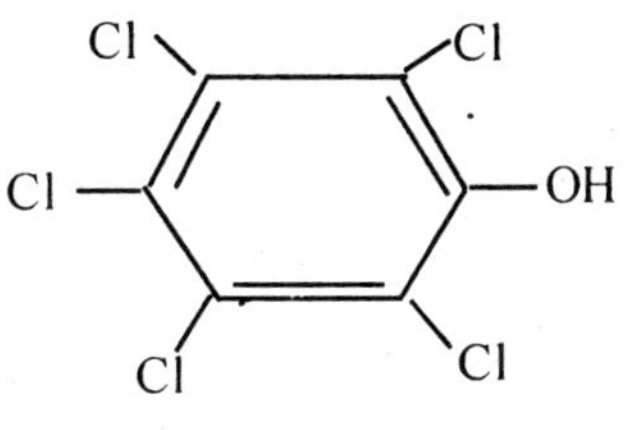 Cl, Cl, Cl–(benzene ring, Cl, Cl)–OH	Some Preemergent and some postemergent with feed crops, cotton; also a defoliant, fungicide, insecticide and wood preservative; suspected of affecting DNA; very toxic to man and animals.
Picloram (4- -Amino-3,5,6-tricloro pi colinic acid)	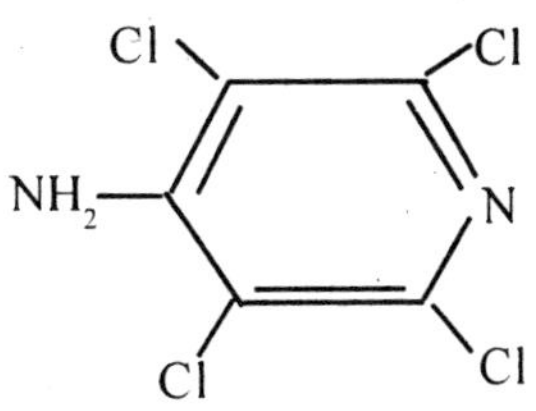 Cl, Cl, NH_2–(pyridine ring N, Cl, Cl)	Used as salts; extremely persistent—perhaps the most persistent and most active herbicideknown; will kill trees if applied to bark at the base.

(Contd.)

TABLE- 7 (Contd.)

1	2	3
Dicamba (Banvel-D)	Cl, OCH_3, $CO_2^{-}\,^{+}NH_2(CH_3)_2$, Cl	Selective; mostly postemergent good against mannual broadlea weeds and some perennia. weeds; used on golf courses and noncrop areas.
Dichlorprop [2-(2,4-Dichloro-phenoxy) propionic acid]; 2,4-DP	Cl, CH_3, $OCHCO_2H$, Cl	Against brush in rangeland clearance; on established lawns to kill emerged, broad leaf weeds.
Mecoprop [2-(2-Methyl-4-chloro-phenoxy) propionic acid; MCPP]	Cl, CH_3, $OCHCO_2H$, CH_3	Selective; Postemergent; found in some home lawn fertilizers; persists in soil several weeks; safer than 2,4-D on sensitive turf.
DCPA (Dimethyl tetrachloroterephthlate: Dacthal)	CO_2CH_3, Cl, Cl, Cl, Cl, CO_2CH_3	Selective; mostly preemergent; applied at time of seeding of alfalfa, vegetables, cotton; probably most versatile for home use: good against seeds of crabgrass and purslane; does not kill emerged broadleaf weeds on lawns.
Paraquat	$[CH_3—N$—(ring)—(ring)—$N—CH_3]^{2+}$ $[SO_4CH_3]_2$	Nonselective; very toxic; contact; noncrop weed control; postmergent; known to induce large chromosome alterations (but not point mutations) in test species.
Atrazine (AAtrex)	$NHCH_2CH_3$, Cl, N, N, $NHCH(CH_3)^2$	Selective; mostly preemergent or postharvest; used for wheat, sugercane, and sorghum.
Diuron	Cl, Cl, O, $NHCN(CH_3)_3$	One of the best for longterm vegetation control.

Contact Herbicides

The contact herbicides are quick-acting substances that kill plants by direct contact with their leaves. Examples are DNOC (dinitrocresol) and PCP (pentachlorophenol). They are not effective against perennials with roots that can form new shoots. DNOC is quite toxic, and tends to accumulate in an animal system. It can be absorbed in toxic amounts either directly through the skin or through the food chain.

Systemic Herbicides

Systemic herbicides diffuse through leaves or roots into the plants system for distributing its fluids. This group includes more selective herbicides. One such group of compounds have hormone —initiating action. Herbicides such as **2,4-D, 2,4,5-T, silvex, dicamba, mecoprop** and **picloram** mimic the plants own hormones but not exactly in the manner in which the plant needs. Inspite of much research to determine how they kill only certain weeds, their mechanism of action is generally unknown. Part of their selectivity against broad-leaf weeds results from greater absorption and translocation than in grasses, but more important factors are involved. It is sometimes stated that they cause a plant to "grow itself to death", but this is misleading. Modern hypotheses suggest that these compounds alter DNA transcription and RNA translation, so that the proper enzymes needed for coordinated growth are not produced properly, but how this is accomplished is unknown.

2,4-D and 2,4,5-T are probably the most widely used herbicides. They are relatively cheap and affect dicoets more than monocots. Because of their selectivity they are often used to kill broad leaf dicot weeds in cereal grains and lawns. **Agent orange,** which was used in the Vietnam war as a defoliant, is an effective mixture of 2,4-D and 2,4,5-T. A common popular misconception seems to be that, 2,4,5-T is a very toxic substance to animals, whereas there is overwhelming scientific evidence that the pure compound is of low toxicity compared with many chemicals in daily use in agriculture.

The toxicity of commercial samples of 2,4,5,-T is due to its contamination with a by-product 2,3,7,8-tetrachlorodibenzo *para*-dioxin (TCDD)

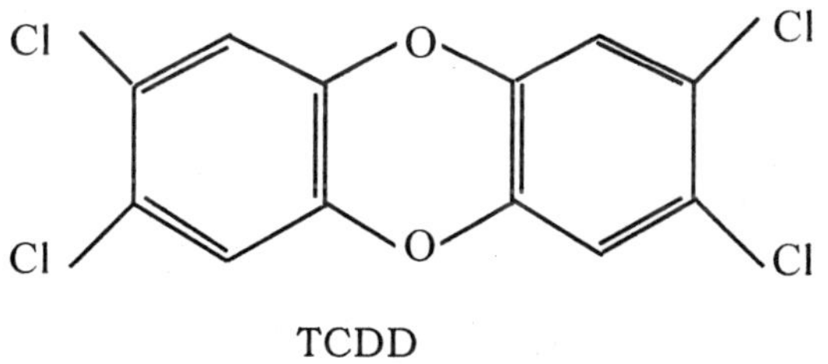

Cl Cl O O Cl Cl

TCDD

TCDD is an extremely toxic substance and the concern over 2,4,5-T has arisen from the use of TCDD—contaminated samples in some areas. Normal application rates of 2,4,5-T approximate 1 kg/acre, and commercially available formulation must contain less than 0.1 ppm of TCDD. Thus, normal applications should disperse a maximum of 1 ppm TCDD per acre.

LINDANE

A persistent broad-spectrum insecticide of **chlorinated hydrocarbon** class; used in agriculture, veterinary medicine, indoor pest control and for killing human head lice. Highly toxic to fish, moderately toxic to rats and birds. Banned in the USA because of suspected carcinogenicity.

Lindane is an active component of a complex mixture called **benzene hexachloride (BHC)**. When this component is removed and purified, it is marketed under the name lindane.

Lindane is toxic to non-target insect species, and is persistant in the environment. Classified by WHO as 'moderately hazardous'.

NEMATICIDES

A class of chemicals that will kill nematodes. Most leave no residues on plants or higher animals that are harmful. Many of these are also insecticides.

Nematodes or round works are thread—like, invertabrate animals, free-living and parasitic. Those parasitic in animals include intestinal round worms, hook worms, whipworms, filarial worms and pinworms. Several hundred species of nematodes are plant parasites, most of them feeding on roots, stems, leaves and flower parts. When they attack the roots of crops, the plants are weakened and are more susceptible to attack by fungi and other pests.

Against Plant Parasitic Nematodes

Most of the nematicides are volatile soil fumigants and are active against not only nematodes but also insects, fungi, bacteria, weed seeds, and almost anything else living in the soil. Several newer chemicals are nonfumigant granular or liquid substances active mostly against nematodes and insects.

The four main groups of nematicides are :

1. Halogenated hydrocarbons, 2. organophosphates, 3 isothiocyanates, and 4. carbamates.

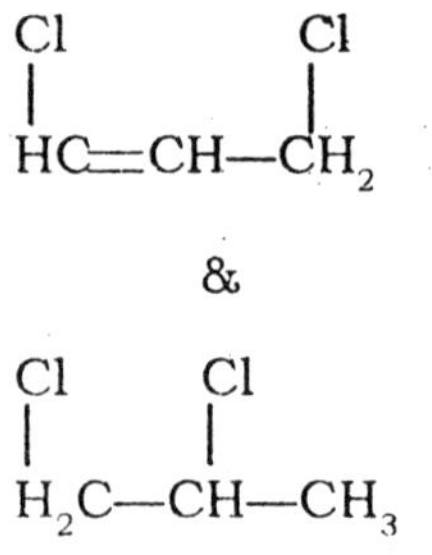

$$\begin{array}{ccc} Cl & & Cl \\ | & & | \\ HC & = CH - & CH_2 \end{array}$$

&

$$\begin{array}{ccc} Cl & Cl & \\ | & | & \\ H_2C - & CH - & CH_3 \end{array}$$

Dichloropropene—dichloropropane mixture

$Br—CH_2—CH_2—Br$

Ethylene dibromide

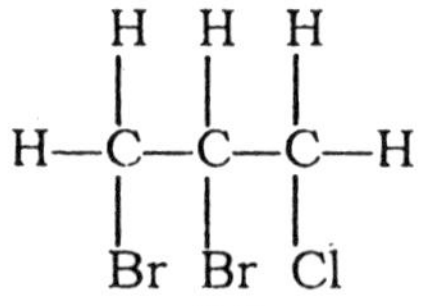

$$\begin{array}{ccccccc} & H & & H & & H & \\ & | & & | & & | & \\ H - & C & - & C & - & C & - H \\ & | & & | & & | & \\ & Br & & Br & & Cl & \end{array}$$

Dibromochloropropane

CH_3Br
Methyl bromide

1. Halogenated Hydrocarbons

Halogenated hydrocarbons include **dichloropropene-dichlore propane (D-D), ethylene dibromide (EDB), dibromochloropropane (DBCP) and methyl bromide (MB).** They are excellent nematicides. Because of their high toxicity, contamination of produce and ground water, and because one of them (DBCP) has caused sterility in male workers in the manufacturing plant, the use of most of them as nematicides has been banned in the United States. A small amount (1-2-per cent of chloropicrin is added to these chemicals to serve as a warning agent. They are applied to the soil by injection at least two weeks before planting. They all kill nematodes and insects, and at the higher dosages most kill soil-borne pathogens and weed seeds. Methyl bromide is a broad spectrum fumigant against soil-borne pathogens and is also used for above-ground control of dry wood termites and for fumigation of agricultural produce.

Halogenated hydrocarbons affect organisms because they are soluble in lipids and disrupt the function of membranes and nervous systems.

2. Organophosphates

Organophosphates include **prorate (Thimet) disulfoton (Disyston), ethoprop (Mocap) fensulfothion (Dasanit), fenamiphos (Numacur), fosthietan (Nem A-Tak),** and a few others. Many of the organophosphates were developed initially as insecticides but are taken up and are distributed systemically through the plant and are effective nematicides. They are available as water-soluble liquids or granules, have low volatility, can be applied before or after planting, and are effective only against nematodes and soil fungi through contact or by digestion. Like the organophosphate insecticides, thee nematicides inhibit the nerve-transmitter enzyme cholinesterase and result in paralysis and ultimately death of affected nematodes.

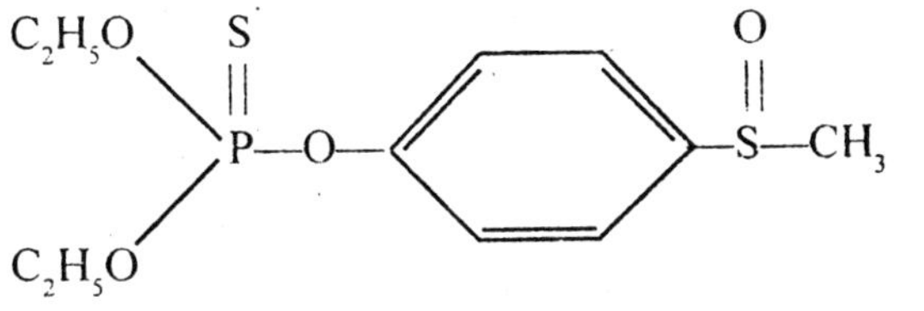

$(C_2H_5O)_2P(=S)-O-C_6H_4-S(=O)-CH_3$

Fensulfothion

$C_2H_5-O-P(=O)(NH-CH(CH_3)_2)-O-C_6H_3(CH_3)-S-CH_3$

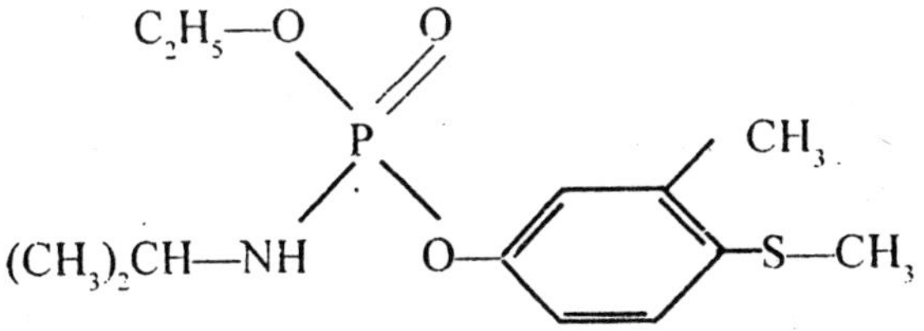

Fenamiphos

Isothiocyanates

The isothiocyanates include **metam-sodium (Vapam), vorlex (Vorlex),and dazomet (Mylone)**. They are active against nema-todes, soil insects, weeds, and most soil fungi. Applied by injection into the soil at least two weeks before planting, they act by releasing either isothiocyanate (metam-sodium) or methylisothiocyanate (Vorlex and dazomet), which inactivates the —SH group in enzymes.

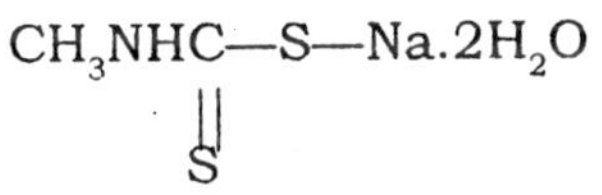

$CH_3NHC(=S)-S-Na.2H_2O$

Metam-sodium

$CH_3-N=C=S$

Vorlex ingredient

$S=C$ ring: $S=C(-S-CH_2-N(CH_3)-CH_2-N(CH_3)-)$

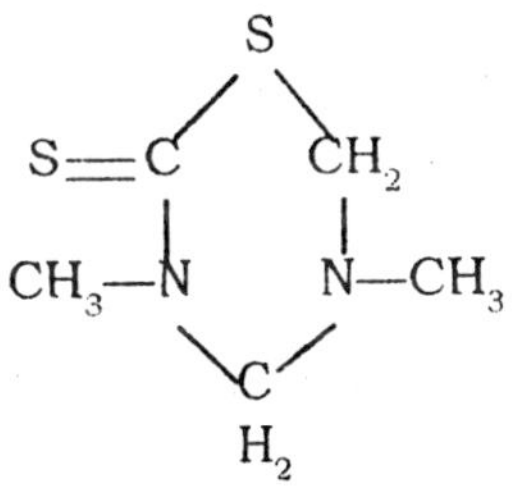

Dazomet

4. Carbamates

The carbamates include **aldicarb (Temik) carbofuran (Furadan), oxamyl (Vydate), and carbosulfan (Advantage)**. They are active against nematodes and soil insects, as well as some foliage insects. Available as granules or liquids of low volatility, they are easily soluble in water and can be taken up and translocated systemically by the plant. They are either drilled into the soil or are broadcast and disked into the soil before or after planting. They act by inhibiting the enzyme cholinesterase and causing paralysis and death of affected nematodes and insects.

Miscellaneous Nematicides

Chloropicrin (Cl_3 CNO_2). the common tear gas, is highly volatile and effective against nematodes, insects, fungi, and weed seeds. It can be used alone or mixed with other nematicides.

Avermectins are a new class of natural compounds that are obtained as fermentation products of *Streptomyces avermitilis* and exhibit nematicidal properties. Avermectins are still used only for experiments.

Against Mammalian Parasitic Nematodes

Diethylcarbanazine is the most effective drug against **filarial nematodes**. It is not highly toxic but it can cause nausea, vomiting and headache in humans, and muscle tremors and convulsions in logs. Thiabendazole is the most widely used drug for the control of gastro-intestinal nematodes, other than *Trichuris* spp., in ruminants. It is also used on citrus fruit and fungicide. Its acute toxicity to mammals is low. **Phenothiazine** is often used against intestinal nematodes in sheep and chickens. It is not used on cattle, horses or humans because of its many toxic effects on these species. *Piperazine* is the drug of choice for the treatment of oxyuriasis and ascariasis in both humans and other mammals, since it is highly active against the nematodes responsible but almost nontoxic to mammals at the therapeutic dose levels.

Whether the drugs used in the treatment of nematode infection in human beings and domesticated animals consitute an environmental hazard does not seem to have been assessed. However, at least 200 million people throughout the world are throught to be infected with nematodes, and presumably a similar proportion of the domesticated animal population is at risk. In consequence, appreciable amounts of these compounds of their derivatives enter the environment.

PCBs

'PCB' is scientific short-hand for polychlorinated biphenyl, a substance obtained by adding chlorine atoms to biphenyl. Products containing a high proportion of chlorine, is known as a highly chlorinated PCB. Products with smaller proportions of chlorine are called lower chlorinated PCB.

Any of the positions marked x can hold a chlorine atom. Depending upon the number of chlorine atoms held and their positions, 210 possible PCBs exist.

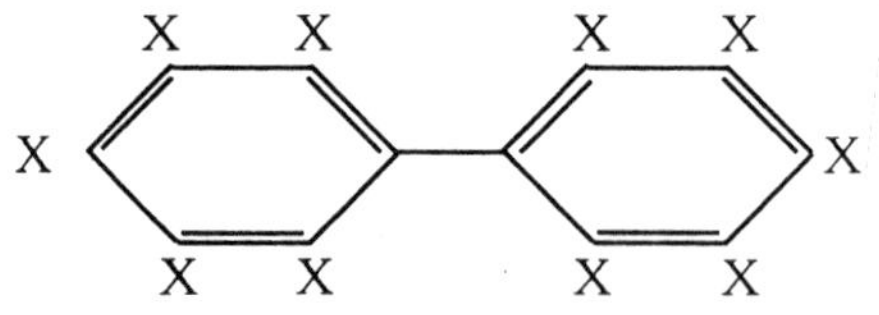

X X X X
X X
X X X X

Polychlorinated Biphenyls

PCB was invented in Germany in 1881 and was first manufactured by the Swan Co., USA, in 1929. At room temperature the PCBs are oils, viscous liquids or sticky resins. Fire-retardant and very stable, with a high boiling point, they soon come to be used in a number of different processes, as lubricants, softeners, laminating agents and hydraulic fluid. While PCBs were considered safe, they were in widespread and daily use. They were added to cement, plaster, sealing liquid, adhesives, plastics and carbonless copy paper. They had applications in heavy industry, being used in large quantities in electric capacitors, transformers and heat exchange units (in which a hot liquid enclosed in a metal pipe is used to warm another liquid).

PCB Disasters

Some scientists had warned of the possible toxic effects of the industrial wonder-liquid, but the warnings were ignored throughout the 1950s and 1960s. In 1958, however, a heat-exchanger was being used in the production of rice-oil in western Japan. There was a pin-hole leak in the exchanger, resulting in a PCB being transmitted into the oil, which was then used for human consumptions; 13,000 people were subsequently found to have symptoms of poisoning which ranged from skin disorders to respiratory problems.

The outbreak became known as the **Yusho epidemic**. Later research into Yusho victims revealed reproductive difficulties, including stillbirths, underweight birth, spontaneous abortions and toxaemia (*i.e.*, blood-poisoning) of pregnancy. Tests on animals indicated that PCB is carcinogenic (*i.e.*, cancer-forming). Post-mortems on deceased Yusho patients have revealed an excess of malignant tumours.

Other outbreaks have occurred, mainly as a result of contaminated food. In 1971, problems arose in the poultry industry in the USA when large numbers of chicken eggs failed to hatch. The cause was eventually traced to the fish-meal they were eating: PCBs in a heat-transfer fluid in the feed plant had leaked into the chickenfeed. In 1979, poultry and eggs in Idaho, USA, were found to be contaminated with PCBs, through leakage of insulating fluid from a transformer. The source was traced to a meat-packing plant in Montana, but not before the contamination had spread to 16 others states. A further outbreak of poisoning by PCBs, involving 2,000 individuals in Taiwan and again attributable to contaminated rice-oil, has recently been reported.

Environmental Hazards of PCBs

One problem with PCBs is that they have been widely dispersed, and do not easily distintegrate in the environment. They are capable of staying in mud or water for long periods of time. They are then absorbed by water insects and small animals, which are eaten by small fish, which is turn are digested by large predators. In 1969, the deaths of over 12,000 birds in the Irish Sea was attributed in part to high level of PCBs in their livers.

As the pollutant is passed along the food chain, concentrations of PCBs increase up to a hundred fold each time it is assimilated by a new predator. The length of the food chain means that final concentrations is an animal's liver can be 10 million times as great as the PCB levels in the water. PCBs have been found to be ubiquitous contaminants of freshwater fish in English and Scottish rivers. Pregnant and lactating women who consume large quantities of fish will pass PCB traces to their infants in their milk. People who consume fish liver run the risk of high intakes of PCBs.

PCBs are extremely difficult to handle or destroy. Special incinerators are used, capable of burning the material at 1,100 to 1,200 degrees C : but before this happen, the material has to be extracted from the transformer or other machine in which it is located. The pollutant can be inhaled or absorbed through the skin, and should be handled only by people wearing gloves, goggles and protective clothing.

According to estimates of the UK Department of Agriculture between 1 and 2 million tonnes of the substance have been produced worldwide since the 1930s.

Once located, the PCBs have to be adequately disposed of. In the past, the pollutant has been dumped on landfill sites, in the UK in quarries at Maendy and Brofiskin in Wales, though this is no longer regarded as accepted. Incineration or expensive long-term storage appear to be the only option.

2,4,5—T

(2, 4, 5,-Trichlorophenoxyacetic Acid)

A herbicide of the chlorophenoxy acid type, (See Herbicide); Acting as a hormone (auxin) in broadleaf weeds, brush, and certain trees and shrubs 2,4,5-T disturbs their patterns of growth and the plants die.

In the late 1960s 2,4,5-T became the focus of almost as much dispute as DDT because of the adverse reactions of many scientists and others over the extensive use of 2,4,5-T (and other herbicides) in defoliation programmes in Vietnam. It was reported that 2,4,5-T cause birth defects.

The use of 2,4,5-T was suspended in the US on 15th April 1970. The toxicity of 2,4,5-T is, however, due to the presence of dioxin in commercial samples. Dioxin is a sideproduct in the synthesis of 2,4,5-T.

Part V

Ecology and Environment

- Ecology
- Ecosystem
- Environment
- Biosphere
- Biological Oxygen Demand
- Biogeosphere
- Biogeochemical Cycles
- Plant Nutrients
- Nitrogen
- Nitrogen in the Biosphere
- Nitrogen Cycle
- Phosphorus Cycle
- Carbon Cycle
- Photosynthesis
- Parasetism & Predation
- Population
- Food Chain
- Food Webs
- Ecological Niche
- Symbiosis
- Algal Blooms
- Allergens
- Acid Rain & Forests
- Petroleum
- Oil Spills
- Toxicity
- Air Pollution & Human Health
- Air Pollution & Plants

ECOLOGY

Ecology is the scientific study of the relationships of living organisms with each other and with their environments. It is the science of interactions among individuals, populations and communities. Autecology concerns itself with the study of individual species, population ecology with populations of each species and synecology with the plant communities. When reference is made to the sum total of communities of a place, the term vegetation is applied. The scope of ecology is more clearly defined by making a study of total integration of living and nonliving components of a defined habitat, the trophic structure, the operation of food chain, cycling of various nonliving components and the net productivity. All these aspects are covered under the term ecosystem.

Every habitat experiences a series of vegetational changes with modification of habitat condition. This process of community development culminates in a climax community which forms an equilibrium with existing climate of the area. This process of community development is termed succession and in unique for each type of habitat.

Vegetation of each climatic region is characterised by life form (appearance of a plant, its method of perennation) of dominant species, which contribute to physiognomy of a vegetation. The relative proportion of different life forms is an area depicts its biological spectrum.

Ecology is related to biogeography as both are concerned with organisms and their relationship with the habitat. Whereas ecology makes a more intensive study of a smaller area, the biogeography makes study of a larger area, the ranges of the species in different parts of the world.

An ecologist has much to share with environmentalist because both are concerned with organisms and their environment, both are bothered about quality of environment, and how to improve it. The environmentalist places primary emphasis on maintaining a clean, undisturbed environment. An ecologist attempts to find a balance between human needs the rest of organisms and the quality of environment. Overlapping with other disciplines—from chemistry to mathematics, from anthropology to zoology—ecology violates traditional academic boundaries. It attempts to provide a balance between human needs and the environment.

ECOSYSTEM

An ecosystem is an ecological system, a natural unit of living and nonliving components which interact to form a stable system in which cycling interchange of materials takes place between living and nonliving components. Although our biosphere forms a complete ecosystem it is much too complex to study at one time. Even a single forest may cover thousands of square kilometres and contain thousands of different types of plants and animals living together. In order to separate these large systems into smaller areas that can be studied more precisely, ecologists work with manageable units such as a hillside, a forested valley, a lake, or a field. Because different components interact with each other and change one another the study of an ecosystem is a complex undertaking. We usually regard an ecosystem as an isolated unit, but this is merely a useful simplification: In reality, ecosystems always interact with one another. For instance, plants, animals, and materials move from one ecosystem to another, as when soil and leaves wash from a forest into a lake, or birds migrate between their summer and winter homes. The different biotic components of an ecosystem belong to a particular trophic level. Green plants which can fix solar energy constitute producers. Animals which depend on food stored in green plants are consumers. Herbivores which feed on plants are primary consumers carnivores as secondary consumers and those who eat smaller carnivores as tertiary consumers. The saprophytic organisms such as bacteria and fungi which decompose dead remains of plants and animals and release nutrients for recycling in the ecosystem are termed decomposers.

Not all ecosystems are natural. A farm may be considered an ecosystem because, in order to manage the farm effectively, all the interactions between crop plants, fertilizers, soil, climate and the natural animal and plant life of the area must be recognized and controlled. Similarly space stations, aquaria, and pots of houseplants are all artificial ecosystems.

Whereas the various mineral components of an ecosystem are continuously recycled, the flow of energy is always unidirectional a fraction getting lost from the ecosystem at each trophic level. The solar energy is thus the single main propeller of any natural ecosystem.

ENVIRONMENT

The word "Environment" encompasses everything that is around us. To the layman it means rather all things that can be damaged by man's activities.

Although it is more convenient to be concerned about the environment of the man, nevertheless, man cannot exist or be understood in isolation from other forms of animal life and from plant life. Therefore, we must deal with the environment of all life forms within the life bearing zone, called biosphere of the planet earth. This vital life zone lies at interfaces between atmosphere, hydrosphere and lithosphere. A deep understanding of all these is vital for any environmental study. The word environment may thus be more objectively defined as all conditions and influences surrounding and effecting an organism of group of organisms of the biosphere. The impact of man in shaping the bringing environment to its present state is, however, too dominant to be over looked.

The role of environment in shaping the form and cultural destiny of different communities of man was first indicated by Aristotle, 4th century B.C., who while dividing the earth into three climatic regions: Torrid, Frigid and Temperate argued that white-skinned people of cold, snowy frigid zone would never amount to much because of severe climate. Black-skinned people of torrid zone, were likewise not expected to be very progressive because of easy life. He predicted that olive-skinned people of Mediterranean, particularly the Greeks, because of ideal climate of the region were destined to rule the world. This idea, commonly known as the concept of environmental determinism was accepted and supported, through many centuries, but has almost been rejected during the course of the present century. Though, it is generally accepted that physical environment does exert significant influence, but it can't control the cultural destiny of man. Man has learnt to migrate to areas of his choice; and developed ways and means to live comfortably in any climatic condition. But in doing so, perhaps, he has fiddled so much with this environment, that its deterioration has reached alarming proportions.

The real impact of Man on the environment began when mining acquired a definite importance, that is only 40 to 50 centuries ago. And environmental effects took a significant importance only since the explosive industrial development began, around 1780, starting the exponential growth of world population and consumption.

The first environmental laws were enacted nearly at the same time. Napoleon ordered in 1810 that measures be taken by all factories to prevent damage or any kind of trouble to neighbours, crops, buildings, etc. Similar regulations were soon imposed in all industrial countries. Nevertheless it was only during the past thirty years that our world has really become "environment-conscious". The wide-ranging character of pollution, of the irreversible damage it can cause in some cases, is a very recent stage of our conscious knowledge. And because it is still very young, the field of environmental protection remains incompletely defined in a number of domains.

Classification of the Environment

The surface of the earth consists of three elements; land, sea and air. Since the land surface has undergone a process of disintegration of rocks scientific considerations call for a somewhat different breakdown, however into : the **lithosphere**, which consists of rock formations; the pedosphere, composed of disintegrating compounds of rock and stone forming soil; the **hydrosphere**, which is a layer of water penetrating the lithosphere and the pedosphere; and the **atmosphere**, which is a layer of air embracing all three above. These are covered by the branches of scientific research designated as geology (study of the earth, especially rocks), pedology (study of soil), hydrology (study of water) and meteorology (study of weather). Biology, the study of life, is concerned with the biosphere (living space), which pervades all the inorganic spheres, and lies at their interface.

All the components of environment mentioned above are interdependent and all interact with one another. The atmosphere contains water which it has absorbed from the evaporation of open waters, transpiration by plants and from exhalations by animals. On attaining a certain saturation point, the atmosphere emits the water it has absorbed and restores it to the other components of the earth in the form of rain, snow, etc. Solid particles are suspended in the earth's atmosphere in the form of dust emanating from soil, stone and microorganisms. In the hydrosphere, gases such as life-giving oxygen are dissolved, rocks are decomposed, their eroded particles are carried away, and they are deposited in new formations. This is the original breeding-ground for all organisms. The lithosphere is impregnated with water and gases. It also contains organic compounds, such as coal and oil, and special forms of life which have adapted to this region. In pores in the pedosphere, the loose erosion products of rock and dead organic substances, there is a constant interchange of water and air. In the soil are the mineral deposits of many forms of life, which are released from the lithosphere with frequently changing intensity. The biosphere, which embraces plants, animals and microorganisms, is contained in the cushion of air nearest to the earth, probably penetrates the entire hydrosphere, and is present in the upper layer of the earth's mantle.

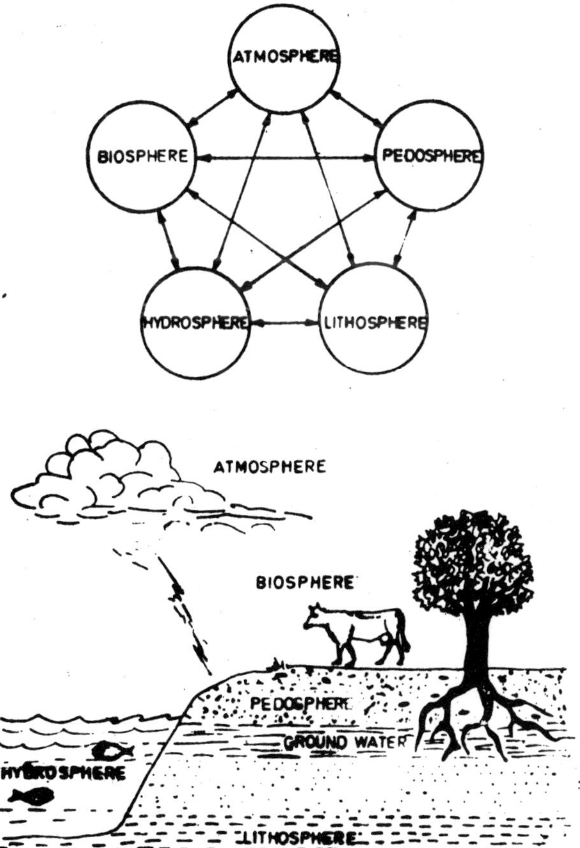

Diagrammatic Representation of the Fundamental Regions of the Environment.

BIOSPHERE

Biosphere is that part of the earth in which the life exists. It may be thought of as a biochemical system capable of capturing, converting, storing and utilizing solar energy. Nearly 300,000 species of green plants and microorganisms, capable of fixing solar energy constitute the producers. This energy is used in combining elements into complex organic compounds. Nearly 1 million species of animals consume energy stored in plants and convert it into animals form and constitute consumers. Bacteria and fungi, constituting decomposers recycle the organic remain of plants and animal and mineral deposited into soil or water for absorption by the producers.

The extent of biosphere is very small considering the size of the planet earth and its atmosphere. No where it is more than 13 km thick. No oceanic organism has been reported below 5 km depth in oceans. Although dormant forms of bacteria and fungal spores may be encountered at higher altitudes, constituting **parabiosphere**, the active life forms do not extend beyond 8 km altitude, which forms the upper limits of biosphere. The producers, the life line of biosphere extend in the form of a narrow layer, not more than 100 m thick close to the surface of land and water, broken up by deserts and snow covered mountain peaks.

The natural environment of living things is a complex, though fragile, membrane around the earth. All living organisms, together with the physical and chemical components of the total environment, make up the **biosphere** (often called the **ecosphere**). It involves all the interactions

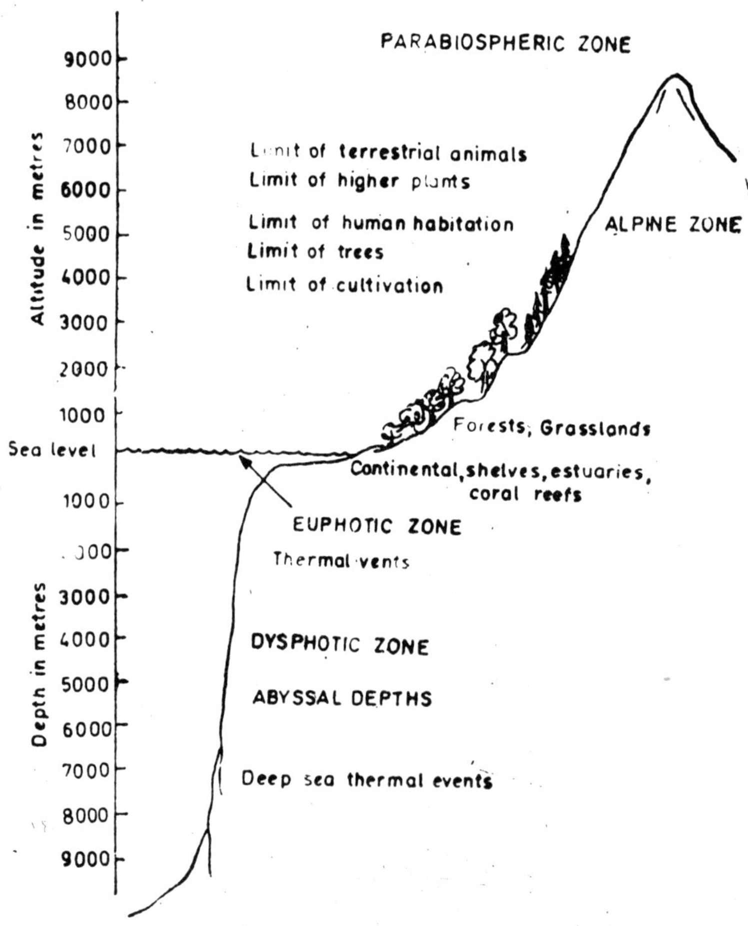

The vertical Extent of the Biosphere.

between the living and nonliving components that produce, support, and affect life. Thus, the biosphere is more than the abode of living things—it is a self-sufficient biological system in which the thread of life is sustained in a dynamic state of equilibrium by a flow of energy from the nearest star, the sun. The energy input from the sun is captured, used over and over again by the web of life, and eventually deposited in the earth or emitted back into space.

The biosphere—is remarkably constant in physical and chemical characteristics. So special are these conditions that perhaps nowhere are they exactly duplicated on any satellite among all the stars in the Universe. However, there may be many that are-similar. The temperatures of bodies in space range over millions of degrees, from intense coldness to the hottest star—the Sun. On the contrary, the biosphere of the planet earth is an island of tranquillity, where the temperature is always within a few degrees of that necessary to maintain water, the giver of life, in its liquid state. The size of the earth and its gravity, rotation, and distance from the sun, as well as the chemical composition of the rocks, the ocean, and the atmosphere, hold within the membrane a delicate balance in which life can flourish. Only slightly beyond these limits, life, as we know, it would perish.

BIOCHEMICAL OXYGEN DEMAND, BOD

The index of pollution by nutrients is called the biochemical oxygen demand, or BOD. The BOD is defined as the amount of oxygen that will be consumed when a biodegradable substance is oxidized (decomposed) in an aquatic system. It is also termed Biological oxygen demand.

Some organic materials, such as chlorinated hydrocarbons, (manufactured by industrial processes) cannot be used as food by bacteria and therefore do not contribute to the BOD.

A biodegradable material (a material that can be utilized by living organisms) if added to a lake or stream, this material by itself, may be harmless. But once it is spilled into an aquatic ecosystem, it will be consumed and oxidized by a variety of organisms. If the organisms are aerobic, the process will also consume dissolved oxygen. If consumption is great enough, oxygen levels in the water will be depleted and aerobic organisms will die. Thus, the quality of the water is degraded and the water has been polluted.

BOD values are expressed in milligrams of oxygen per liter of water. Pure water saturated with air at 25°C contains 0.0084 g, or 8.4 mg of oxygen per litre. Some typical BOD values are given below :

Type of Water	*BOD (mg/L)*
Pure water	0
Fresh natural water	2 to 5
Domestic sewage	100 to 900
Sewage after primary and secondary purification	10 to 20

Thus domestic sewage requires more than 10 times as much oxygen as avilable in pure aerated water. Even treated sewage requires up to twice as much oxygen as is available.

Measurement of Biochemical Oxygen Demand (BOD)

The rate of biochemical oxidation depends on the temperature of the environment and on the particular kinds of microorganisms and nutrients present. If the first two factors are controlled, then the rate of oxidation depends only on the amount of nutrient. After five days, under typical conditions, almost all the nutrient is gone and almost all the oxygen that is going to be used has been used. A standard test is carried out by saturating a sample of the polluted water with oxygen at 25°C and determining how much oxygen has been used up after five days. The amount of oxygen thus consumed per litre of contaminated water is the practical measure of the BOD.

BIOGEOSPHERE

Biogeosphere commonly refers to that part of the land including lithosphere and pedosphere, within which the living organisms can live, sometimes synonymous with true biosphere.

THE BIOGEOPHYSICAL ENVIRONMENT

The nature of the biosphere and its relationship to living organisms are determined in large measure by the physical characteristic of the earth. The earth is the fifth largest planet in the solar system, having a diameter of slightly less than 12,800 km. The surface of the earth is slightly irregular, although this would be hard to discern by an observer in space.

The land part of the earth is called the **lithosphere** from the Greek word ***lithos***, meaning "stone". The solid part, literally "stonesphere" extends 30 or 50 km beneath the surface. Beneath that, a zone called the **mantle** consists of heavy rock like material that flows like sticky gum when under pressure. The deeper the rock, the hotter it is. The core at the center is either molten or, as believed by some geophysicists, so highly compressed that it consists of solid metal, mainly iron and nickel.

The outer portion of the earth is sometimes called the crust. It consists of two layers, the **subcontinental** or inner layer and the **continental** or outer layer also known by the names **oceanic basins** and **land masses** respectively. The subcontinental layer lies beneath the continents and makes up the ocean floor. It consists mainly of a type of rock called basalt. The continental layer forms all the continents. It is mostly a type of rock called granite. As basalt rock is dominated by elements silicone and magnesium, the subcontinental layer is also known as **sima**, formed from first two alphabets of each element. Continental layer is similarly also known as **sial**, formed from two dominant elements silicone and aluminium in granite rock.

About 70 per cent of the earth's surface is water and about 30 per cent is land. Slightly less than 10 percent of the land area is in Antarctica, too cold to be populated by man. The soils of the crust were made from erosion of the rocks by the action of water, ice, and wind. Glaciers tear away pieces of rock, winds literally sand-blast the surfaces, and rivers wear down rock and soil to deposit millions of tons of rocks, mud, sand and silt in the oceans and other bodies of water every day.

The earth is constantly changing not only by the erosion of mountains but also by internal distortions that lift parts of the crust to higher levels. Thus, while the material of the mountains is slowly ground away and deposited in the oceans, some of the same material again becomes part of the dry land of mountains and plains.

The history of the earth is written in its crust. Twisting, tilting, upheaval, and distortion clearly show the release of tensions in the earth that changed the surface drastically over millions of years. Volcanic eruptions repeatedly laid new rock over the old. The rock show the record of glaciers that advanced and receded several times and covered large parts of North America and Europe. There is strong evidence that entire continents, once part of the same land mass, drifted apart over a long period of time. This is called *continental drift.* The history of life on earth from the emergence of primitive plants and animals to the appearance of advanced forms contemporary with man is imbedded in the rocks that were formed by sediments containing the remains of living forms.

The Biogeochemical Environment

The earth's crust contains all of the 88 natural elements in various combinations. The most abundant elements are oxygen and silicon. They comprise 46 per cent and 28 per cent, respectively, of the total composition of the crust. The other elements are present in smaller amounts (Table 1.) Analysis of the average granite rock shows that it consists of 67 per cent silicon dioxide (SiO_2), 15 per cent aluminium oxide, and about 14 per cent oxides of iron, calcium, sodium and potassium. The remaining elements make up less than 4 per cent of the rock.

The soils of the earth are created by the weathering of rocks. They furnish much of the nourishment for all life on land. Most of the green plants are dependent on the soil for the primary nutrients : nitrogen, phosphorus, and potassium. The abundance of these minerals in any particular place determines in large measure the productivity of the soil. A critical factor in the growth of plants is the availability of the so-called micronutrients, such as zinc, iron, and manganese. When even one of these essential elements is present in insufficient amounts or in a form that is not readily available, plants do not thrive. When deficiencies exist, all the living components of the ecosystem are affected.

Table - 1: Major Elements in the Earth's Crust

(besides oxygen and silicon)

Elements	*Uses*	*Abundance in* Earth's Crust *parts per million (ppm)*
Aluminium		81,300
Iron		50,000
Magnesium		20,900
Chromium	Stainless Steel	200
Nickel	Stainless Steel	80
Tungsten	Abrasion resistance	69
Copper	Electric wire	70
Lead	Storage batteries	16
Zinc	Galvanized Steel	220
Tin	Tinplate for cans	40
Molybdenum	Hardening steel	15
Mercury	Chlorine production	0.5
Silver	Photographic film	0.1

Organisms not only depend on the geochemical environment; they modify it is many ways. For example, plants growing in a square metre plot of soil build it into a microcosm entirely different from its original state. Small plants grow in the shelter of the larger ones; fungi and insects proliferate in the discarded plant parts; worms find a congenial home in the organically impregnated soil and bacteria and other microorganisms multiply. The soil changes in structure and in nutrient content. The decomposed rock becomes a biological metropolis teeming with visible and invisible life. Similar changes take place in forests and valleys but on a grander scale. Conversely, the removal of plant growth by whatever means produces drastic changes. The soil becomes compacted and hard. Its capacity to absorb moisture is reduced, increasing run-off and erosion. Nutrients are lost, some of them irretrievably. Plants can be reestablished only with great difficulty. Having no plants, the land is shunned by animals and microorganisms that depend on plant life for sustenance.

BIOGEOCHEMICAL CYCLES

Although the productivity of an ecosystem may be limited by the supply of sunlight or availability of water, in many cases productivity is limited instead by the availability of important inorganic nutrients that plants need.

Plants and animals require six elements in relatively large quantities—carbon, hydrogen, oxygen, nitrogen, phosphorus, and sulphur. Some of these elements are present in the form of compounds in rocks. They are released by erosion and weathering into soil, rivers, lakes and oceans. Nitrogen (as N_2), carbon (as CO_2) and oxygen (as O_2) are also present in the atmosphere. The movements of nutrient elements through an ecosystem are called biogeochemical cycles. The nutrient elements, unlike solar energy, may be used over and over again by living systems. On the average, every breath we inhale contains several million atoms once inhaled by Alexander or any other person in history we care to choose.

Nutrients are sometimes recycled rapidly through ecosystems, as in grasslands, where the above-ground vegetation dies back each year and its nutrients are made available again the following season. In other cases nutrients may spend many years away from the activities of the biological world. For example, remains of marine organisms may sink to the bottom of the ocean and be incorporated into rocks that are uplifted and exposed to the erosion that releases nutrients only after millions of years. Fossil fuels present similar example. Every nutrient element has a somewhat different fate, depending on its physical and chemical properties and its role in living organisms.

TRACE ELEMENTS

Trace elements are plant nutrients needed by plants in every small quantities. Plant nutrients are classified in the traditional manner into the principal ingredients such as nitrogen, phosphorus and potassium, and into trace elements such as boron, zinc and manganese. This classification is graded into the quantitative requirements of plants for the individual elements. Thus the higher plants require some thousand times more potassium than boron, for example.

More practical is a subclassification from a chemical point of view, based on the effect the nutrients have on the plant organisms as worked out in recent years. On this basis the following groups of plant nutrients have been established :

1. The nonmetals comprise *inter alia* the basic elements of organic matter, including carbon, hydrogen, oxygen, phosphorus, sulphur, silicon and nitrogen.
2. In the alkali-alkaline earth group, which includes potassium, calcium, magnesium and manganese, the elements are present in the ground in the form of ions and are absorbed by the plants in that form. Their primary task is the neutralization of charges of organic ions which, because of their size, cannot migrate through the membrane system of the cells.
3. Among the heavy metals, special mention may be made of copper, molybdenum, zinc and iron.

The physiologically most active trace elements are to be found in groups 2 and 3. Although the trace elements are needed by the plants in only the very smallest quantities, they are still of great importance to the nutrition of cultivated plants, since they play a vital role components of enzymes at essential stages of their metabolism.

Plant Nutrients

Elements that are essential to the growth and normal development of plants and whose function in the plants cannot be performed by any other element.

Nitrogen

Nitrogen is an essential element in the formation of protein material in the plants.

Occurrence in Soil

Nitrogen occurs as nitrate, nitrite, ammonium and free ammonia dissolved in soil solution. Most of the nitrogen in the soil, however, occurs in organic form partly in nitrogenous plant and animal remains (humus), partly in microorganisms, and partly in soluble organic compounds, e.g., amino acids and amides.

In the upper soil, nitrogen is mainly combined as organic nitrogen with animate and inanimate matter. It gets there from the almost inexhaustible nitrogen reserves in the atmosphere with the help of microorganisms, which absorb atmospheric nitrogen in the organic composition of their own body.

As rocks contain little or no nitrogen, replenishment of the soil solution depends on breakdown of humus and nitrogen fixation. Gaseous nitrogen in the air spaces of the soil is unavailable to many plants, but microorganisms can fix it to form nitrogen compounds such as nitrate and ammonia.

On a farm, nitrogen is as a rule introduced into the soil as mineral fertilizer in the form of nitrate or ammonia. Either form can be converted into the other in the soil by the metabolic activities of microorganisms. When there is a high oxygen content in the ground air, the nitrogen is mainly deposited as nitrate, and when the oxygen content is low it assumes the form of an ammonia compound.

The nitrogen is absorbed by plants from the soil by elutriation, by the erosion of surface material or by the release of gaseous ammonia. Most higher plants absorb nitrogen from the soil mainly in oxidised form (NO_3-). Reduced nitrogen (NH_3-, NH_4+) is also taken up quite readily, as are soluble organic nitrogen compounds such as urea, $CO(NH_2)_2$, and amino acids. The quantity of nitrogen received from the soil depends to a large extent on the nature of the cultivated plants. For example, crops like potatoes, turnips, cabbage, beans and peas, draw much nitrogen from the soil, while corn crops need very much less. If insufficient nitrogen is present in the soil, the plants will suffer from the shortage which will be reflected in stunted growth and in faded, pale yellow colouring of the leaves; stems may become red or purplish due to excessive anthocyanim production.

All these symptoms are observable because nitrogen is a constituent of a variety of organic compounds like nucleic acids, proteins, chlorophyll, and various coenzymes, including the nicotinamide adenine dinucleotides (NAD and NADP), and all these compounds are essential to the structure and metabolism of plants.

Overfertilization with nitrogen can also have unfortunate consequences for both plants and consumers. If too much nitrogen fertilizer is applied, leaf vegetables like spinach can accumulate nitrate, which during cooking is converted into nitrite, and this, if present in large quantities, has toxic effects on humans.

An important factor as regards the environment is the elutriation of nitrogen from the soil, particularly nitrate. As nitrate is highly soluble in water, it soon gets washed by rainwater into layers of soil well below the surface. This is particularly prone to happen a light soils, which precipitation water inflitrates rapidly. But loamy soil can also be affected in the same way. The effects of summertime precipitation are not so severe as in winter, as in summer the evaporation is heavy, and at times of high evaporation the penetrating seepage water rises by capillary action and carries the nitrate anions with it. In winter, on the other hand, especially when the snow begins to melt, the water movement in the ground is mostly vertically downward, so that the nitrate washed out of the upper soil gets carried away. In lakes and rivers in the neighbourhood of intense agricultural activity (vegetable cultivation), heavy concentrations of nitrate and consequently of nutrients can thus be expected.

A similar result is also to be expected from the nitrogen of eroded ground matter passing into the outflow system. This nitrogen, too, is eutrophic in its effects; that is to say it increases the nutrient material in the water, has a stimulating effect on the growth of algae, and disturbs the suspended biological equilibrium of flow.

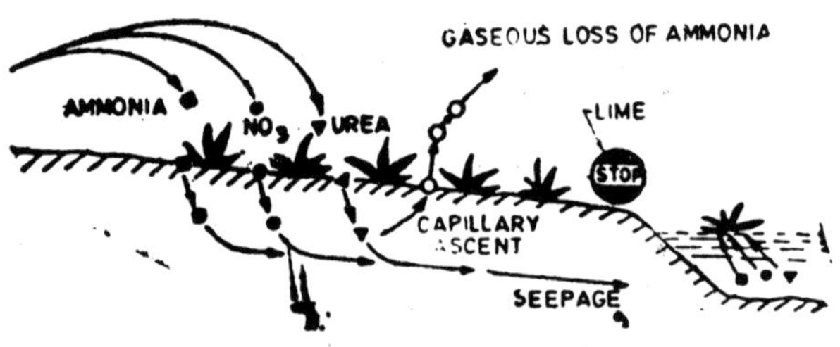

Nitrogen is fed into the soil as a mineral fertilizer.

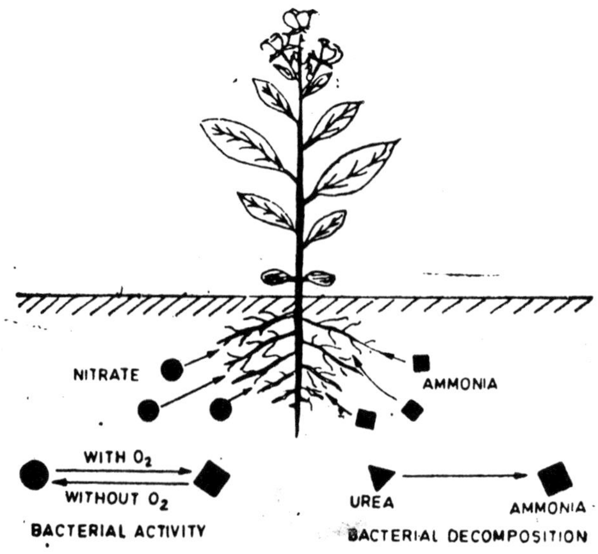

Transformation of Nitrogen Compounds by Microorganisms in the Soil.

The losses of nitrogen occasioned by the gaseous release of ammonium nitride from the soil are of relatively minor importance in their effects on the environment.

Phosphorus

Like nitrogen, phosphorus too belongs to the range of nutrients (macronutrients) that are essential to the normal development of plants and for which no substitute exists. It is required in quantities about one tenth as great as nitrogen. But unlike nitrogen, whose content in the bedrock of the earth plays no role in the nitrogen nutrition of plants, the plants must meet their phosphorus requirements from the mineral phosphorus elements in the ground, which are made available to plants by the chemical weathering processes of rock (in particular apatite). In addition to the mineral form, phosphorus is also found in organic combination in plant remains, humus substances and ground organisms. The univalent form of phosphate is absorbed less readily than the highly oxidised bi and trivalent ($HPO_4{}^{2-}$, $PO_4{}^{3-}$) phosphate.

Phosphorus is a common component of organic compounds in plants. It occurs as a component of nucleic acids, of phospholipids and of activated carbohydrates (*e.g.*, glucose 6-P, fructose-6-P, phosphoglycerate, etc.). There are also phosphoamino acids (*e.g.*, phosphoserine) that may occur in the phosphoproteins.

The central role of phosphorus is in energy transfer. Carbohydrates are phosphorylated prior to metabolism. The presence of phosphoros in the molecular structure of sugars make them more reactive. The highly reactive nucleotides—adenosine triphosphate (ATP), adenosine diphosphate (ADP), guanosine triphosphate (GTP), guanosine diphosphate (GDP), uridine triphosphate (UTP), uridine diphosphate (UDP), cytosine triphosphate (CTP), and cytosine diphosphate $(CDP)_2$— play an essential role in energy transfer of phosphorylation.

In agriculture phosphorus is used as a mineral fertilizer, especially in the form of Thomas meal (basic slag), a by-product of steel manufacture, or in the form of superphosphate (phosphate incorporated with sulphuric acid), or as raw phosphate (ground apatite). The phosphate applied to the soil is rapidly transformed there; in soil with a low pH value, the phosphate is established as an iron-aluminium phosphate aggregate, and in soil rich in lime as calcium phosphate. This phosphate deposit necessitates supplying phosphate fertilizers at regular intervals. It also explains why, after over-fertilization, no decline in growth occurs (as is the case with nitrogen or potassium), but rather that each application of phosphate exercises a positive effect.

Following intensive cultivation of land, the constant phosphorus fertilization leads to a gradual phosphate enrichment of the upper soil. This can be produced by what is known as humate effect : the accretion of organic substances in the ground improves the availability of phosphate for the plants. This effect probably stems primarily from an increase in micro-organism activity. By means of complex-forming metabolic products, the microorganisms are able to combine large quantities of calcium ions, so that phosphate ions are not transformed into insoluble phosphates. The phosphorus thus remains available in the soil solution for the plants. Owing to the exudation of hydrogen ions in the immediate vicinity of plant roots, it is also possible for the plant roots themselves to convert the phosphates of low solubility into soluble form by pH reduction.

The processes just described also depend directly on the water content of the soil. In dry years less phosphate is removed from the plants than in wet ones. In Intensively fertilized areas particularly, the erosion of upper soil material may lead to the removal of considerable quantities of phosphate by drainage into streams, rivers, etc. Here, too phosphorus has a similar stimulating effect to that of nitrogen on the growth of water plants, especially algae, by which the biological balance of ditch and drainage systems is considerably disturbed. This can be countered to some extent by erosion-preventive measures such as reduction of fallow stretches or introduction of organic substances into the soil and measures to control water flow.

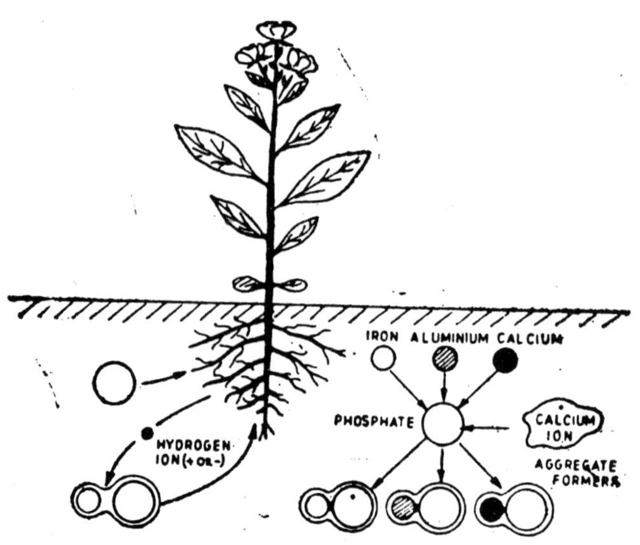

Complex Formation of Phosphate in the Ground.

Despite the central role of phosphorus in energetics and protein metabolism, the deficiency symptoms are rather nondescript. If there is a marked deficiency, there will be reduced growth, and fruits mature slowly.

Small phosphorus deficiency leads to the development of brown necrotic areas on leaves and petioles, and a dark, blue green colouration of the leaves often occurs. It is the older leaves which show deficiency symptoms as phosphorus (as well nitrogen) is highly mobile and moves progressively into younger leaves.

Potassium

Like nitrogen and phosphorus, potassium is one of the most important plant nutrients. However, unlike the other macroelements, potassium is not known to occur in plants in organic form or in an organic complex. The average potassium content of mineral soil is about 3 per cent, but in organic (*e.g.*, moorland) soil and in sandy soil poor in clay, the potassium content is considerably lower. Potassium is present in the soil solution as fixed potassium. It occurs between the layer packs of clay minerals, as building stone in the molecular lattice of various minerals and in natural organic and mineral ion exchangers.

Under natural conditions, potassium is transmitted by chemical erosion processes into the ground solution and converted into a form available to plants. From primary clay minerals (mica, feldspar), illite is formed, which after a further loss of potassium is transformed into vermiculite and mont morillonite. On drying out, the latter can again absorb the potassium ions and retain them between the layer packs. When they humidify again, the layer packs will again expand so that the potassium returns to the ground solution.

The process of potassium absorption by clay minerals is known as ***potassium fixation.*** In potassium fertilization this plays an important role, as potassium is fixed as a result of mineral fertilization. According to the particular circumstances, potassium fixation will have a promoting or a restrictive effect on the nourishment of the cultures. In ground rich in mont-morillonite, the potassium fertilizer will pass by preference into the intermediate layers, where after a drying-out period it will be retained and so abstracted from the plants. On the other hand, the potassium fixation serves as a defense against erosion.

Eroded potassium (fertilizer) passes through ground or surface water into water outlets (streams, rivers, etc.), in which it acts as a fertilizer on the water plants. If heavy leaching of the potassium occurs, however, there will be a high salt loading in the water outflow, which eventually results in salt damage to the flora and fauna in the waterways.

Besides the loss by erosion, a reduction in the potassium content of the soil will occur in particular as a result of absorption of potassium by the plants. In this respect, potatoes, turnips and cabbages, together with alfalfa (lucerne) and red clover, make the heaviest demands on the potassium supplies in the soil. Garden crops such as carrots, cucumbers and cabbages attain extraction rates as high as 350 kg of potassium oxide per hectare.

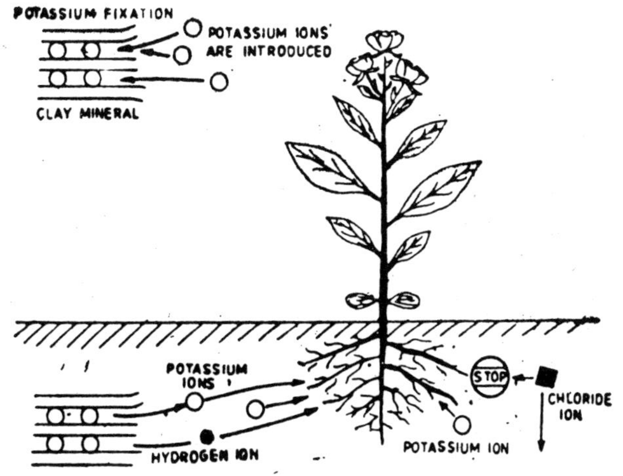

Potassium Fixation by Clay Minerals and Potassium Absorption by the Plants.

Increasing acidity of the soil solution tends to decrease the availability of potassium because a high proportion of the cation exchange sites in an acid soil are occupied by hydrogen ions. Hydrogen ions also compete with potassium cations for carrier sites in the root.

Under intensive soil cultivation this removal of potassium from the soil must be offset by applying potassium fertilizers. Here the choice lies between two types of fertilizer; potassium chloride and potassium sulphate. As both these compounds are highly soluble in water, the potassium ions are easily absorbed by the plants. Both the accompanying ions, the chloride ion and the sulphate ion, have their own effects on the plants, however, and these have to be considered in selecting the fertilizer. In humid conditions, the chloride ion is rapidly eroded. There are certain cultivated plants, such as tobacco, potatoes and vines, which react unfavorably to chloride, so that for these plants a potassium sulphate fertilizer is indicated.

Since, it is readily translocated throughout the plant body, deficiency symptoms occur first on older leaves as a slight yellowing. Leaf edges subsequently become brown and necrotic. In grasses, interveinal areas turn brown. These brown patches have been found to contain large amounts of an unusual nitrogenous compound, putrescine, which is produced as a result of disturbed metabolism and eventually kills the cells. The leaves sometimes develop a metallic sheen (bronzing) prior to the development of others symptoms and later they may curve downwards and roll inwards towards the upper surface.

In the absence of potassium the whole structure of the plants withers and fades. The reason for this is that potassium retains the water in the cell sap of the vacuole, as it has a swelling effect. Plants with an insufficient potassium reserve release the water easily, so that the cell pressure (turgor pressure) declines and the plant withers.

Maintenance of the tumefying effect of the plasma colloids, however, is not the only function of potassium in plants. It is abundant in plant vacuoles and is bound through ionic bonds to proteins and other components with negative functional groups. Potassium functions as a cofactor or activator for many enzymes of carbohydrate and protein metabolism. One of the most important, pyruvate kinase, is a central enzyme of glycolysis and respiration.

Over forty enzymes are known which require a univalent cation for maximum activity; potassium is usually the most effective. Transport ATPases from animal sources require sodium as well as potassium ions for maximum activity. No such synergistic effect of the two cations has been found in plants although often plant ATPases are activated *in vitro* by sodium as well as by potassium.

If the function of the potassium is only to activate enzymes it is at first sight surprising that it is required by plants in such large amounts. An explanation of this anomaly may be that because potassium has a very low affinity for protein a high concentration is needed to maintain the potassium-protein complexes essential for optimal enzyme activity. Plants seem to have an almost unlimited capacity to accumulate potassium but above a certain level growth is not improved by additional absorption ('luxury' consumption). Potassium salts of organic acids accumulate in guard cells when stomata open and are released when they close. This contributes to a change in osmotic potential upon which the turgor of the guard cells depends. When leaves wilt there is an increase in the concentration of an inhibitor, abscissic acid, and this seems to cause the stomata to close, apparently by facilitating release of potassium ions from the guard cells.

NITROGEN

Colourless, odourless and tasteless diatomic gas constituting about four-fifths of atmospheric air (about 78% by volume). In some bacteria and blue-green algae nitrogen fixation is an important means of augmenting the nutrient supply, especially for higher plants when they grow in association with the nitrogen-fixing micro-organisms.

Although nitrogen comprises the major bulk of the air, it is almost inert insofar as animals are concerned, except when one breathes air under increased pressure, such as during deep-sea driving, at which time the effects of this gas can be catastrophic. Nitrogen is a deadly threat to divers, who may fall victim to absorption of the gas into the blood stream. This affliction is known to divers as the "**bends**", or *Caisson's disease.*

The high partial pressure of nitrogen during dives forces the gas readily through the alveolar membranes. Under a pressure sometimes several times the normal atmosphere pressure, depending on depth, nitrogen diffuses rapidly into the blood stream. But the tissues have no mechanism for chemically absorbing it as with oxygen and carbon dioxide, Nitrogen can be eliminated only through the lungs. Under high pressure it remains in solution in the plasma and during deep dives the nitrogen in the blood is thus greatly increased. If ascent is too rapid, diffusion through the alveolar membrane does not take place quickly enough to allow the nitrogen to leave blood and return to the air in the lungs. Instead, the nitrogen may bubble out in the blood vessels like gas from a bottle of soda water when the pressure is suddenly released by removing the cap. The "bends" is a serious and sometimes fatal condition. It can be prevented by rigidly observing the rules for slow return to the surface or which instruction and practice are required.

Nitrogen Narcosis

Breathing nitrogen under pressure has an intoxicating effect. At depths of about 10 metres and below, sufficient to increase the pressure to 4 atmospheres or more, the amount of nitrogen forced into the blood stream through the alveolar membrane reaches a concentration at which the nitrogen has a narcotic or anaesthetic action. It is a delightful but dangerous sensation. The condition is called **nitrogen narcosis** or "rapture of the deep."

Nitrogen narcosis is similar in effect to alcohol intoxication. The diver becomes uncoordinated and lacks judgement. The most dangerous part is that under the influence of nitrogen narcosis, the diver may want to go deeper, which could be fatal. Nitrogen narcosis becomes so severe at a depth of 300 feet that a man may be unable to perform even simple task. But the affected diver recovers quickly upon return to the surface.

Nitrogen Fixation

All living organisms need nitrogen in large amounts. Nitrogen is a key element in the amino acids as well as in many other dynamic molecules of the protoplasm. The use of nitrogen compounds in the metabolic manufacture of nutrients within the tissues and the fate of nitrogen products during and after decomposition are part of a complicated series of processes referred to as the *nitrogen cycle*. Much of the nitrogen in animals, plants, and their waste products is returned to the soil and reused by plants as part of the cycle. However, microorganisms called *denitrifying bacteria* act on soil nitrates and nitrites and in doing so produce molecular nitrogen (nitrogen gas), which returns to the atmosphere. Probably all of the nitrogen gas available for food in the soil would be lost if it were not for another part of the nitrogen cycle in which atmospheric nitrogen is eventually returned to the soil.

Nitrogen gas is a chemical substance of moderate chemical stability and does not readily enter into chemical reactions. But its abundance in the atmosphere provides an almost unlimited source of nitrogen. In nature, there are primarily two ways in which nitrogen is returned to the soil. One is by the energy of lighting which forms small amounts of nitrogen oxides from molecular nitrogen. The nitrogen oxides drop to the earth in moisture and become available for further modification by microorganisms as plant nutrients. A more important way in which nitrogen becomes "fixed", is by the action of nitrogen fixing organisms among the algae and bacteria. It is one of the most crucial links in the nitrogen cycle.

Two groups of bacteria are largely responsible for nitrogen flxation : (1) those which live in the soil independently of higher plants, and (2) those which grow in association with higher plants and have a mutual relationship with them. The relationship of two species of organisms living together in an intimate association is called **symbiosis**. The particular kind of symbiosis in which different kinds of organisms living together are of benefit to each is called **mutualism**.

There are two types of nonsymbiotic nitrogen-fixing bacteria, those which require oxygen and those which do not. Aerobic (oxygen-requiring) bacteria called. *Azotobacter* do the work in alkaline soil, while anaerobic bacteria with the formidable name *Clostridium posteuranium* work in acid soil. (Anaerobic means that it does not depend on oxygen for survival).

The symbiotic bacteria are the most important nitrogen fixing organisms. Several kinds of bacteria called *Rhizobium* have a cooperative arrangement with a group of plants called legumes, such as peas, beans, and alfalfa, whereby the two dissimilar kinds of organisms jointly manufacture nitrate from nitrogen gas. To accomplish this *Rhizobium* invades the roots of the pea plant or other legume, where it causes innumerable swellings or nodules packed with bacteria. Without *Rhizobium*, the root tissue of the pea plant could not produce nitrate.

Man has devised a way for nitrogen fixation that is probably as important, at least in agriculture, as the sum total of electrical discharges in the atmosphere and biological activity. The most widely used industrial process for nitrogen fixation use only nitrogen and hydrogen as raw materials. Large amounts of energy are needed to compress the gas mixture to 5,000 pounds per square inch. The crucial step produces ammonia :

$$\underset{\text{Nitrogen}}{N_2} + \underset{\text{Hydrogen}}{3H_2} \rightarrow \underset{\text{Ammonia}}{2NH_3}$$

The ammonia is used in anhydrous form as a commercial fertilizer and as a starting point for the manufacture of other nitrogen-containing products for industry and agriculture aqua ammonia, ammonium salts, urea, and nitrates. Urea is used as a fertilizer, a cattle feed additive. and an industrial chemical.

NITROGEN IN THE BIOSPHERE

Without nitrogen, life on this planet as we know it today could not exist. This is because nitrogen is a major component of a number of compounds which are essential for the structure and functioning of biological organisms, from the primitive prokaryotes to the highly evolved and sophisticated eukaryotes. The intense interest of the biologist in this remarkable element is because of a series of biological processes involving this element namely—the vital importance of the nucleic acids in perpetuating the characteristics of an organism and directing its growth, development and metabolism; the central position of proteins (as enzymes) in executing the instructions of the nucleic acids, or building the mobile framework of the protoplasm and other essential structures of the organism; and the fundamental roles of compounds such as adenosine triphosphate (ATP) is energy transfers, or nicotinamide adenine dinucleotide (NAD) and the flavoproteins in the redox reactions so crucial in biological metabolism.

It is a curious fact that, apart from water, nitrogen is the key substance that limits the primary productivity of the plant kingdom in most parts of the world. We know that nitrogen is a constituent of many important plant compounds, but we also know that nitrogen is one of the commonest elements an Earth 78% gaseous N_2 in the atmosphere.) Why then, should it occupy this growth-limiting role in the biosphere ?

The answer to this enigma lies in the inert dinitrogen (N_2) molecule, which most biological systems are unable to utilize in their metabolism. The N≡N molecule is highly stable (inert due to very high bond energy. The normal assimilation processes of the plants are unable to convert N_2 into useful nitrogenous compounds. Thus, the plant has to rely on nitrogen occurring in more reactive combination, *e.g.*, with hydrogen (as in ammonia) or oxygen (as in nitrate or nitrite), to build its organic nitrogen-containing molecules. Such nitrogen is referred to as 'combined nitrogen' and is not as common as the dinitrogen state. Geochemists have calculated that of earth's total supply of nitrogen (some 57.4×10^{10} kg), only 0.0025 per cent (1.5×10^{15} kg) is available in the biosphere for plant growth. The vast bulk (93%) is locked away in the rocks of the earth's mantle, while 7 per cent (3.8×10^{18} kg) is found as dinitrogen in the earth's atmosphere. Of the 0.0025 per cent combined nitrogen present and available for biological use, 57 per cent occurs in inorganic molecules while 43 per cent occurs in the organic form.

Biologically available nitrogen, therefore, forms only a minute fraction of the total present on earth Furthermore, much of this nitrogen is temporarily inaccessible to plans as it is present in the form of dead organic material from which it is only slowly released by microbial action.

Thus, in the biosphere there is a real shortage of the nitrogen form that can be made use of by living organisms and a real necessity exists for the conservation of this material. Nature responds to this necessity by recycling combined nitrogen in a complex process known as the Nitrogen Cycle.

NITROGEN CYCLE

The nitrogen cycle involves the following :

1. The scavenging of combined nitrogen from dead plants and animals by microbial action.

2. Its passage through the soil in various forms.

3. Its final re-absorption by plants to form the proteins, nucleic acids and other nitrogenous compounds of the living cells of plants and animals.

In this cycle combined nitrogen can be lost to the nitrogen pool of the biosphere, the earth's atmosphere, by a process called denitrification and gained from it by natural and artificial nitrogen fixation. The major processes of the nitrogen cycle in which various soil and symbotic microbes play a dominant role are described below.

Mineralization of Nitrogen

When plants and animals die, their nitrogenous remains are converted into inorganic forms of nitrogen (*mineralized*), mainly ammonia, by a host of bacteria, fungi and actinomycetes. In the degradation of nitrogen-rich material (*e.g.*, animal excreta, flesh) anaerobic bacteria are often involved, resulting in the production of foul-smelling products such as amines, mercaptans (thioalcohols such as C_2H_5SH) and hydrogen sulphide as well as ammonia. This process is known as *putrefaction*. Aerobic degradation results in the production of less obnoxious end-products, particularly ammonia.

Typical mineralizing organisms are :

Ammonifying bacteria	*Actinomycetes*	*Fungi*
Pseudomonas spp.	Streptomyces spp.	Alternaria spp.
Bacillus spp.		Aspergillus spp.
Clostridium spp.		Mucor spp.
Serratia spp.		Penicillium spp.
Micrococcus spp.		Rhizopus sp.

Many ammonifying organisms release protein-hydrolysing enzymes (proteases) when invading decomposing organic material. This results in a rapid hydrolysis of the protein into amino acids which can be absorbed by the invading organisms and further hydrolysed inside their cells. Some microorganisms do not, however, secrete enzymes and rely on other organisms to do the initial protein breakdown for them. Many microorganisms also secrete nucleic acid hydrolysing enzymes (ribonucleases and deoxyribonucleases) which degrade deoxyrihonucleic acid (DNA) and ribonucleic acid (RNA), the nitrogen of which ends up as ammonia.

The nitrogenous compounds of plants, particularly those which from complexes with materials such as polyphenols, may take months or years to mineralize (for example, many forms of plant litter). This slowly mineralizing material forms the humus content of soils.

Microbial Immobilization of Nitrogen

The final product of mineralization, ammonia, does not necessarily become immediately available for the nutrition of the higher plant. It is often immobilized by the soil microflora which use it for their own growth and multiplication. The extent to which the ammonia released by mineralization can become immobilized in this way depends largely on the carbon : nitrogen (C:N) ratio of the decomposing material. It may be many weeks before the newly arrived nitrogen becomes available for plant growth. This fellows mineralization of the microbial population when it dies off after the exhaustion of the carbohydrate supply.

Nitrification

In many soils ammonium, produced by mineralization, does not accumulate because of the activity of the ***nitrifying bacteria*** which rapidly convert ammonium into nitrite and then into nitrate. These bacteria are ***chemotrophs*** and are completely dependent on this oxidation process for their energy supply.

There are two genera of bacteria which are important in this process (although others, *e.g. Nitrosolobus* and *Nitrocystis*, do exist).

(a) *Nitrosomonas.* which converts ammonia to nitrite (one species, N. *europaes* forming ellipsoidal or short rods)

$$NH_3 + \frac{1}{2}O_2 \rightarrow NH_2OH \qquad +15 \text{ kJ}$$

$$NH_2OH + O_2 \rightarrow NO_2 + H_2O + H^+ \qquad -289 \text{ kJ}$$

(b) *Nitrobacter* which converts nitrite into nitrate (one species, N. *winogradskyi*, a short rod bacterium)

$$NO_2^- + \frac{1}{2}O_2 \rightarrow NO_3 \qquad -77 \text{ kJ}$$

Most crop plants usually receive nitrogen in the form of nitrate, as the fertilizer applied is either in this form or, if in the ammonia or urea from, is rapidly oxidized to nitrate.

The advantage of nitrate to the plant is that it is non-toxic (ammonium is, except at low concentrations). The disadvantage is that it is a negatively charged ion and consequently it is easily leached from the soil (which is also negatively charged) by rain and washed into rivers which it then pollutes. Also, the higher plant has to use large quantities of energy in reducing it back to ammonia for use by the plant. In some parts of the world, chemicals (*e.g.* nitrapyrin) are added with fertilizer to retard the process of nitrification by poisoning *Nitrosomonas.*

Factors Affecting Nitrification

The main environmental factors that influence the process of nitrification are :

(i) *pH* In acid environments nitrification proceeds very slowly and the responsible bacteria are rare or totally absent. Nitrification is acid soils is markedly enhanced by adding lime. Thus in acid soils, ammonium is usually the predominant form of inorganic N, whereas in neutral or alkaline soils, nitrate is the dominant form.

(ii) *Aeration.* Oxygen is absolutely essential for nitrification and in poorly aerated soils (e.g. waterlogged soils) nitrification is heavily suppressed.

(iii) *Soil moisture.* Excess water as found in waterlogged soils suppresses nitrification because of lack of oxygen. In dry soils, however, there is not enough moisture for bacterial metabolism and the moistening of such soils rapidly increases the biosynthesis of nitrate.

(iv) *Temperature.* Nitrification is markedly affected by temperature. Below 5°C and above 40°C the rate is very slow with an optimum temperature between 30° and 35°C. So in winter the rate of nitrification is much lower than in summer.

Losses of Combined Nitrogen from the Nitrogen Cycle

The major loss of combined nitrogen to the atmospheric dinitrogen (N_2) pool is by the process of denitrification. Strangely enough, there are no microorganisms that are *specifically* engaged in carrying out this process. Instead, a number of ammonifying bacteria which under normal aerobic conditions use O_2 as their electron acceptor in respiration, switch to using nitrate when O_2 is in short supply, for example, when the soil becomes waterlogged. The nitrate becomes reduced to either N_2 gas, or the gaseous oxides of N, nitrous oxide (N_2O), nitrite oxide (NO) or nitrogen dioxide (NO_2), which are volatilized and lost to the soil. The commonest denitrifiers are *Pseudomonas spp., Paracoccus* spp. and *Thiobacillus denitrificans.*

Fungi are not often associated with denitrification. Denitrification is an extremely serious form of combined nitrogen loss from soils.

Factors affecting Dentrification

The main environmental factors which affect the process of denitrification are :

(i) *Aeration and soil water content.* Denitrification only proceeds when O_2 supply is insufficient to satisfy the microbial demand.

(ii) *Nitrate content of soil* If this is high and poor aeration conditions exist in the soil, the rate of denitrification is high. Thus there can be an enormous loss of combined N from heavily fertilized fields after heavy rain.

(iii) *Carbohydrate supply.* A high soil carbohydrate level will promote denitrification because it supplies energy for microbial respiration.

(iv) *Temperature.* Denitrification is markedly affected by temperature; it proceeds very slowly at 2°C with an optimum above 25°C and is still rapid at high temperatures up to 60-65°C. Thus, the organisms responsible are mainly thermophilic.

Denitrification, represents a permanent loss of combined nitrogen from the biosphere. There may also be temporary losses which can neverthless last for a considerable time; for example, when nitrogenous material is deposited from large bodies of water to form sediments below the euphotic regions, thus becoming largely unavailable for reuse by biological systems. Much sewage nitrogen discharged into lakes or the sea, is thought to be temporarily 'lost' in this manner.

Gains of Combined Nitrogen in the Nitrogen Cycle

There are several avenues whereby the depleted supplies of combined nitrogen can be replenished.

Gains from the Earth's Mantle

Weathering of rocks can release small quantities of combined nitrogen for use by living organisms. The source of this ammonia depends on the type of rock being weathered.

Gains from Atmospheric Deposition

Nitrogen compounds scrubbed from the atmosphere by rain (wet deposition) or deposited as wind-blown dry organic material (dry deposition *e.g.* pollen grains), can provide significant additions of combined nitrogen to the soil.

The bulk of this nitrogen comes from volatilization of nitrogen as a result of grass and forest fires and, in recent years, industrial combustion of fossil fuels, with volcanic activity also providing an input. Thus only about 2 per cent of the combined nitrogen acquired by the soil as a result of atmospheric deposition can be considered as true accession, the rest being simply recycled combined nitrogen derived mainly from the burning of organic matter.

Gains from Biological Nitrogen Fixation

By far the greatest accession of newly combined nitrogen to the biosphere comes from the fixation of atmospheric N_2 by certain prokaryotic organisms; there is little doubt that the largest proportion of the world's organic nitrogen has arisen from this source. Why the genes for this tremendously important nitrogen fixing ability possessed by these primitive organisms have not been transferred to higher life forms during the course of evolution, is still one of the enigmas of biology. It is estimated that over 100,000,000 tonnes of combined nitrogen are made available to the biosphere each year by this process.

The organisms responsible for biological nitrogen fixation may be classified according to their mode of nutrition. The two primary groups are the symbiotic and non-symbiotic classes.

Non-symbiotic nitrogen fixation : The organisms responsible for this type of fixation are found free-living in the soil. The energy required for the fixation of nitrogen in these soil organisms comes from either photosynthesis or the oxidation of organic materials absorbed from the soil.

The *photosynthetic organisms* (phototrophs) are of two types. The most important are certain genera of the Cyanophyta (blue-green algae) such as *Nostoc* and *Anabaena* which fix their nitrogen in special cells called heterocysts. These algae are particularly important in the wetland cultivation of rice (the staple diet of a large proportion of mankind), where as much as 50 kg/ha/annual crop can be fixed and eventually made available for crop growth, thus rendering nitrogen fertilization largely unnecessary.

Certain green photosynthetic bacteria and sulphur photosynthetic bacteria (which derive their electrons from reduced sulphur compounds), and a number of non-sulphur photosynthetic bacteria (*e.g. Chlorobium* spp., *Rhodospirillum* spp.) can fix nitrogen using light energy.

The *non-photosynthetic* soil organisms (heterotrophs) which fix nitrogen, rely on the oxidation of organic compounds present in the soil to provide the energy for the fixation process. These are bacteria from a number of genera, the better known of which are *Clostridium* (anaerobic), *Azotobacter, Azotococcus* and *Beijerinckia* (all aerobic).

Symbolic nitrogen fixation. The most effective nitrogen-fixing organisms are those which form symbiotic relationships with higher plants and are able to draw on the carbohydrates produced by these plants to fuel their nitrogen fixing activity and provide carbon skeletons for the production of nitrogenous compounds.

The most important of these relationships (from the agricultural point of view) is the symbiosis that exists between a large number of members of the legume family, Fabaceae (Leguminosae), and several species of the bacterial genus *Rhizobium*. The nitrogen fixation was brought about by bacteria living in nodules on the roots of these. This was confirmed by experimentation with ^{15}N.

It is now established that these bacteria occur free-living in the soil (where they do not fix nitrogen) and are attracted to the roots of young legumes by secretion from these roots. A certain degree of specificity is considered to exist between the legume host species and the bacterial symbiont. For example, *Rhizobium leguminosarum infects* the roots of *Pisum* spp., *Lathyrus* spp. and *Vicia* spp., *R. japonicum*, the roots of the soya bean (*Glycine max*) and *R. phaseoli* the roots of *Phoseolus* spp. Host recognition is apparently assisted by lectins. These are proteins produced by the bacteria which recognize certain sugar residues in the root hair wall of the 'correct' host and allow the bacteria to bind to it (shown in fig). The rhizobia synthesize compounds which deform the root hairs of the host, causing them to curl and allowing them to be penetrated by the bacteria. Following infection the root hair wall invaginates and extends into a tube-like structure, thus initiating the 'infection thread' which penetrates the walls of cortical cells and allows the bacteria to migrate into the inner cortex of the root. They are finally released into the cytoplasm of these cells where they multiply and enlarge into swollen, irregularly shaped *bacterioids*, Groups (4 to 6) of these bacteroids become surrounded by membranes produced by the cell to form a number of isolated nitrogen fixing colonies within the cell.

Just prior to the liberation of bacteria from the infection threads, the cortical cells of the host, in the region of infection, multiply rapidly to form the root nodule. Its final structure consists of a central region containing nitrogen-fixing bacteroids, surrounded by a rhizobia-free area into which the vascular tissue of the root has developed (see fig.). The cells of the central core of the nodule possess twice the chromosome number of the normal cells of the host plant; the significance of the phenomenon is still a matter for debate.

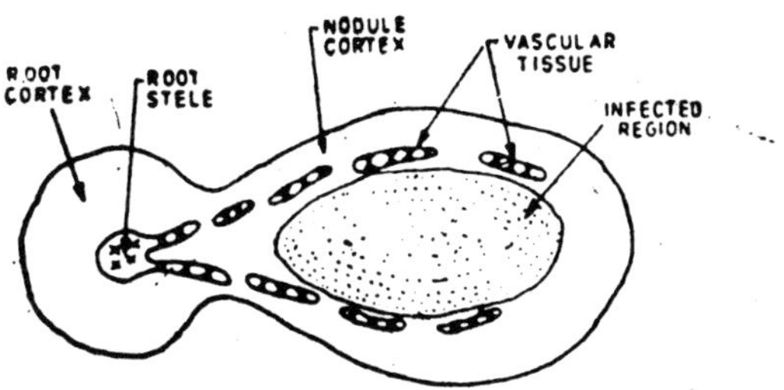

Plan of a Typical Root Nodule.

On cutting open a functional root nodule, the central region will be seen to have a light pink colour due to the presence of *leghaemoglobin* in the membranes surrounding each bacteroid group. The haem protein is thought to regulate the supply of oxygen to the bacteroids whose nitrogen fixing enzymes, nitrogenase, is readily inactivated by free oxygen.

The bacteroids are able to fix nitrogen, using energy supplied as carbohydrate produced by the host and delivered via the nodule's phloem link with the root vascular system. The bacteroids in turn supply their host (via the xylem link) with nitrogenous products which are produced in excess of their own needs. Thus the symbiosis is a true one, both organisms benefiting from the relationship. It is also a highly specialized one with the host demonstrating a number of features which accommodate the succour its symbiont (*e.g.* nodule structure, diverted vascular strands, leghaemoglobin, and rhizobium attracting root exudates).

Although ammonia is the first product of nitrogen fixation in the bacteroids it is rapidly converted to glutamine by the enzyme glutamine synthetase. Fixed nitrogen may be exported in this form by the nodule to the host plant. Usually it is converted into compounds such as asparagine, citrulline or the ureides for loading onto the xylem connection to the host.

In addition to the *Rhizobium* legume symbiosis a number of other nitrogen fixing symbioses are important in the accession of combined nitrogen to the global nitrogen cycle. Some significant examples are :

(a) *Actinomycetes with Alnus spp. (alder), Casuarine spp. and Myrica* gale *(bog myrtle).* The actinomycetes, a transitional group of micro-organisms between simple bacteria and fungi, are capable of fixing between 12 and 200 kg N ha^{-1} y^{-1} in the symbioses they form with *Alnus*, *Casuarina* and Myrica.

(b) *Blue green algae with gymnosperms, ferns, lichens and liverworts. Macrozamia, Encephalartos* and *Cycas* have certain of their roots infected with blue-green algae of the genera *Nostoc* or *Anabaena*. These roots adopt negative geotropism and emerge above the ground where the algae, present in a ring of cortical cells near the epidermis, are able to fix nitrogen using photo-synthetically derived energy. The free-floating fern, *Azolla*, also forms a symbiotic relationship with blue-green algae and can fix nitrogen. In the paddy fields of Indonesia, decomposing *Azolla* contributes a major part of the nitrogen required for the growth of rice crops. Certain lichens and liverworts are also involved in nitrogen fixing symbioses with blue-green algae, but the ecological significance of these associations is considered to be small.

(c) *Nitrogen fixing bacteria with plant leaves.* A number of plants (*e.g. Pavetta* spp. *Psychotria* spp., *Ardisia* spp. and *Gunnera* spp.) produce leaf nodules which become infected with nitrogen-fixing bacteria or blue-green algae. Only in the case of *Gunnera*, however, does the host plant appear to derive any benefit from the association.

(d) *Rhizosphere Associations (Rhizocoenoses).* Increasing attention is being paid to the loose association that exists between heterotrophic, non-symbiotic, nitrogen-fixing organisms and the roots of a number of species of higher plants. The contribution of such free living bacteria to, global nitrogen fixation is considered to be relatively small because of the constraint placed on their assimilatory activity through lack of adequate energy resources. Bacteria living in the rhizosphere are, however, able to draw on the considerable energy supply provided by root exudates for their nitrogen-fixing activity. Certain nitrogen-fixing bacteria have been shown to be present in large concentrations on or near the root surface of plants such as maize, wheat, sorghum, rice, sugar cane and a fairly wide range of grasses. There seems to be little doubt that the higher plants concerned benefit greatly from this loose association with rhizosphere nitrogen-fixing bacteria as they grow surprisingly well in the absence of fertilizer nitrogen. The bacteria are special strains of *Azotobacter, Spirillum, Azospirillum* and *Beijerinckia* which appear to be adapted to life in the rhizosphere.

All terrestrial ecosystems continuously lose some nitrogen when nitrates and organic matter are washed, or leached, out of the soil into groundwater and streams. In most cases, the nutrients are then eventually deposited in the oceans. In a few places on Earth, leached nutrients have collected in natural geological deposits. For example, certain streams flowing out of the Andes Mountains in Chile travel across one of the driest deserts in the world. Much of this water evaporates before it completes its journey to the sea, leaving behind deposits of nitrate rocks.

Today, large quantities of fixed nitrogen are artificially added to agricultural soils in the form of fertilizers. The production of nitrogen fertilizers is accomplished by combining nitrogen and hydrogen in the presence of heat and pressure. This process has increased the fertility of agricultural systems but consumes large quantities of energy.

PHOSPHORUS CYCLE

Phosphorus is a major constituent of biological membranes. Many animals also need large quantities of this element to make shells, bones and teeth. Phosphorus probably is the limiting nutrient in more circumstances than and other element because of its scarcity in accessible form in the biosphere.

Two chemical properties of phosphorus are responsible for this natural scarcity. One is that phosphorus does not form any important gaseous compounds under conditions encountered in the environment. The second is the insolubility of the salts formed by the phosphate anion PO_4^{3-} and the common cation Ca^{2+}, Fe^{2+} and Al^{3+}. The lack of gaseous compounds deprives the phosphorus cycle of an atmospheric pathway linking land and sea. Its cycle unlike those of carbon and nitrogen is a **sedimentary cycle** : That phosphate forms insoluble compounds with constituents of most soils retards its uptake by plants and slows its removal and transport by surface water and ground water.

Bulk of the phosphorus present on the earth is integrated in rocks, soil or sediments and distributed fairly uniformly over the entire earth. In the course of the earth's history, deposits have formed at various points of the phosphorus rich-mineral apatite, and these have been used as natural reserves for the steadily rising phosphorus requirements of man.

Energy is required in connection with all the processes of life in the biosphere, and is converted by green plants from solar radiation into chemical energy. Of fundamental importance in the conversion process is adenosine triphosphate (ATP), a nucleotide consisting of one molecule of the base adenine, one molecule of the sugar ribose, and three molecules of phosphoric acid. As ATP is formed readily by the supply of energy from adenosine monophosphate (AMP) and adenosine diphosphate (ADP) and phosphoric acid, and decomposes again easily upon the release of the energy, it is a suitable compound for transferring energy for metabolic reactions or for storing energy released by decomposition reactions. In its combination with ATP, phosphorus is to be regarded as a key substance in life.

Phosphorus is obtained by plants out of the ground as a phosphate ion. Owing to the poor solubility of phosphorus in water, only a few phosphate ions are available in solution in the ground. The availability of phosphorus is heavily dependent on the soil acidity (pH value) and the form of phosphorus in the soil. In acid soils, iron and acid phosphates dissolve less easily as the decline in the pH value increases. If the ground is basic (alkaline), calcium phosphates are formed, the solubility of which declines as the pH value rises. Ground phosphates are available to plant life in large quantities when the soil reaction is neutral.

If the plants or parts of them die, the organic substance is returned to the soil and decomposed there by phosphatising bacteria and fungi. In the process, part of the phosphorus incorporated in the plants will be absorbed by the micro organisms, and the rest will be released and pass into the ground. According to the soil acidity and the supply of aluminium and calcium ions, the phosphorus will again be established in the form of iron, aluminium or calcium phosphate. If the plants are consumed by animals, phosphorus contained in them will be ingested by the animal organism and ultimately be returned to the soil in excrement or with the carcass. Thus, phosphate cycles round and round on land from plants to animals and back again. Land ecosystems preserve phosphorus efficiently, since both organic and inorganic soil particles absorb phosphate, providing a local reservoir of this element.

In an undisturbed ecosystem, the intake and loss of phosphorus are small compared with the amounts of phosphorus that are internally recycled in the day-to-day exchange among plants and animals. Some phosphorus is inevitably lost by leaching and erosion of the soil into streams and rivers. When an ecosystem is disturbed, for instance by mining or farming, erosion can become so significant that large quantities of phosphorus and other minerals are washed away. When phosphorus reaches the ocean, it encounters other minerals in the water and reacts with some of them. Since most phosphates are not very soluble in water, they form insoluble compounds and eventually settle to the ocean floor. One reason for the infertility of open oceans is their low phosphorus content.

In the atmosphere, phosphorus is largely combined with aerosols. Its displacement in this form is of consequence only in dust-and sand-storms and volcanic eruptions. When the air turbulence subsides, the dust particles settle down again. Phosphorus normally gets into stream and river currents only through erosion of soil particles containing phosphorus.

The links between the terrestrial and oceanic parts of the phosphorus cycle are very weak. Little dissolved phosphorus is carried by rivers, because of the low solubility of phosphorus salt (the amount carried in suspended soil particles by erosion in about ten times as large). Aside from the slow link provided by the sedimentary cycle, the only sea-to-land transport is in fish and shellfish harvested from the sea and consumed on land, and in the excrement that fish-eating sea birds deposit on land. The only other way that phosphorus can return to land is by way of extremely slow geological processes in which sea floor sediments rise from the sea to form islands or continents. Soil erosion robs an ecosystem of its phosphorus, and it may take thousands or tens of thousands of years to recoup this loss through the weathering of rocks. Phosphate is much in demand as a fertilizer, but the richest and most easily mined deposits of phosphate are being depleted rapidly.

Phosphorus enters the water by several anthropogenic pathways. It is applied in fields as fertilizer and is present in sewage and in the wastes from livestock feeding. Phosphates added to detergents to improve their "cleaning power" become the ingredient of municipal waste water and are not removed effectively from sewage by primary or secondary treatment plants.

Phosphorus is the limiting nutrient in the growth of aquatic plants due to the less-soluble nature of phosphorus compounds and thus it does not readily move through soil into the water ways so its concentration is less in most natural bodies of water.

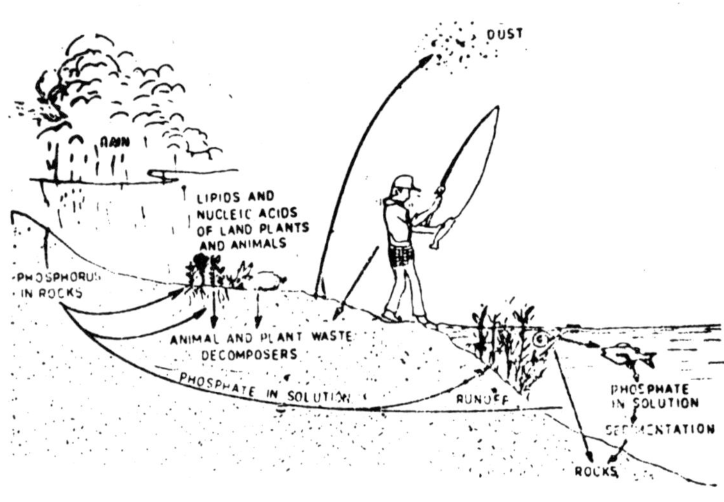

The Phosphorous Cycle

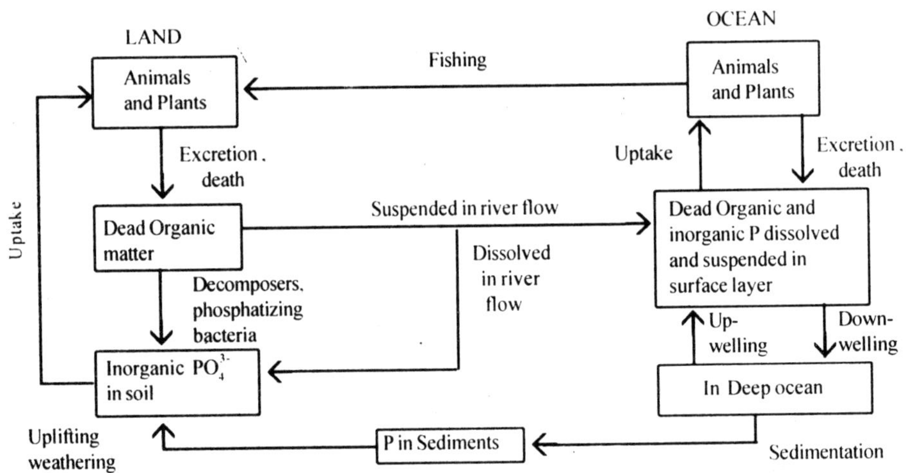

Global Phosphorous Cycle.

Because phosphorus compounds are so tightly bound to the soil, even very heavy fertilization does not lead to the leaching of phosphorus into ground water and surface water. Fertilizer phosphorus that reaches water ways does so almost entirely by being carried along in eroded soil particles. Even in the water, most of that phosphorus remains in suspension (not in solution) so it is not readily accessible for use by aquatic plants. By far the largest source of soluble phosphorus in waterways is municipal sewage, which contains phosphorus both from excrement and from detergents. Many examples of cultural eutrophication of lakes, due mainly to phosphorus addition in sewage, have been documented.

CARBON CYCLE

The main sources of carbon dioxide to the atmosphere are the decomposition of organic matter by microorganisms, gas exchange in the ocean, respiration by animals, and the burning of wood, coal and petroleum (Shown in Fig.). Photosynthesis converts CO_2 into the structural material for plants. When plants shed their leaves or upon death the structural material is decomposed by microroganisms and is then either released as CO_2 or incorporated into soil organic matter. The organic matter may remain in the soil or it may be leached into rivers and eventually carried to the sea.

In aquatic ecosystems, the corresponding exchange is between living organisms and carbon dioxide (CO_2) or bicarbonate ion (HCO_3^-) dissolved in the surrounding water. The cycling of carbon occurs quite rapidly, completing itself in a matter of minutes, hours, days, or week. In part, the rapid movement of carbon in ecosystems is made possible because the large reservoir of carbon dioxide is directly available to nearly all organisms as a gas in the atmosphere or dissolved in the ocean water.

In some ecosystems, however, carbon may accumulate as undecomposed peat in bogs and moorlands. Such carbon has entered a cycle of longer duration, and it may be millions of years before it is released. Deposits of coal, oil, and natural gas are organic compounds that were buried, millions of years ago before they could decompose and were subsequently transformed by time and geological processes. When they are burned, their carbon is finally released back into the atmosphere, largely as carbon dioxide.

The carbon cycle is complicated by the fact that carbon can also be bound into mineral formations, thus entering sedimentary cycling, which is usually much slower than the cycling of atmospheric gases. Certain acquatic animals absorb carbon dioxide and convert it to insoluble calcium carbonate ($CaCO_3$) which is then used to construct hard protective shells. When the animals die, the seashells accumulate as bottom sediments, which may eventually turn into sedimentary rocks such as limestone and dolomite. After millions of years, these rocks may be lifted above sea level and exposed to erosion, releasing their carbon as carbonate and bicarbonate which act as a global carbon dioxide sink.

Because carbon dioxide is one of the atmospheric gases important in determining the Earth's climate, a doubling of CO_2 might cause a global surface air warming. These increases are predicted to vary considerably over latitude and over the seasons. The warming of the polar regions could be two or three times as great as that in the equatorial regions, and the warming of polar winters will probably be greater than that of polar summers.

In addition to CO_2, the atmospheric concentration of several other "greenhouse" gases produced by human activity also appears to be increasing. These gases are nitrous oxide (N_2O), methane, and ozone. If increases of all these gases are forecast, their synergistic effect will probably cause climate changes to occur significantly earlier than if CO_2 were acting alone.

The annual release of carbon dioxide to the atmosphere from the burning of coal and petroleum is equal to about 10 per cent of the carbon dioxide used during natural photosynthesis by terrestrial plants. This release has increased the amount of carbon dioxide in the global atmosphere from early nineteenth century values of approximately 265 $\pm$ 30 parts per million (ppm) by volume to 352 ppm in 1989. Estimates of the future combustion of fossil fuels based on an energy growth of between 2 and 3 per cent each year imply that the atmospheric CO_2 concentration will be double sometime in the next century.

PHOTOSYNTHESIS

The process by which chlorophyll-bearing plants use energy from the Sun to convert carbon dioxide and water to sugars.

The sugars and their more complex relatives, the starches, collectively called

carbohydrates, are produced in green plants from carbon dioxide and water. The reaction is driven by the sun which pumps energy into space in the form of electromagnetic radiation. Packets of light energy called **photons**, penetrate to the earth through the "optical window" of the atmosphere. Energy equal to 100 million Hiroshima bombs reaches the earth every day. Part of the radiation is absorbed by green plants, which use it as the energy for the reaction of photosynthesis.

Photosynthesis takes place in specialized pigments of the plant. The green pigments, in which the reaction mainly takes place, are collectively called **chlorophyll**, usually packed in granules called **chloroplasts**, which can be seen easily under the microscope and are visible in the green colouration of plants. It is a curious fact that the chlorophylls are a type of structure called **porphyrins**, related in structure to the porphyrin in haemoglobin. Instead of iron as in haemoglobin, the porphyrin structure of the chlorophylls contains magnesium.

Light in the green part of the spectrum can be used only slightly by chlorophyll, for most of it is reflected and comprises a large part of the "colour" of foliage. The light rays that energize chlorophyll most effectively are in the blue, orange, and red parts of the visible spectrum.

The capture of solar radiation by plants and its utilization to energize the synthesis of sugar from carbon dioxide and water by photosynthesis is the most important biological exploit of nature. The biological mechanism of the process has fascinated scientists since the beginnings of plant studies. Yet the process remained one of the most puzzling problems in biological science until a series of brilliant laboratory experiments, made possible by the new techniques of radioactive tracers, revealed the biochemistry of photosynthesis.

In 1949 Melvin Calvin, a biochemist at the University of California at Berkeley, investigated the processes by which the green plant manufactures carbohydrates. Calvin and his co-workers used the radioactive isotope of carbon, carbon-14 (usually written ^{14}C), which they incorporated in molecules of carbon dioxide. Their experimental plants were placed in an atmosphere in which part of the carbon dioxide contained carbon-14. Calvin found that the reactions of photosynthesis occurred in split seconds. By stopping the process after exposure of the plants to the radioactive environment for seconds at a time, Calvin was able to follow the very first steps of photosynthesis. For this work Calvin recieved the 1961 Nobel Prize in chemistry. Subsequent work with both radio active carbon and a heavy isotope of oxygen, oxygen-18 (usually written ^{18}O)- has further elucidated the complicated steps in photosynthesis. The following overall equation summarizes the basic process :

$$\underset{\text{Carbon dioxide}}{CO_2} + \underset{\text{water}}{H_2O} \xrightarrow{+\text{ Light}} \underset{\text{Carbohydrate}}{C_6H_{12}O_6} + \underset{\text{Oxygen}}{O_2}$$

The oxygen that goes into the makeup of the carbohydrate comes from the carbon dioxide, while the oxygen that is liberated as free gas comes from the 'splitting' of water. A more complete overall equation is frequently given, although even this is a gross simplification of an intricate biochemical mechanisms :

$$\underset{\text{Carbon dioxide}}{CO_2} + \underset{\text{water}}{12H_2O} \xrightarrow{+\text{ Light}} \underset{\text{glucose}}{C_6H_{12}O_6} + \underset{\text{oxygen}}{6O_2} + \underset{\text{water}}{6H_2O}$$

The equation represents the conversion of electromagnetic energy (in the form of light rays) to chemical energy (stored in the form of a sugar). Glucose (dextrose) the simplest sugar, contains 670 calories of available energy per 180 grams.

The process of photosynthesis cannot be duplicated artificially. However, changes in the carbon dioxide content of the air can profoundly modify the natural process. Experiments by James Riley and Carl Hodges at the University of Arizona to determine the response of plants to enrichment of the air with carbon dioxide showed that plant growth increased with increasing carbon dioxide concentration. Plant yields were greatly increased at 2,500 ppm (parts per million). The optimum concentration for acceleration of photosynthesis with the species of plants used in the study was around 5 times the normal concentration of carbon dioxide in air. But above an optimum concentration, further increases caused a reduction in growth. The effectiveness of higher carbon dioxide concentrations in bringing about an increase in growth was largely dependent on the degree of turbulence at the surface of the leaves. It is the increase in ventilation resulting from turbulence, rather than greater exposure to light, that causes the "border effect," an increase in growth often observed in plants growing along the border of a field.

PARASITISM AND PREDATION

Parasites are organisms that live in or on the body of other organisms, called the host, and derive nutrition from it. It may be a temporary parasite such as a wood tick, or it may be a permanent resident, such as a tapeworm. It may weaken, debilitate, or eventually kill the host or it may cause little harm to the host. The boundaries between parasites and predatory organisms, which devour or suck nutrients from other organisms, are not sharp. A distinction is made between facultative parasites (casual parasites), which normally live on a substance that is decomposing by a natural process but can also penetrate from the alimentary tract or from an incision into a living organism, and obligate parasites, which have so adapted to a host that without it they are no longer capable of independent life. There are harmless parasites, which inflict only minor damage on their hosts, and others that injure them severely and sometimes even kill them.

Parasitism is virtually universal in all plants and animals. In vertebrate animals according to their location, a distinction is made among endoparasites and ectoparasites. Endoparasites are found within many of the organ systems of the body, most commonly in the digestive, circulatory, respiratory and urogenital systems. Ectoparasites occur on or within the skin and its appendages, such as hair and scales. Man is beset with many parasitic organisms, including intestinal worms (*e.g.*, tape worms, liver flukes, round worms); intestinal protozoa (*e.g.*, amoeba, ciliates and flegellates) abdominal cavity parasites, which live in the abdominal cavity between organs (*e.g.*, ichneumon flies, whose larvae are parasitic in or on other insects); blood parasites, which get into the bloodstream or blood corpuscles are propagate there (*e.g.*, the protozoan malaria which is transmitted by a blood-sucking mosquito and micro filaria); and muscle parasites (trichina and the bladder worm, which is an intermediate stage of the tapework); and a variety of ectoparasites (lice, mites; ticks, mosquitos).

Many infectious microorganisms, including bacteria and viruses, are also parasites in the sense that they extract nutrients from the host. Some of these are pathogenic *i.e.*, they disturb the health and normal functioning of the host *e.g.*, *Shigella*, a bacterium responsible for bacillary dysentery, others, such as some intestinal flagellates, are non-pathogenic and do not harm the host. Many parasitic organisms may be pathogenic in one individual and non-pathogenic in another. For example, most individuals infected with *Endamoeba histolytica* in the digestive tract do not cause amoebic dysentery while in other individuals, the same organism may produce severe disease and even death.

The major microbial predators are the protozoa which may engulf bacteria and more rarely algae and other protozoa. In the simplest form the protozoan population *e.g. Tetrahymena* is limited by its bacterial food *e.g.*, *Klebsiella*. Some protozoa prey on others and the most studied example of this is the *Didinium* which attacks the engulfs *Paramecium*. Predatory fungi have been considered as possible biocontrol agents for some diseases of plants caused by soil animals. Nematodes and protozoa may be trapped by a variety of net like hyphae, sticky surfaces and nooses. The animal are digested by hyphae. Predatory bacteria are much more in dispute and the only extensively studied example is *Bdellovibrio*. It is apparently qúite widespread in aquatic habitats and attacks other bacteria, by boring a hole in the well, entering the bacterium and causing lysis with the eventual release of many small vibrio-shaped bacteria. *Bdellovibrio* is sometimes considered a parasite rather than a predator.

Parasitism among microorganisms is very common. Many microbes are attacked by viruses or bacteriophages and these have been reported to play a part in limiting populations in cyanobacterial blooms. Some fungi especially chytrids parasitize algae and protozoa commonly in aquatic habitats. There are fungal parasites on other fungi, so called mycoparasites, where the host hypha may be intertwined in or penetrated by, the parasite.

Many microbes are, of course parasites on higher plants and animals. Protozoa and bacteria are most important in diseases of animals, and fungi are the major causes of plant disease.

In the majority of cases, a parasite is associated quite specifically with a particular host. The infestation of a host by a parasite is often a complicated interplay. When a parasite has invaded the host, the host may turn out to be the stronger and overpower the parasite by its defensive reaction. When the host is weaker than the parasite, mass reproduction by the parasite may result sooner or later in the death of the host. When the balance of power between the two is more or less even, the parasite may remain a long time in a host that can sustain the burden without serious damage. A working arrangement between parasite and host of this order holds out the best prospects for the parasite of constant propagation and dissemination of its species. For example pinworm infections in children are relatively harmless, though they may cause a minor irritation around the anus. In West Bengal over 75 per cent of the local population harbours hookworm (*Necator* sp. and *Ancylostoma* sp.), but there is no demonstrable pathology or effect on the people. Most wild animals have parasites and are able to maintain excellent health. Zebras in Kenya are intensively infected with internal and external parasites, but are in good health.

Most parasites tend to adjust to their living conditions, such as by deterioration of mobility and sense organs which are no longer needed; increase in numbers of eggs (which in the case of the mawworm may be as high as 50 million), since owing to the often very complicated development process heavy losses can occur; self-fertilization of the parasites by parthenogenesis, since a change of location for intercourse between male and female is often not possible; loss of pigmentation, as in the case of cave dwellers; and deterioration of the mandibles and digestive organs insofar as the nutrient is already chemically decomposed or in liquid form, as in the case of parasites in the alimentary canal or blood ducts.

POPULATION

The world human population has shown an unprecedented increase during the present century, so as to become a matter of great concern, especially in tropical Asia and Africa. The man, as he is today, goes back thirty thousand years. Eight thousand years ago, world human population was less than five million, that would mean 1 person for every 30 square kilometre of the land surface. The population reaches nearly 200 million around the birth of Christ. It took nearly 1600 years for this population to double, and is touched only 500 million during the 17th century. This slow rise of human population was, however, mainly due to natural checks like disease, famine and wars during the period. In the fourteenth century, at least 25 per cent of adult population died in epidemics of bubonic plague, the population of England was reduced by almost 50 per cent between 1348 and 1379 A.D. Famines, malaria and yellow fever have been the major checks in Asia and Africa. Some 1800 separate famines were recorded in China alone. The world population next doubled in only 200 years reaching 1,000 million mark in 1830; the next doubling in only 100 years touching 2,000 million in 1930; next in only 45 years touching 4,000 million in 1975. The population crossed 5,000 million in 1987 and according to UN estimates it may cross 6,000 million in 2000 A.D. Each human being today, thus can make claim for less than one-thirtieth of one square kilometre; precisely the world human density today is 34 per square kilometre. In India, where according to 1991 census, the population has crossed 840 million, as compared to 685 million in 1981, there are nearly 250 persons for every square kilometres of space today. China has a population density of 106 and Japan 318 per square kilometre.

The world is becoming an increasingly crowded place, and this trend will continue through the next century. Currently the world population is increasing at the rate of 2 per cent per year—a net increase of over 70,000,000 people per year, or more than 1,400,000 per week. This is mainly the result of better health care resulting in fewer deaths and longer life span, less epidemic and famine deaths as compared to earlier centuries.

The resulting stress on environmental resources will be enormous. People need to eat; where there is competition for land, this will result in over-intensive use. They need heat, which can only be supplied by cutting down trees or consuming fossil fuels. They need land, which will be achieved by taking over previously uninhabited areas. There will be an enormous impact in terms of human misery. It is estimated that, at the moment, 450 million people are acutely malnourished, with 10 million of these on the verge of starvation. The number of acutely malnourished people is expected to rise by the end of the twentieth century.

The disproportionate number of young people being born in Asia and Africa means that there will be land shortages; 42 per cent of Africa's population will be suffering shortage of land by the year 2000, with similar problems being experienced by 82 per cent in the Far East and 87 per cent in the Near East. The economic problems of the less-developed countries will be aggravated by the demands of ever-increasing numbers for water, food accommodation and a minimal standard of living.

The lack of land, and pressures to find some sort of income, will continue to force people off the land into the cities. This migration will drastically change urban environments, with a new race of 'super cities' whose population will be larger than that of some medium sized countries. It is impossible to predict how all these changes will be accommodated, and what the effects on the global environment will be.

One suggested remedy for population pressure is often suggested to be the birth control. While this is necessary other aspects are more important.

Similarly, Western arguments about there not being enough food for everyone have been questioned. To meet human nutritional requirements, the world needs to produce an average of

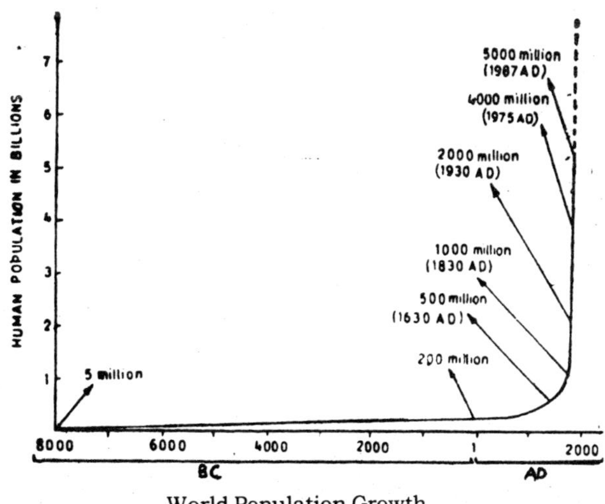

World Population Growth.

2,300 calories per person per day, roughly equal to 250 kg. of grain per person per year. On average, the amount of grain sold annually in the world is about 1,300 million tonnes—enough to provide the necessary 250 kg. for almost 6 billion people. A quarter of this, however, is fed to livestock to produce meat for the developed world. More grain is consumed by the world's livestock than by the entire population of China and India together.

There is enough energy to supply all the world's people; enough wood for everybody's fuel needs, provided that the developed countries modify their consumption; and the money needed to supply contraceptives to all the people who want them could be found by diverting ten hours' worth of global expenditure on armaments. Povery-stimulated population growth will stop when the citizens of the developing world feel secure in their future prospects; this happens when the developed world makes this possible.

FOOD CHAIN

Sunlight is the ultimate source of all life in our biosphere. Green plants and some microorganisms can fix this radiant energy, convert this into chemical energy and utilize this for building up organic substances from CO_2 and H_2O. These organic substances in turn are used in releasing energy for metabolic processes and for building up more complex organic molecules, using minerals absorbed from soil or aquatic habitat. These plants are designated as **producers** and form the first link in the food chain.

Unlike green plants, animals do not have the capacity to build up organic substances out of inorganic matter by the direct utilization of sunlight. They rather have to convert their energy and structural requirements by eating organic compounds synthesized by the green plants. Animal life, including humans, is completely dependent on the vital activity of green plants and thus constitute consumers.

The following members of the food chain represent the different animal consumers : first come the plant eaters (**herbivores**), followed in second place by flesh eaters (**carnivores**). Whereas the herbivores are designated **primary consumers**, the carnivores, which for their part consume herbivores or other carnivores, are classified as secondary, tertiary, etc., consumers according to their position in the food chain. **Omnivores**, which are classifiable as consumers at various points in the food chain, consume both animal and vegetable foodstuffs. Organisms with a wide food range are also described as polyphagous, as opposed to those that specialize in a particular diet, which are designated **monophagous**. An intermediate position is occupied by **oligophagous** creatures, which vary in the selection of their diet within certain limits.

The human being, a polyphagous animal, consumes anything that is palatable. The oligophagous species include many insects that live on a small selection of food plants not related to one another. Many parasites that are only found on a particular type of host plant or animal and often only on certain of its organs are monophagous. Head and body lice, for instance, concentrate exclusively on human beings.

The final link in the food chain is formed by decomposing organisms called decomposers. This groups consists of bacteria, fungi and many ground-dwelling animals that live as saprophytes or saprozoa on decaying organic substances. They live in humus and convert organic matter into materials required in return by the plants for their sustenance.

In the food chain, the organisms consumed are normally smaller than those that live on them. Grain for instance, is smaller than the mouse that consumes it, and the mouse in turn is smaller than a cat. The successive increases in size need not be pronounced, however, as in the sequence plant sap—plant house—ladybug—songbird. On the other hand, the prey can be disproportionately large, like the victims of snakes, or it can be diminutive in size, as in the case of the giant bearded whale, which lives entirely on plankton.

Shorter the food chain, greater is the biomass accumulation since at each trophic level only one tenth of the energy absorbed is used for increasing weight, rest is wasted or used up in respiration. This is the reason that antarctic seas, during the polar summers are among the most productive areas. For one thee is 24 hour energy input in polar summer and secondly phytoplankton to baleen whale it is just a 2-link food chain. Also organisms in cold polar water have a low respiratory rate to ensure less loss of energy due to respiration. In longer food chains of temperate and tropical climates, the food chains of different ecosystems get inter-connected to form food webs.

FOOD WEBS

For an ecosystem is to be self-sustaining, it must at the very least contain some source of food, usually produced by autotrophs, and decomposers to recycle nutrients.

An ecosystem is usually much more complicated than the basic chain of life and death of producers and decomposers. The energy and nutrients in the bodies of producers are resources that may be exploited by a variety of hetrotrophic *consumers*. Consumers include **herbivores** and **carnivores**. A few insect-catching plants are both producers and consumers, and many animals, such as pigs, bears, rats, and humans, are **omnivores**, animals that eat both plants and animals.

The flow of food energy in an ecosystem progresses through a **food chain** in which one step follows another : Primary consumers eat producers, secondary consumers eat primary producers, and so on. In nearly all natural ecosystems, particularly in temperate and tropical climate the patterns of consumption are so complicated that the term **food web** is more descriptive because there are many cross-links connecting the various organisms.

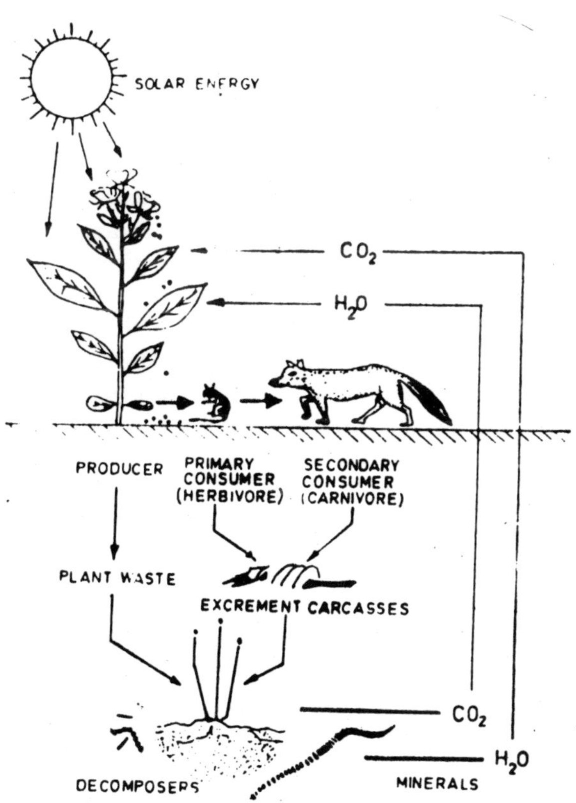

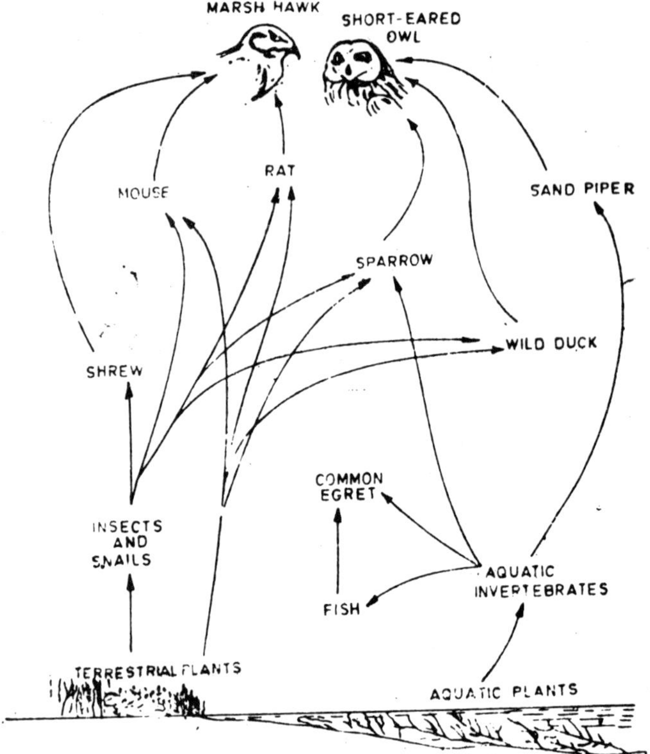

Fig. A simplified diagram of a Food Web

Food webs are so complex that human interference, particularly with systems that are incompletely understood, can have unforeseen results. Ecologist Lamont Cole investigated one such situation in the 1950s. The World Health Organization tried to eliminate malaria from Borneo by spraying the environment with the insecticide DDT. The spray killed the mosquitoes that carry malaria. DDT was consumed by cockroaches, which are larger than mosquitoes and being more resistant to DDT, did not die immediately. Geckoes (insect-eating lizards) that ate the cockroaches ingested the insecticide in turn suffered nerve damage due to the insecticide. Their reflexes became slower, were caught and eaten by cats in larger numbers. Because most of the gecko predators were now dead, their prey caterpillars eating the thatched roofs of local houses multiplied unchecked, and the roofs started to collapse. The cats were soon dying of DDT poisoning, rats moved in from the forest, and with them came rat fleas carrying the bacteria that cause plague. If untreated, plague is more immediately fatal than malaria. Thus the problem escalated, and as a consequence the World Health Organization stopped spraying DDT and, in an attempt to remedy the damage already done, sent a large number of cats into the jungle. The whole experience was an expensive lesson on the importance of understanding a food web before trying to alter or improve a complex system.

ECOLOGICAL NICHE

There are about 2 million species of living organisms on earth, of which one and a half million are animals and half a million are plants. Each species performs unique functions and occupies specific habitats. The combination of function and habitat is called an **ecological niche**. For every species **habitat** is its address, the place where it lives; its niche is analogous to a human profession, the way in which it makes a living. An organism's niche includes all its interactions with the physical environment and with other organisms that share its habitat. To describe a niche, one would first have to describe all the physical characteristics of a species' home. For plants and the less mobile animals, one would describe the preferred micro-environment, such as the water salinity for species living at the interfaces of rivers and oceans, or the soil acidity for plants. An animal's trophic level, its exact diet within that trophic structure, and is major predators are also importance in the description of its niche. Mobile animals generally have a more or less clearly defined food-gathering territory, or **home range**, which is another factor in establishing the physical niche.

A niche is not an inherent property of a species, because it is governed by factors other than genetic ones. The niche of a given species in a given ecosystem is not a set of conditions that would be best suited to the genetic makeup of the organism but rather the set of conditions in which it can actually survive. The organisms that occupy same or similar ecological niches in different geographical regions are known as **ecological equivalents**. When reference of an ecological niche is concerning its habitat it is termed **spatial or habitat niche**, when functional role is considered, it is termed **trophic niche**. Since, each species is controlled by several factors, the ecological niche is more appropriately represented as a hypervolume and the term **hypervolume niche** is used. The hypervolume (constructed by tolerance range of each factor) that a species can fill in absence of competition is termed **fundamental niche**. If the niche of two species overlap partially, they can coexist by one species occupying the whole fundamental niche, excluding the other from overlapped part and the other species occupying only non-overlapped part of its fundamental niche or what is termed *realised niche*. Alternately both species will be excluded from the overlapped part, so as to occupy only their realised niche. If the fundamental niche of the two species completely overlap, the competition would result in exclusion of one from the area.

SYMBIOSIS

Symbiosis is an association of two or more different organisms in which each partner, unlike the case of parasites, benefits by the relationship. This includes fleeting relationships as well as long lasting ones. A mycologist named Anton deBary first coined this word in 1876.

Symbiosis is often a very convenient mode of life; if two species by living together can help each other feed, or protect one another, then both have a better chance or survival. Some species have become so reliant on another that they stop functioning as wholly separate entities. One species may totally depend on the other for its food supply, for example termites and their gut microorganisms or they may be unable to function properly in some way without the other. On the other end some individuals meet only for a short time and help one another with one specific problem, for example yucca moth which assist in the pollination of the yucca plant. These are therefore many different degrees of symbiosis ranging from the casual meeting to total interdependence. Some associations are between two animals, some between two plants and some between animals and plants. Some association are between two individuals and some are between while populations.

It is therefore true that the range and depths of these associations is vast and covers every conceivable environment and condition. Symbiosis is thus of considerable ecological importance, and several of these relationships are important to the well-being of all living things as described below :

1. Corals

One of the most significant symbiotic associations is where partnership between tiny unicellular algae-dinoflagellates (*Gymnodinium*) and various marine animals together enable the formation of coral reefs. Coral is made up of the calcareous skeletons of millions upon millions of tiny marine organisms. The lower cell layers of the organisms lay down calcium so that the reef grows beneath the cell layer, building the coral continuously. Coral needs water that is shallow and free from pollution. Unless the tiny algae can receive sunlight through the water above they are unable to photosynthesize. So any depth greater than 50 m and pollution would filter away the necessary light. Coral forming animals are filter feeders, sifting through the water and retaining animal plankton. In the course of this sifting they take in dinoflagellates along with their food. By some means, the tiny plants are separated out from all other food and pass into the body of the polyp. There they begin to photosynthesize and provide their host with oxygen and carbohydrates surplus to their own needs. The dinoflagellates are also able to utilise their host's excretory products or carbon dioxide and nitrogen. They convert these waste products and make them usable again for the host. Some coral forming animals are totally dependent on the photosynthetic by-product from the plants, and also these photosynthetic activities increase the deposition of calcium in their host animals which significantly increases reef formation. The dinoflagellates help to oxygenate the water thus making the area more suitable for other forms of life. The surplus organic matter produced by the corals support large colonies of other animals that live within or close to the reef.

Crabs and Sea Anemones

A crab called *Lybia tesselata* lives within coral formations in the Indian Ocean near the Seychelles. It associates with small sea anemones and together they form a fairly close symbiotic relationship. The anemones are small, with tentacles around the 'mouth' end of their bodies. Along the tentacles are tiny stinging cells called nematocysts. These cells can stun or kill prey species which are then drawn into the anemone's body as food. *L. tesselata* uses these anemones as an additional weapon in its armoury. It already has the usual exoskeleton making it pretty well defended against predators, but feels the need for extra defences. The crab carries an anemone on each claw which it thrusts into the 'face' of any enemy that dares to attack. The stinging tentacles brush into the attacker and convince it that other prey would perhaps make an easier meal.

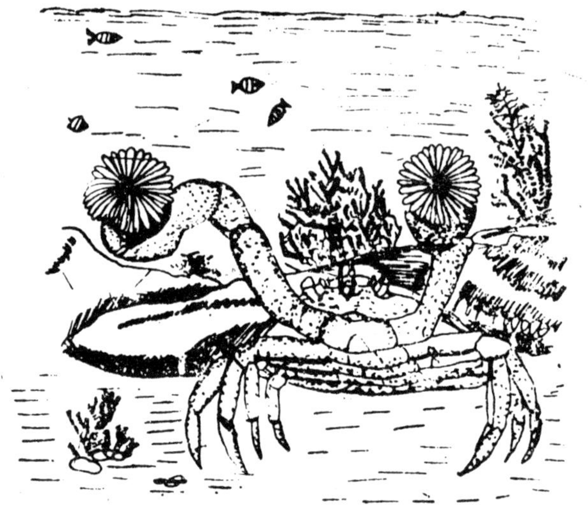

Lybia tesselatia with anemones on each claw. The claws are no longer used for their normal function of feeding.

Both species benefit from this relationship. The crab is better able to defend itself from predation and the anemone also gets protection from the crab, as few creatures attempt to eat the two species when they are working together. Normally the anemones are slow-moving creatures and consequently have a small range, but when living on a crab the anemone is able to move around a much wider habitat. This increases the feeding area and adds variety to the food sources available to the anemone.

2. Milkers

Several species effectively 'milk' other species for a foodstuff that they desire. The milker strokes the other creature, thus stimulating it to ooze a substance usually called honeydew. This substance is neither honey nor dew, but it is a sweet solution.

Mealy bugs and aphids exude this sticky substance over the leaves of the bushes on which they live, and then the fierce heat of the sun dehydrates it, and a white crystalline formation is left. This glistens in the sun like frost on leaves and is often collected by native peoples as food. The collector can obtain a kilogram of honeydew a day and sell it as a delicacy in the local markets.

Three milking relationships are known, which are all quite different from one another and indicate the range of these associations.

Alphids and Ants

Aphids the tiny green, white or black flies suck the highly concentrated sugar solution from the plant and extract the tiny amount of useful nutrients that are contained in this solution. Because such a small amount of the liquid is useful the aphid has to consume large quantities of the plant juices and has therefore to get rid of considerable amounts of unwanted sugar and water. Thus is the sticky waste substance called honeydew, which is exuded from the aphid's anus.

Ants Construct Shelters from Mud for the Protection of their Domesticated Aphid populations.

Wherever colonies of aphids are found ants of various species often abound as well. This is no coincidence, as the ants are interested in the aphids' honeydew. They find this substance almost irresistible, although it is not a very useful food. The ants approach the aphid from the rear and stroke its hind end; this stimulates the aphid to release honeydew, which the ants lap up.

Various species of ant have extended this basic relationship and have developed different methods of colonising or domesticating the aphids. This adds the element of symbiosis to an association that would otherwise be only one-sided. In one species, the ants take fine earth up to the leaves and stems of plants and, using their own saliva, cement together tiny shelters, shaped like mud huts, for their aphid partners. These shelters help to protect the aphids from severe weather and to some extent from predators.

Other ant species will go out searching for aphid colonies and gently bring them back to a convenient plant, close to the ant nest. This makes the aphids constantly 'on tap' for milking by the ants. Some ants are left on duty to protect the aphids from predation. As the ants have a vicious bite, their presence keeps away may species of aphid predator. It is not only mealy bugs and aphids that produce honeydew; some larval caterpillars produce a very similar substance, and the ubiquitous ants are always around to take advantage of the situation.

3. Cleaners

Cleaning symbioses are found in the sea, in freshwater, on land and in the air, but the greatest number of examples concern marine species. Some creatures are physiologically unable to keep themselves clean and free from parasites, as they are shaped in such a way that it is impossible for them to remove any foreign matter from their bodies. Fish are obvious examples of this. They are not able to bend in two, and although some can reach their tails with their mouths it is not possible for them to reach any other part of their bodies. There are other animals that need assistance with cleaning, such as hippopotami. Their mouths are so huge and ungainly that even if they could reach the parasites they would be totally incapable of removing them; the same applies to crocodiles, alligators, some iguanas, elephants, rhinoceroses and turtles.

Other animals are able to groom most of their bodies but find small areas difficult. These creatures often engage in mutual grooming with their own kind; apes and monkeys are good examples.

It is essential for all creatures to have some method of keeping themselves clean and free from parasites. If they do not, they will probably fall ill from infected wounds or the effect of disease and blood loss from parasites. For these species that are unable to clean themselves is obviously vital to find some other animal to perform this cleaning function.

Cleaners are found in all the oceans, and sometimes the associations are between closely related species but they are often between distant ones. The vast majority of cleaners are fish. Not only that, shrimps of various species crabs marine iguanas, seagulls are known to pick up parasites. Cleaner symibiotic relationships are also observed in lizards and birds, whales and birds, birds and herbivorous mammal, fish and hippopotami, birds and crocodile, birds and ants, etc.

The last cleaning symbiosis is quite different. Here Blue bottles and Blowfles, which are found around dustbins and animals are found to lay her eggs on decaying matter of all kinds e.g. animal corpses, tainted meat or fish. Larvae feed on only necrotic tissue and pus and not on healthy tissue. The excretions of larvae act to clean out the wound. These grubs may often save host's life by preventing the infection spreading all over the body.

4. Protective Symbiosis

Small creatures are vulnerable all the time, day and night. They protect themselves as much as possible by sheltering in rock crevices or coral by burrowing in sand or by living in

Ants Living within the Hollow Thorns of Acacia Plants.

enormous shoals. Nevertheless, even these forms of self protection cannot eliminate the extreme danger of predation that they constantly live with. Only the very large and predatory fish are safe from natural predation. Some examples of protective symbioses are association between Blind Shrimp (*Alphens djiboutensis*) and Gobid Fish (*Cryptocentrus lutheri*). Here the shrimp is allowed to feed in relative safety by the presence of the fish, and the fish has belt-hole for daytime and a safe resting place at night in the burrow which is dug by the shrimp. Similar examples of mutual protection can be observed in the association of fish and jelly fish, crabs and sea anemones, crabs and sponges, and antelopes and baboons.

The final example of protective symbiosis is quite different. It concerns a plant and an insect, and there is a kind of role reversal. The plant 'uses' the insect to protect it, just as many animals and insects use plant material to protect them. The Acacia (*Acacia sphaerocephala*) is a bush that grows in the African savannah. It has large vicious-looking black thorns 3 cm long, which give the plant some degree of protection from leaf-eating animals. The thorns are, however, hollow and therefore brittle; so any really determined animal can break off or flatten the thorns without too much difficulty.

A species of ant (*Pseudomyrma*) has become adapted to live in the hollow thorns and provides the plant with added protection. The ants have a virulent bite and when an animal attempts to browse on the acacia the ants come out of their thorns and attack the soft nose of the animal. This discourages most creatures with the notable exception of the giraffe. Giraffes have long, strong, prehensile tongues with which they strip the leaves off trees. Their sensitive noses do not come into contact with the plant material and therefore they are not badly bitten.

The ants nest in the thorns, going in and out through tiny holes that they make at the base of the thorn. They feed on a diet of sugar, oils albumen provided by the plant. This food supply means that the ants never have to leave the plant, and the thorns make a well-armoured home for them. The substance that the ants feed is specially manufactured by the plant, which has no need of it for its own purposes. It is intended to keep the ant colony and deter it from straying, which means that the ants are always 'on duty' to protect the plant. This is a most unusual example of a plant enlisting the services of an animal. The relationship provided both plant and ant with solid protection, with the added benefit to the ant of a constant food supply.

5. Algal Symbiosis

Various animal monocells, such as the slipper animalcule (*Paramaecium bursaria*), contain in their cells single-cell green alga (*Chlorella*) which, owing to their capacity for photosynthesis, can emit assimilation products and oxygen. The host, in turn, provides protection for the green algae, transports them on the basis of its capacity for progression into favourable light conditions and imparts to them carbon dioxide from its respiratory metabolism, which again helps the green algae with the synthesis of its assimilation products. In general the symbionts are resistant to the assimilation enzymes of the host. It can happen, however, that a surplus of algae, which are constantly increasing through cell division, are reabsorbed or assimilated from the host.

Other examples are association between Green Hydra (*Hydra viridis*) and an alga *Chlorella*. The flatworm *Convoluta roscoffensis* and algae, sea slugs (*Elysia viridis*) and chloroplasts from the seaweed *Codium fragile*.

The last example in this chapter of a relationsh ip with algae is quite different. It does not concern an animal but rather associations between algae and fungi. The algae invade the fungi and the two together are called a *lichen*. This is an example of a third 'species' being formed by the joining of two symbiotic partners.

By examining a cross section through a lichen its symbiotic character instantly emerges.

Its main tissue consists of fungal fibre, which gives the lichen its characteristic form. Among these fungal tissues, especially in the case of thallophytes, there are green algae and blue-green algae, which have a capacity for photosynthesis. A symbiont category is determined for each particular type of lichen. In particular there are single cell or thread shaped multicellular representatives of green algae (Chlorophyta) or blue-green algae (Cyanophyta); in the case of fungi they are mainly from Ascomycetes but often from Basidiomycetes. Symbionts living in association with a lichen can surprisingly also remain separated from one another on an artificial feeding ground. Neverthless they flourish in symbiosis considerably better. In symbiosis the fungus first receives part of the assimilation products of the partner, and attends to the supply of water and salts to the lichen itself. In this way it makes it possible for the algae and blue-green algae, which normally flourish only in moist areas, to live in relatively dry places.

It is interesting to note that algae or blue-green algae have lost the capacity to propagate independently in lichen. The propagation of lichen comes about by the formation of fruit organs on the part of the lichen fungus, in which small young lichens develop, which already possess certain algae or blue-green algae, so that when they are scattered by the wind and settle on suitable terrain, they are able to develop in symbiosis from the outset. The lichen has by this association been able to colonise the most unlikely and inhospitable habitats, and also help in the continuous process of rock erosion.

6. Microbial Symbiosis

Most bacteria found in nature are positively beneficial, and it is only a tiny minority that cause any real trouble. An extraordinary fact is that none of the herbivores can digest grasses without the aid of intestinal bacteria. These symbiotic microorganisms live in various parts of the gut, according to the type of herbivore, but they perform essentially the same function for all.

Herbivores

Domestic cattle provide a typical example of this symbiosis. In cows, the stomach is divided into four chambers. The grass that the cow eats passes into the first chamber on rumen, where it is fermented by the micro-organisms present. This fermentation produces large quantities of methane gas (500 litres of gas in one day for each cow) and so conditions in the rumen are anaerobic (without oxygen). This analrobic environment is essential for the survival of the dense

A Normal Population of Micro-organisms in the Rumen of a Cow.

population of bacteria and protozoans which live in the rumen. The relationship does not end at breaking down plants for digestion by the host, however. The bacteria have a very short life span of about 20 hours and when they die the cow can digest their protein-rich bodies. As the cow's diet of grass is low in protein, this extra nourishment is vital and can provide the cow with 20 per cent of its protein intake. This is therefore a very complete relationship. The cow cannot survive without the bacteria and the bacteria need this type of anaerobic environment with constant supplies of cellulose. The micro-organisms are so adapted to this anaerobic environment that they die in the presence of oxygen. Both parties are thus ideally suited by the arrangement. Other animals e.g., nursing piglets, require micro-organisms to provide them with an enzyme that breaks down their mother's milk sugar from lactose to glucose without this they rapidly get hypoglycemia and decline instead of grow.

Similarly, many blood sucking animals are dependent on gut bacteria for digestion of the blood that they feed on as well for the synthesis of vitamins. Without the synthesis of vitamins the host blood-sucker would not be able to grow and develop.

Man

The gut of man is fairly heavily infested with micro-organisms. Some work has been done on the effects of these organisms, but no real benefit or harm has been established for most of them. Some people have been kept in a germ-free state for a short while and have suffered no noticeable harm.

Termites

Another species that is totally dependent on gut micro-organisms (in this case mainly flagellate protozoans) is the wood-eating termite. In this case, both the termites and the flagellates could not exist without one another. This then is the most extreme symbiosis possible. These termites live on a diet of wood, but, like the herbivores, are unable to break down the cellulose component of the wood. The young newly hatched termites do not have any flagellates in their gut but rapidly become infested with them by eating the faeces of adult termites. By this means, the flagellates eventually reach the hind gut of the young termites. The flagellates act in the same way as the microorganisms in herbivores, by converting cellulose to soluble carbohydrates. In exchange for a constant supply of wood they provide the host with nourishment. Once again an anaerobic environment is provided by the termite's gut.

Leguminous Plants

It is not only animals that rely upon bacteria for their nourishment; there is a family of plants, the Leguminosae, which utilise bacteria to convert nitrogen into nutrients. Nitrogen in the form of a highly stable molecule (N_2) is one of the constituents of the air, and it is washed down into the soil by the action of rain. Most plants require nitrogen for their survival and the legumes utilise the bacterium *Rhizobium* to convert this nitrogen to ammonia which can then be readily absorbed by the plant.

These bacteria are micro-organisms which normally live in the soil. If they are close to the root of a legume the relationship begins. It is thought that the plant's roots secrete a substance which activates the bacteria to change shape and enter the plant's roots via the root hairs (tiny, hair-like branches which take nutrients and water from the soil). Once inside the plant, the bacteria multiply very quickly and activate the plant root cells which divide (multiply) and form tuberous swellings called root nodules. Inside these nodules the plant and bacteria work together to fix the nitrogen to make it into a form usable by the plant. For details see Nitrogen cycle.

Fish and Squid

One more example of animals using bacteria is quite different in that bacteria in this case act a torch for the host and have little to do with feeding. The bacteria receive nourishment from the host organs and a degree of protection from the outside world. The light produced by the bacteria enables the fish to see its path and also act as lure enticing small prey fishes as well as to attract a mate.

7. Fungal Symbiosis

Fungi may be associated with trees. In a dense forest trees grow up in close proximity. There is intense competition for root space and thus the root system is not able to draw enough nutrients and water from the soil to allow them to grow and mature. It is in this situation that the trees' association with various fungi is of paramount importance. The fungi cover the tips of the tree roots with hyphal tissue, forming a cobweb-like net which penetrate the soil, effectively increasing the tree's root system and enhancing root efficiency. This form of association is known as **mycorrhizae**. Any one tree species may form mycorrhizal relationships with many different sorts of fungi and conversely most fungi (*e.g., Cenococcum*) have a wide host range, though some (*e.g., Boltus elegans*) are apparently restricted to a single host genus (*Larix*). Common mycorrhizal genera are basidiomycetes, particularly Agaricales such as *Amanita, Tricholoma, Russula* and *Lactarius*. Many of these fungi, or at least their mycorrhizal strains, do not have ability to break down complex carbohydrates in culture and thus receive their carbohydrates from the host tree and utilise it for its on self. The tree has increased mineral uptake (phosphate in particular) due to the presence of the mycorrhizal fungus. The mycorrhizal roots live longer and less susceptible to disease than uninfected roots. Without the aid of these fungi, most of our enormous, dense forests would not be able to exist. Instead the trees would grow much more sparsely and would therefore not produce anything like the quantity of oxygen that they do now.

Another relationship of the fungi is with the seeds of orchids. The tiny orchid seeds are unable to germinate as such as no nutrients are carried along with the seedings as in other seeds. But if they are infected with various fungi they could be persuaded to germinate and grow. These fungi provide the seeds with an environment enriched with carbohydrates and vitamins which because of their size they do not naturally possess. When the seeding has developed a root system the fungi form a fine web of fungal filaments which extend the plant's roots and help it to survive until it develops shoots and leaves which allow it to photosynthesize and survive unaided. Some of the species of orchids are totally dependent on the fungal 'roots' for their nourishment and water supply so that they d not develop root system of their own. The fungus benefits from a stable 'home' and shares in the surplus sugars produced by the plant.

The Ambrosia Beetle (*Xylerborus ferrugineus*) and its larvae burrow in wood, making the tunnels and galleries in which they live. But they are unable to digest the wood cellulose without the aid of a fungus, which they cultivate. The fungus cannot break down the cellulose from the intact wood, but needs the beetle to chew it up into small pieces. The beetle transfers fungi from tree to tree when necessary by filling a sack on its thorax with fungal spores and planting them in tunnel made in the new wood. Moreover the beetle eats some of the fungi to keep it under control, or otherwise it might smother the galleries. The ambrosia beetle is unable to pupate without the ergosterol which is synthesised by the fungus. The fungus is provided with an even climate and protected from outside world, and it is also 'gardened' by the beetle to keep it healthy and is given chewed-up wood as nourishment.

This review of symbiotic relationships is not comprehensive by any means but may perhaps give an idea of the range of different associations which may be formed by different organisms in many different habitats.

ALGAL BLOOMS

Every lake or pond goes through a life cycle that may take from a few days up to thousands of years for the larger bodies of water. Each lake or pond faces the prospect of becoming more shallow with time as silt and organic matter are swept in by streams and run off and deposited in the water. Shallow ends, often near the outlets of streams, tend to become marshes. Algae and other plants proliferate under these conditions and often appear as a green scum. We can see various stages in the life history of lakes, from those with fresh, clear, sparkling water to the shallow, growth-choked, scum-surfaced swamps on the verge of becoming dead lakes.

In lakes and slowly moving waters that have been enriched, a number of surface water algae accumulate at times in such numbers as to form loose, visible aggregations called **blooms** which may cover very large lakes and reservoirs or even streams. Blue-green algae may form these water blooms particularly during periods of warm, calm weather, when the algae which were previously distributed through the water rise to the surface. Other blooms may be associated with a rapid reproduction of a particular alga. Some water blooms have resulted in fish kills by interfering with reaeration, by excluding light necessary for photosynthesis in the lower areas and thereby preventing release of oxygen into the water, or by depleting the oxygen through decay or respiration within the bloom. Some water blooms release substances extremely toxic to fish, domestic animals, and birds, and extensive kills have resulted in several areas.

A considerable number of species of algae are capable of producing blooms. Some of the genera most frequently involved are the blue-green algae, *Anacystis (Microcystis), Anabaena, Aphanizomenon,* and *Oscillatoria*; the green algae, *Hydrodictyon, Chlorella,* and *Ankistrodesmus*; the diatoms, *Synedra* and *Cyclotella*; and the flagellates *Synura, Euglena,* and *Chlamydomonas.* Blooms of blue green algae are particularly obnoxious Blooms of flagellates and green algae are often encouraged by the addition of fertilizer in farm hatchery ponds in order to increase fish production.

Related to blooms are the **mats or blankets** of filamentous green algae such as *Spirogyra, Zygnema, Oedogonium,* and *Cladophora* which at times may cover large areas of a reservoir or lake. Certain blue green algae such as *Gloeotrichia natans* may also form extensive floating masses. These growths cause unsightly appearance in a community's water supply and serve as breeding places for gnats and midge flies. They may clog intake screens, cause tastes and odours, and interfere with the functioning of multiple purpose reservoirs by collecting as debris on the shores and interfering with fishing and bathing. Many of the mat forming algae are resistant to effective treatment of copper sulphate.

Only species from three Cyanophyceae genera *i.e., Microcystis, Anabaena* and *Aphanizomenon,* have been studied in detail with regard to toxin production. *Microcystis aeruginos* a produces a toxin called Microcystis FDF (**fast death factor**) which is a small cyclic peptide. *Anabaena flosaquae* produces a toxin which seems to be a low molecular weight tertiary amine or alkaloid. The toxin of *Aphanizomenon flos-aquae* has been studied but not characterized. The endotoxins are released in water as the blooms die and animals may be poisoned by drinking the contaminated water. Although these toxins have not caused any human fatalities, they have been blamed for causing dermatitis, gasteroenteritis and respiratory disorders in human beings. Decomposition of the Cyanophyceae not only releases toxins but also leads to a depletion of oxygen, accompanied by production of hydroxylamine and hydrogen sulphide which are themselves highly toxic. The toxicity of contaminated water shows diurnal fluctuations. Around mid day, when photosynthesis is maximal, dissolved oxygen levels are high and the pH of the water may reach 9.5. Since all Cyanophyceae toxins are alkali labile this lowers the toxicity of water, especially in warm waters ($>20^{\circ}C$). At night only respiration occurs, the oxygen level of the water drops and the pH falls to 6.5. The stability of the toxin under these conditions makes for maximum concentrations. Thus any steps to destroy algal blooms should be taken in the middle of the day when the toxin released will break down quickly and so pose a minimal threat to zooplankton and fish.

Algal blooms are sometimes caused by conditions that are not related to added nutrients.

Aritificial warming of the water, such us that from cooling towers and condensers, may stimulate the growth of blue-green algae. The elimination of other aquatic flora, either mechanically or by the presence of some toxic substance, promotes the growth of bloom-forming algae by removing competition for the available nutrient supply. The release of organic nutrients by the decomposition of aquatic plants killed by herbicides may cause an increase in algal growth, and the absence of plankton-feeding organisms may favour excessive production of bloom.

Algae, that are the principal growth products of eutrophication, need at least fifteen elements to sustain growth. If any one of them is in short supply, it will be a limiting factor in the growth and development of the algae. However, many of the essential nutrients, particularly the so-called micro-nutrients, are almost always present in abundance in lake water. Also, algae normally have access to an abundant supply of carbon, especially from carbon dioxide in the air and from carbonates dissolved in the water. (Some scientific investigators believe that a deficiency of carbon dioxide can become a limiting factor). Nutrients, which are utilized by the algal plants in large amounts, are apt to be the most important in determining the rate and degree of growth. The primary plant nutrients are nitrogen, phosphorus, and potassium. Of these three, nitrogen and phosphorus are in short supply naturally, and since some of the blue-green algae have nitrogen-fixing abilities, phosphorus is in many cases the most important added nutrient in bringing about excessive growth.

There are several man-made sources of the nutrients that cause **eutrophication**. They include domestic, industrial, and agricultural wastes. **Sewage** is an important source of nitrogen and phosphorus as well as organic matter. **Detergents** are important contributors of phosphates. **Agricultural fertilizers** and livestock and poultry wastes are also major sources. Wastes from slaughter houses and food processing plants often end up in the waterways. Boats discharge a variety of pollutants including sanitary wastes.

Several lines of attack are needed to solve the problems of water pollution. Better sewage treatment to remove nitrogen and phosphorus and more effective disposal of the organic material as sludge are needed on a large scale. Control of industrial and agricultural wastes would lessen the problems to some extent. Since phosphate-containing detergents are a major contributor of phosphorus; the elimination of phosphates from detergents would go a long way toward retarding eutrophication. Because food and other agricultural products come from the land, the return of agricultural wastes to the land would complete their cycle.

ALLERGENS

Allergens are substances produced by plants that cause contact dermatitis or allergic reactions, the latter are primarily due to pollen and spores. Allergic reactions are a side effect of immunity, caused by the interaction of allergy-producing factors (antigens) with the body's antigen-attacking factors (antibodies). During an antigen-antibody reaction cells are stimulated to produce substances that are actually harmful to the body.

The most important of these substances is histamine, which has simultaneous effects. Histamine causes a buildup of pressure in the small vessels, thus producing extensive "leaking" of fluid into tissues from the capillaries. The result is tissue swelling (edema)—in the sinuses, for example. Histamine can also cause spasms of the small breathing tubes (bronchioles) in the lungs, making breathing difficult. There are drug preparations that contain antihistamines to counteract the effects of histamines.

In allergic response, there a is latent period between the time of contact and the development of symptoms. During the latent period (more than 24 hours, typically 5 days) immunological changes occur within the body. In addition, an individual must be predisposed or sensitized to the substance by inherited tendency, prior exposure, or a combination of both. Sensitivity to allergy-producing factors varies widely among individuals. People usually possess some initial resistance but become sensitized by continued contact with specific allergens. Sensitization is accumulative so that successive exposures increase the potential for a severe reaction.

The development of an allergic response also depends on factors inherent within each plant. Not all parts of the plant may contain equal quantities of the chemical substance. It may be located on plant surfaces, or internally, so that bruising, cutting, or crushing releases it. Hay fever allergens are located on pollen grains or spores. The concentration and distribution of allergens within a plant body depends on stage of maturation, season and environmental factors. Allergic responses may progress into secondary complications, including bacterial infections of scratched skin, drug-treatment side effects, generalized edema, and asthma.

There are hundreds of plant species to which humans can develop allergic reactions. Allergic responses to plants usually take two basic forms; contact dermatitis and hay fever.

Poison ivy (*Rhus radicans*) and poison oak (*Rhus toxicodendron*) are undoubtedly the leaders in causing human allergic contact dermatitis because of an oleoresin (phenol) known chemically as a 3-*n* pentadecylcatechol, or urushiol. The reaction is of a delayed response type. Several hours to days after contact the skin of a sensitized individual becomes reddened and itchy, with water blisters at sites of contact. Contrary to a popular misconception, the blister fluids do not spread the inflammation, for they are formed inside the body. The inflammation is spread by traces of urushiol on the hands, clothing, or other objects (even pets). Droplets of the substance can also be spread in smoke when the plants are burned. As is typical of allergies, individual sensitivity to urushiol varies greatly.

People who work intensively with particular plants are quite likely to develop sensitivities to them—for example, horticulturists and florists working with English ivy (*Hedera helix*), tulips (*Tulipa*), *Narcissus, Philodendron* species and certain lilies; agricultural employees working with hops (*Humulus lupulus*), onions, carrots, tomatoes, potatoes, and garlic; lumbermen and wood-workers with various types of wood, especially poplar (*Populus*), and certain lichens and leafy liverworts that may be covering the tree bark. Usnic and fumarprotocetraric acids appear to be particularly active in this regard, but only a minute proportion of people handling lichens become affected. Usnic acid causes photosensitivity as it converts light to chemical energy with resultant rashes.

Many pollens cause contact dermatitis even though pollen is usually thought of only in terms of hay fever. The two different allergic responses are stimulated by totally different pollen components. Water-soluble substances stimulate hay fever whereas oleoresins cause allergic contact dermatitis. Pollen dermatitis affects exposed skin, such as the face, neck, and arms, forming a sheetlike dermatitis. Ragweeds pollen is the most common culprit which were given the genus name *Ambrosia*— "food of the gods." Some problem pollens are box elder (*Acer negundo*), poplar (*Populus*), maple (*Acer*), ash (*Fraximus*), cocklebur (*Xanthium*), carrot grass (*Parthenium*), mare's tail (*Hippurus tetraphylla*) and marsh elders—but not poison ivy!

Although several hundred plant species cause allergic contact dermatitis, most of the offenders are found in nine plant families (Table 2)

Urticaria, hay fever, asthma, eczema, and allergic gastrointestinal disturbances are examples of atopic allergies, experienced by persons who have abnormal immunity reactions to everyday antigens, such as pollen, spores, fungi horse hair and organic material in dust.

Urticaria—itchy wheals—is characterized by development of large welts all over the body. Itching inside the ears, on roof of the mouth and in throat may be a constant source of irritation. Antihistaminic drugs are very effective in treating urticaria.

Table 2: Nine Plant Families Particularly Significant in Containing Species that Produce Allergic Contact Dermatitis of these, the Anacardiaceae is the most important

Plant Family	*Representative Examples*
Amaryllidaceae	Daffodil, narcissus
Anacardiaceae	Poison ivy, poison oak, poisonwood, mango, Florida holly, cashew
Asteraceae	Dog fennel, chrysanthemums, daisies, carrot grass
Euphorbiaceae	Spurges, castor bean plant, poinsettia, pencil tree or milkbush, crown of thorns, manchineel tree, snow-on-the mountain.
Poaceae	Grasses, cereal grains
Litiaceae	Tulip, hyacinth, garlic, onions
Primulaceae	Primrose
Rutaceae	Citrus fruits
Apiaceae	Carrot, celery

Hay fever is one of the most common and annoying chronic medical problems. What basically happens is that if the genes carry the requisite allergic tendency, the body develop a substance called **reaginic-antibody** or I.g.E. In hay fever this antibody is to be found in the lining of the nose and bronchial tubes, eyes and skin. When an antigen, a pollen, comes into contact with I.g.E. antibody a reaction takes place on the surface of the cell and various substances, including histamine, are released. The surrounding tissue swells full of fluid and there is a watery discharge full of mucous. Antihistaminic drugs are effective in treating the edema but do not prevent nasal irritation, decongestant sprays containing ephedrine make the nose comfortable.

The pollen grains of many plant species cause hay fever, as do the spores of some fungi, especially *Alternaria* and *Cladosporium.* Spores of *Helminthosporium, Aspergillus, Penicillium, Fusarium,* and *Rhizopus* are of secondary importance. Pollens that are lightweight and wind borne cause hay fever—not the heavier insect-borne pollens. Hay fever causing pollens vary seasonally (Table 3).

Asthma occurs when antigen-antibody reactions are localized in the lungs, causing constriction of bronchioles and consequent breathing difficulty. Severe asthmatic reactions ("attacks") can quickly reduce the lungs' ability to provide adequate oxygen and emergency medical treatment is necessary. Anti asthmatic drugs may be helpful.

Table 3. Some Pollens that Cause Hay Fever

Early Spring

Mainly tree pollens, such as box elder (*Acer negundo*), poplar (*Populus*), elm (*Ulmus*), birch (*Betula*), oak (*Quercus*), hickory (*Carya*), and ash (*Fraximus*).

Late Spring and Early Summer

Mainly grass pollens, such as timothy grass (*Pheleum pratense*), Bermuda grass (*Cynodon*

(Contd.)

Table 3. (Contd.)

dactylon), Kentucky bluegrass (*Poa pratensis*), and redtop (*Agrostis gigantea*); also some broad-leaved plants.

Late Summer and Autumn (Especially August and September)

Mainly ragweeds (*Ambrosia* spp.), especially in the Midwest and East, and saltbush (*Atriplex* spp.) and Russian thistle (*Salsola kali*) in the West.

Atopic allergies are often treated successfully by gradual desensitization. Sterile allergens (*e.g.*, specific pollen extracts, house dust, fungi and horse hair are injected into a person. If a weal develops in a few minutes the person is sensitive to the extract. A vaccine, made to suit the sensitivity reaction is injected in increasing strengths every week until the course is completed.

ACID RAIN AND FORESTS

Over the past decade, scientists have amassed considerable evidence that air pollutants from the combustion of fossil fuels, both oil and coal, and the smelting of metallic ores are under -mining sensitive forests and soils. Damage to trees from gaseous sulphur dioxide and ozone is well documented. Recently, **acid deposition**, more commonly called **acid rain**, has emerged as a growing threat to forests in sensitive regions. Acid deposition refers to sulphur and nitrogen oxides that are chemically transformed in the atmosphere and fall to earth as acids in rain, snow, or fog, or as dry acid-forming particles. Acid deposition is known to have killed fish and plants in hundreds of lakes in Scandinavia and eastern North America. However, its links to forest damage remain circumstantial.

Temperate forests have a long history of stress and acidification. Since the end of the last continental glaciation 10,000-15,000 years ago, soils have slowly formed from the sterile layers of gravel, sand, and silt left behind by the retreating ice. Pioneering plants, animals, and micro-organisms aided this soil development, helping form an intricate cycle of nutrient uptake and release. Death and decomposition of these inhabitants generated acids in the soil. Where acids developed faster than the natural processes that could neutralize them, the soils gradually acidified. This process continues even today. Centuries of human use and abuse of forest ecosystems have added to this natural acidification. Many temperate forests in Europe and North America are now recovering from decades of intense burning, grazing, and timber cutting.

Air pollutants and acids generated by industrial activities are now entering forests at an unprecedented scale and rate, greatly adding to these stresses carried over from the past. Many forests Europe and North America now receive as much as 30 times more acidity than they would if rain and snow were falling through a 'pure' atmosphere. Ozone levels in many rural areas of Europe and North America are now in the range known to damage trees. Despite air quality improvements made during the seventies, the average concentration of sulphur dioxide in many areas is high enough to reduce the growth of the trees.

Needles and leaves yellow and drop prematurely from branches, tree crowns progressively become thin, and ultimately, trees die. Even tree that show no visible sign of damage may be declining in growth and productivity. Moreover, the tendency of the acid rain to leach nutrients from sensitive soils may undermine the health and productivity of forests long into the future. Taken together, these direct and indirect effects threaten not only future wood supplies but the integrity of whole ecosystems on which society depends.

The role of acid rain and other forms of air pollution is under intensive investigation. In spite of the dimensions of the forest damage, however, a firm link has not been established. One can get some idea of the difficulties by contrasting the forest decline with clear-cut cases of forest poisoning by air pollutants. Smelters and chemical plants that emit sulphur dioxide, oxides of nitrogen or fluoride compounds are often girdled by dead timber. In such cases there is a clear correlation between tree damage, a specific pollution source and a threshold concentration of the pollutant. The forests that are now dying, in contrast, are far from any source and are exposed to pollutants in concentrations well below the levels previously reported to injure trees. If air pollution and specifically acid rain, plays a part in forest decline, it probably does so less as a lethal agent than as a stress.

Many stresses, both biotic and abiotic, combine to affect the vigour of a forest. The tree's genetic endowment or age can be a source of stress: a stand may be genetically weak or senescent. Other stresses may take such forms a diseases, insects, parasitic fungi, a shortage of light, water or essential nutrients and sporadic injury from events such as floods, high winds and ice storms. Stresses easily withstood in isolation can combine with debilitating or fatal effects. A fatal sequence of stresses may begin with a "predisposing" stress, such as a shortage of nutrients. The tree may then be seriously weakened by an "inciting" stress, such as a severe winter. It is then defenseless against a final, "contributing' stress—the actual cause of death—such as disease or insect attack.

Acid and other pollutants could add to the high level of abiotic stresses, including thin soil, low temperature and desiccating winds, present in a high-elevation forest. That is, the pollutants might handicap the trees with one more predisposing stress as they face subsequent stresses. But what is the nature of the added stress?

Investigators, most of them in Europe, have put forward a number of hypothetical mechanisms (around 180) many of which would ultimately lead to nutrient deficiency in the tree.

The most commonly accepted is the '**cup of stress**' theory, in which a tree is represented as a cup full of environmental stress, composed of acid pollution, unacceptable ozone levels, nutrient deficiency caused by acid soils, weather stress, etc. Once a tree reaches this threshold level, any additional stress—a hot summer, a cold winter, an attack of beetles—will kill it. All species can be affected, although within a species some individuals will be more resistant than others.

Sulphates and nitrates raining down as acids have drastically different effects on different forests and even on different tree species in the same forest. Incoming acids affect interactions between the soil and living biomass of an ecosystem in complex and varying ways. Soil structure and composition, vegetation type, climate, and elevation are only some of the natural determining variables. Yet, research over the last decade has uncovered some common effects of acidity that point to several pathways by which acid deposition can threaten forests.

Trees derive their nutrition primarily from elements such as calcium, magnesium, and potassium that are weathered from minerals in the soil. Acid deposition adds hydrogen ions to the soil, which displace these important nutrients from their sites bound to soil particles. Soils with a pH of 5 or more are seldom in danger since they have plentiful calcium carbonate (the constituents of lime) or silicates (which have abundant calcium, potassium, and/or magnesium) that effectively neutralize the acid ions. Yet soils at lower pH levels have fewer of these **buffering agents**. Competition from incoming acids causes the leaching of calcium and magnesium from the soil. Large areas of the southeastern United States, New England mountain ranges, eastern Canada, and extensive areas in Scandinavia, for example, are underlain by slightly acidic, poorly buffered soils that are especially susceptible to the effect. Although soil changes generally take place over a long period of time, soil studies in Sweden suggest that substantial leaching of nutrients from sensitive soils can occur in just a decade.

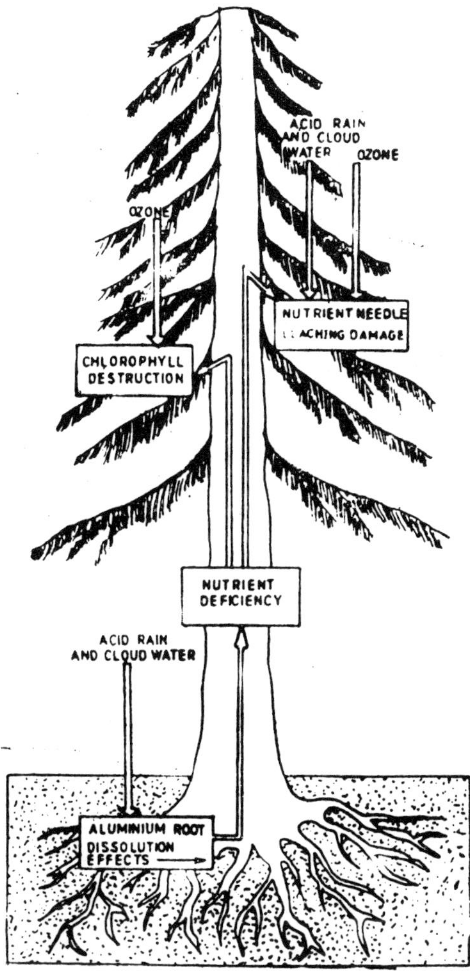

Acid Rain and Ozone Damage to Trees.

Sulphates and nitrates, the other key constituents of acid deposition along with hydrogen ions, can initially have a fertilizing effect on many soils and for a time may actually boost tree growth. Yet this enhanced growth is short-lived, for eventually these "fertilizer" supplies will exceed the forests' capacity to use them. **Sulphate saturation** usually precedes **nitrate saturation**, but excess quantities of either or both simply pass through the soil carrying vital nutrients with them. With forest productivity closely tied to nutrient availability, this leaching of soils by H^+, SO_4^{2-} or NO_3^- ions eventually reduces forest growth.

Research has also shown that heavy metals—either mobilized in the soil or introduced from the atmosphere—may be involved in the forest damage. Bernhard Ulrich, a soil scientist who has studied damaged beech and spruce forests in the Solling Plateau of West Germany for nearly two decades, has proposed that as soils become increasingly acidic, aluminium which is normally harmlessly bound in soil minerals, becomes soluble and toxic. The free aluminium attacks the tree's root system, making the tree less able to take up moisture and nutrients and to protect itself from insect attacks and droughts.

Fig. 1 illustrates the mechanism by which acid rain and Ozone together could lead to nutrient deficiency in a coniferous tree, according to a currently favoured mechanism for their possible role in forest decline. The ozone might act both by destroying chlorophyll and by degrading the waxy coating of the needles, the former being essential to photosynthesis. Acid rain or cloud water could get absorbed into needle tissue and leach out nutrients. In the soil the acid might increase the nutrient deficiency by mobilizing aluminium, which could displace calcium from the tree's fine roots. The tree would be stressed by lack of nutrients, thus it will be accessible to destruction by insects, disease or some other stresses.

PETROLEUM

A liquid mixture of hundreds of organic compounds, mostly hydrocarbons produced millions of years ago from the remains of plants and animals.

Petroleum is perhaps the most versatile fossil fuel. Crude oil, as it is pumped from the ground, is a heavy, sticky, vicious, dark liquid. The oil is refined to produce many different materials such as propane, gasoline, jet fuel, heating oil, motor oil, and road tar. Some of the chemicals in the oil are extracted and used for the manufacture of plastics, medicines, and many other products. It is difficult to imagine what would happen to our civilization if the supply of liquid fuels ran out. Automobiles, air-planes, and many appliances could not operate. Many industries would have to redesign their factories. Yet reliable estimates indicate that before the year 2000 there will not be enough petroleum to meet worldwide demand, many of the richest fields will be depleted, and production will slow down. Yet the need for oil will continue to increase. People will want more oil than is available. Therefore, a real and permanent shortage will result.

Composition

Petroleum comprises a mixture of several hundred or even thousand hydrocarbon constituents. In many cases these are the only constituents, while in others a large number of other organic compounds are present, such as oleic acid, asphaltene, gums and a variety of sulphur and nitrogen compounds. The vast proliferation of hydrocarbons stems from the tetravalence of carbon and its many possible compounds—in the form of chains of different lengths with single, double and triple bonds, of rings of different bond types and sizes and of the most varied bonding with one another. Gums and asphaltenes are high-molecular-weight organic compounds, which give to the oils their dark colour and increase their viscosity. The sulphur and nitrogen compounds are in the form of chains or rings, in which carbon atoms are replaced by sulphur or nitrogen atoms.

Origin

The explanation of how petroleum is formed is one of the most difficult problems encountered in geological research. Almost all the facts point to its organic development from decomposition products of plant and animal organisms that were deposited in the mud beds of waterways, were preserved there from immediate putrefaction by the lack of oxygen, and then formed rock bitumen. One problem regarding this explanation of the origin is that organisms are composed to a large extent of oxidized compounds, whereas petroleum consists of reduced (oxygen-free) compounds. Reduction is conceivable, however, only as a result of energy input. In the first phase of transformation at least, the necessary energy could not have been derived from the heat of the earth, as the petroleum would then still contain elements from the chlorophyll and blood pigment, which will not sustain a higher temperature than 200°C and will persist only in an oxygen-free environment. One possible reaction, which also occurs in lower temperatures and is anaerobic (with out free oxygen) is biological reduction by bacteria. In the sediment in water and in oil a whole series of anaerobic bacteria can be recognized that are able to reduce organic substances and to transform them into fatty acids, methane, and ethane, and also into other hydrocarbons. The hydrocarbons is question do not form petroleum however.

The further transformation of the long-chain compounds (fatty acids) into petroleum with its light paraffins (5-7 carbon atoms) can occur only at lower temperatures and in the absence of active clays as catalyzers or at great depths under high pressures by thermal decomposition. Under this thermal decomposition or fission, long-chain hydrocarbons will be split up into methane and other light gaseous hydrocarbons.

Natural gas is produced in the same way, but contains carbon dioxide, nitrogen, hydrogen sulphide, and mixtures of rare gases in addition to hydrocarbons.

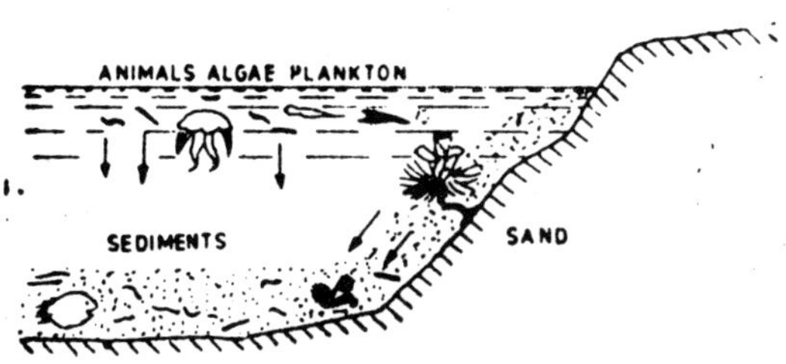

(a) Sedimentation.

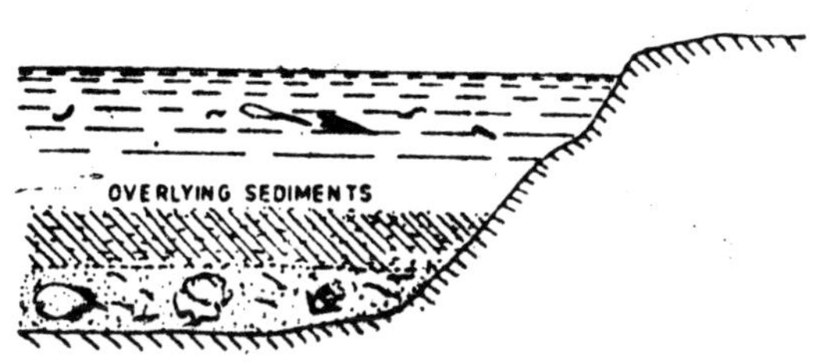

(b) Anaerobic Bacteria Decompose the Organisms Deposited.

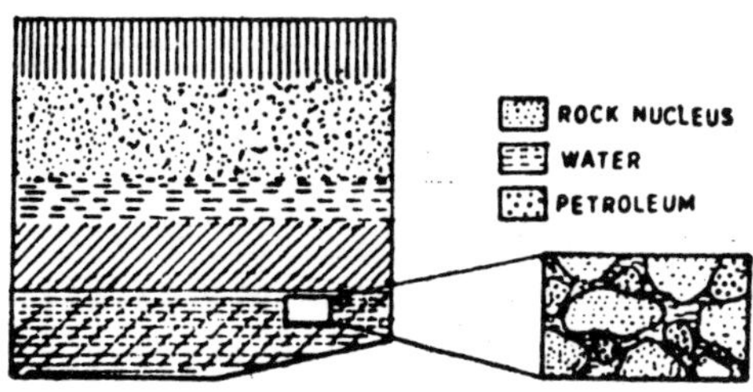

(c) Different Deposits Intensify the Pressure.

Formation of Petroleum.

The formation of petroleum (Shown in Fig.) thus possible represents a process that embraces some very different reactions (bacterial) and thermal decomposition or fission) and extends over long periods. The formation of the most recent petroleum fields dates back roughly one million years.

OIL SPILLS

Accidental or deliberate dumping of oil or other petroleum products onto the ocean and its coastal waters, bays and harbours, or onto land, into lakes or rivers.

The components of crude oil are mostly hydrocarbons; many are not biodegradable. Much of the environmental damage due to oil spills is caused because of the physical properties of oil. The most obvious effects are due to its hydrophobic nature and its impermeability to oxygen. Thus when it coats organisms, they die from asphyxiation (lack of oxygen). Similarly contaminated areas of the environment suffer from oxygen depletion. Oil is chemically stable therefore, environmental damage may occur long after the input to the environment has ceased.

The harmful effects of oil on living organisms may be divided into two (i) that are primarily physical and (ii) that are primarily chemical. Physical effects are caused by oil coating the organisms or their immediate environment. This is very clearly seen when water birds become covered with oil. By matting the feathers, the oil destroys their insulative capacity, reduces buoyahcy in the water and prevents flight. In other organisms, oil coating may cause death by asphyxiation. Oil films on the surface of natural water reduce light transmission and, hence, photosynthetic primary production. Such films also retard oxygen uptake by water and so cause a lower dissolved oxygen concentration and the death of many organisms.

Chemical effects of oil can be related to the components involved. Low boiling point saturated hydrocarbons, at least up to octane, can produce anaesthesia and narcosis in many lower animals. Low boiling point aromatic hydrocarbons are even more toxic, and their greater water solubility tends to enhance their distribution and uptake by aquatic organisms. Benzene, toluene, naphthalene and phenanthrene are amongst the compounds in this group; being fat soluble, they accumulate in fatty tissues and hence can be concentrated as they move up the food chain. Benzene characteristically inhibits blood cell formation in bone marrow. All cause local irritation of the respiratory system and excitation or depression of the central nervous system. Many may be mutagenic, carcinogenic or teratogenic. Polycyclic aromatic hydrocarbons are especially dangerous in this respect. High boiling point saturated and aromatic hydrocarbons may not exert much direct toxicity but may interfere with the responses of aquatic organisms to chemical stimuli, such as, sex attractants, with equally serious consequences.

Water birds, especially diving birds, are hard hit by oil slicks. An oily coat breaks down the natural oils and waxes that insulate and protect the bird from loss of body heat, feathers are damaged; nesting, mating or migrating activities may be seriously disrupted. The most serious long-term effect or marine life from oil pollution may be the confusion caused by odourous materials in oil among marine predators and prey which use their sense of smell to attack or avoid other species. Fish can smell very small concentrations (ppb) of some substances. Migratory fish are believed to depend upon their sense of smell to find their way into particular channels and rivers.

Sources of soil pollution are tanker disasters, ballast water and bilge washings. Most oil spills in the past have occurred usually within 15 km of shore and 35 km from the nearest port. Hose failures and pumping errors account for most oil spills in ports and harbours. The global marine transport of oil is more than 1300 million tonnes, of which 579 million tons (43%) is shipped from the Gulf Countries. Main routes are across the Arabian Sea. One of these is through the Mozambique channel round South Africa to the Western Hemisphere, while the other is round Sri Lanka across the Southern Bay of Bengal through the Malacca Strait to the Far East. This, with the increasing emphasis on offshore oil exploration in many countries of the region, makes the northern Indian Ocean vulnerable to oil pollution. The most recent oil slick and probably the worst in history occurred in the Persian Gulf on 26th January 1991 during the Gulf war between Iraq and U.S. led multinational forces. Several million barrels of oil was allegedly dumped into the persion Gulf by Iraq. Iraq, however, maintained that the oil slick had been caused by bombing of two of its oil tankers by the Allied Forces. According to reports Iraq pumped out

100,000 barrels of oil a day into the sea, killing marine life and threatening desalination plants at Jubail on the western coasts of the Gulf that provide water to the Arban States of Saudi Arabia, Qatar, Baharin and UAE. The oil slick estimated to be about 140 km long and 40 km wide forced Saudi Arabia to shut its desalination plant at Safaniya on the Saudi coast.

Oil Spill Clean up

The procedures available for reducing or removing oil slicks fall into three categories. The use of cleaning agents, the establishment of control barriers, and oil skimming.

Not only are there hazards associated directly with oil spills, there are others associated with oil clearance from natural waters and from beaches. Containment with booms followed by removal by skimming devices causes the least damage but is frequently impossible. As an alternative, detergents may be used to disperse the oil. This may give aesthetically satisfactory results but may enhance the biological destruction done by the spill. All detergents have appreciable toxicity and can facilitate uptake of oil by aquatic organisms. Further, they may cause oil on beaches to penetrate the sand more deeply, thereby prolonging harmful effects on intertidal animals and plants.

TOXICITY

A measure (often called a dose) of how much quantity of a given chemical substance or physical agent (such as radioactive radiation) will kill an individual or cause disabling harm. Acute toxicity is used to describe overwhelming effects—serious illness or death—from one dose given at one time. Chronic toxicity designates a condition causing an illness resulting from exposure to small amounts repeatedly or constant exposure to a low concentration of the toxic substance over a long period of time.

Assessment of Toxicity

Attempts to regulate the emission of atmospheric toxicants have necessitated the determination of safe atmospheric concentrations. In relation to the industrial environment, these may be established by reference to **threshold limiting values (TLVs), formerly known as maximum allowable concentrations, (MACs)**. The threshold limiting value of a toxicant, or mixture of toxicants, is defined as the maximum concentration to which it is believed healthy workers may be repeatedly exposed without ill effect, on the basis of an eight hour working day. In relation to the atmosphere at large, the threshold limiting value is replaced as a reference point by the 'three minute mean concentration'. This is defined as the maximum permissible concentration of a toxicant, or mixture of toxicants, averaged over three minutes, under the conditions most likely to favour high concentrations at ground level beside a stationary source. The three minute mean concentration is often calculated as a fraction of the threshold limiting value, usually one-thirtieth or one fortieth. Where there is a risk of accumulation of a toxicant *e.g.*, a heavy metal, an absolute mass emission limit may be established. The absolute mass emission limit is the total amount of toxicant that may be emitted from a stationary source per hour, and which may not legally be exceeded.

The most commonly used assessment of toxicity is the measurement of short term lethality. For a given substance, this involves determining the median concentration which is lethal to a fixed proportion, usually 50 per cent of a test population of organisms after continuous exposure for a fixed time, usually 48 or 96 hrs. This is interpolated from the sigmoidal dose response curve (see fig. on p. 248) which is obtained by plotting percentage mortality at each dose against dose. The **lethal concentration (LC)** for 50 per cent of a population after continuous exposure for 48h is referred to as the 48h **LC_{50} or LD_{50} (LD=lethal dose).** This abbreviation may be altered appropriately to correspond to different exposure times and proportions of the population. Usually the concentration referred to is that in aqueous solution but it may also be a concentration in air.

If the test compound is insoluble for sparingly soluble in water, it must be uniformly dispersed for a repeatable LC_{50} to be obtained. If an emulsifier or other solubilizing agent is used for this purpose, it should be chosen carefully to ensure a minimal contribution to the toxicity of the system.

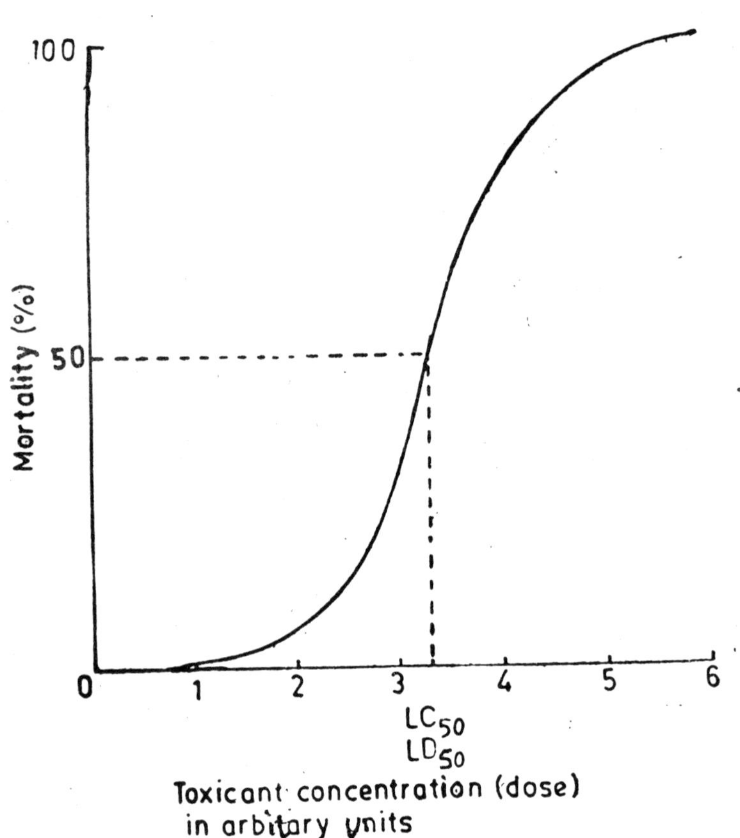

Graph showing the relationship between mortality and toxicant concentration, or dose, following continuous exposure of organisms for a short fixed time, usually 48h or 96h.

In medical usage the LC_{50} frequently refers to an injected concentration, and this must be borne in mind when attempting to extrapolate from such studies. Lethal concentrations have been expressed in a variety of units, most frequently in milligrams per litre or per kilogram body weight, but it is desirable that a more uniform approach should be adopted. Where possible, molarity (moles per litre) or molality (moles per kilogram) should be used as this will give a uniform chemical basis for comparing toxicity. The value of the LC_{50} as a measure of toxicity has been questioned and it is important to realize that it may be readily misinterpreted. Firstly, it is not constant. Considerable variation is observed, even in a given laboratory with the same population under the same conditions and the same investigator. Secondly, there is no simple correlation between organisms permitting direct extrapolation, even between organisms of the same Class such as rats and man. Thirdly, it gives no indication of the concentrations at which sublethal harm may occur. It is quite possible that a toxicant may be lethal only at very high doses but produce significant ill effects at relatively low concentrations. The most information the LC_{50} gives is an idea of the order of magnitude of the lethal dose under specified conditions. However, it is relatively quick and cheap to determine and provides a basis for initial assessment of the likely hazard from a toxicant and of the effects of various parameters on its toxicity.

Apart from the LC_{50} a number of other figures may be derived from studies of short term lethality. For example, the minimum lethal dose (MLD) is the dose which will kill at least one organism in the test population over the test period. This may be described as the LD_1. Another possibility, is that the concept of **potential toxicity (pT)** be introduced on an analogy with pH, i.e., pT= —log(T) where (T) is the molal concentration of the toxicant and the logarithm is to the base 10. A toxicant with a 24h LD_{50} of 0.001 mol kg^{-1} would have a 24h (T_{50}) of 10^{-3} and a 24h pT_{50} of 3. The calculated pT corresponds directly to the effect of toxicants on the test population. A small pT_{50} should indicate a relatively harmless substance while a large pT_{50} will indicate a highly toxic substance.

Measurement of sort term lethality is one aspect of assessment of acute toxicity. Acute toxicity has been defined as 'the total adverse effects produced by a toxicant when administered as a single dose. However, alternative definitions are used, *e.g.*, the total adverse effects produced by a toxicant when given in a single dose or multiple doses over periods of 24 hours or less'. Another parameter may be defined, *i.e.*, the EC_{50} or ED_{50} which is the concentration or dose which produces a specific ill effect in 50 per cent of the organisms tested. The EC_{50} is subject to the same provisions given for the LC_{50}.

AIR POLLUTION AND HUMAN HEALTH

The air pollution and human health studies involve a large number of variables. Consequently it becomes difficult to prove that air pollution has a clearly demonstrable effect on human health at 'tolerable' urban concentrations of air pollutants. It is well established that at the elevated levels air pollutants can cause severe changes in human physiology and a quick death. It is also clear that air pollution has adverse effects on those who already have respiratory problems. However, it is difficult to show that air pollution is the basic cause of the disease. Some adverse effects of ambient levels of air pollutants on human health are discussed here.

Numerous studies of the effects of gaseous pollutants and particulates on animals have been carried out. Most of such studies have been done at very high concentration and thus have no direct significance for humans since the concentrations of pollutants in human habitats are lower. Furthermore, there is a wide variety of response by different animal species to the same concentration of pollutant. Thus, any extrapolation of these effects is a complicated exercise. For example, the guinea pig, expected to be most susceptible to sulphuric acid aerosols, can withstand concentrations which would be intolerable to man.

Although one cannot yet implicate most ambient air pollution level as a source of illness, it is generally acceptable that increasingly smaller dose can be shown to have an effect.

Pollutants may attack the lungs through five main sites of action or mechanisms :

1. In the respiratory system, which reacts by the initiation of a constructive reaction of a bronchi reflex;
2. In the blood vessels of the bronchus and its branches, which try to reduce the absorption or harmful substances through the bronchial mucosa;
3. In the blood vessels of the lungs, where they react by decreasing absorption from the alveolar (pulmonary) capillaries;

4. In the heart and large blood vessels taking part in the transport of toxic substances;
5. By penetration into organs, tissues or cells and by affecting metabolic processes.

In order to understand this let us first give a brief description of the anatomy of the human respiratory system and the mechanisms by which it cleans itself. This is followed by the effects of air pollutants on human health as determined by epidemiological studies. Some terms used in this connection are given in Table 4.

Table 4. Some Important Terms

Alveoli (pl.) Alveolus (sing.)	Small, sac-like dilations at the innermost end of the airway, through whose walls gaseous exchange takes place.
Bronchiole	One of the finer subdivisions of the bronchial tree.
Bronchitis, chronic	A long standing inflammation of the bronchi, manifested by cough and the production of sputum.
Bronchi (pl.)	One of the larger air passages in the lung.
Bronchus (sing.)	
Cilia	Small, hairlike structures attached to the free surface of a cell, capable of rhythmic movement.
Emphysema	A condition in which there is overdistention of air spaces and resultant destruction of alveoli and loss of functioning lung tissue.
Epidemiology	A science dealing with the factors involved in the distribution of a disease in a population.
Morbidity	The occurrence of a disease state.
Mortality	The ratio of the total number of deaths to the total population, or the ratio of the number of deaths from a given disease to the total number of people having that disease.
Synergism	A situation in which the combined action of two or most agents acting to gether is greater than the sum of the action of these agents acting separately, also called potentiation.
In vivo	Within the living body.
In vitro	Within a test tube or artificial environment.

The Respiratory System

The human respiratory system functions to take in oxygen needed by the metabolic processes and to remove carbon dioxide produced in the body. The exchange of these two gases takes place in the lungs. Blood acts as the carrier to transport oxygen to the tissue and to bring carbon dioxide to the lungs. We breathe about 15-18 times per minute at rest and take in about 0.5 litres of air per breath. Most of this air is mixed in the lungs with air that is already there. However, a portion remains in the "**dead space**", which is the name given to the large air spaces in the respiratory system where no gaseous exchange takes place.

Most of the oxygen and carbon dioxide transported by the blood are not dissolved in the plasma but instead combine with the haemoglobin. **Oxyhaemoglobin** is the name given to haemoglobin to which oxygen is attached. Haemoglobin in which carbon monoxide has become attached is known as **carboxy haemoglobin**, COHb. Since the affinity of CO for haemoglobin is about 200 times more than that of O_2 it is easy to understand why low levels of CO can still produce significant amounts of carboxyhaemoglobin.

The respiratory system is shown in Fig. 1. The nasopharyngeal section, the tracheobronchial system, and the pulmonary structure are the three main regions. The nasal passages lead through the nasopharyngeal structure (throat) to the trachea and bronchi. The **bronchi** break up into smaller **bronchioles** which then terminate in the **alveolar sacs**. The exchange of oxygen and carbon dioxide between the air and the blood, occurs in the alveolar sacs. Presence of a large number of these tiny sacs provide a very large surface area for this exchange.

The bronchial and alveolar structure of this lungs, are shown in Fig.

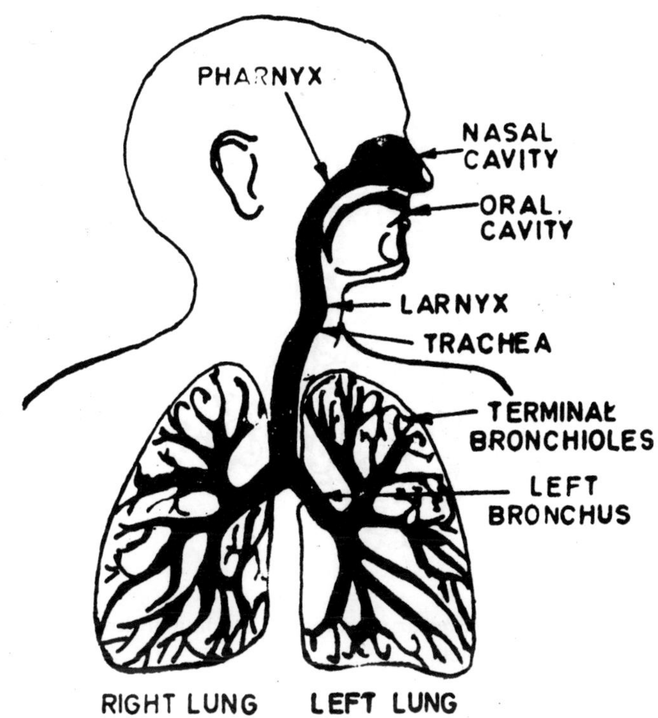

The major anatomical features of the human respiratory system. (The diagram shows the major divisions of the human respiratory tract into nasopharyngeal, tracheobronchial, and pulmonary compartments).

The nasopharyngeal and the tracheobronchial tubes have a layer of cells which have many small hairlike projections on them called the cilia. These beat in a rhythmic movement to produce a wave motion upward which carries mucus and entrapped particulates, germs, and dissolved gases away from the lungs. This material is then either swallowed or coughed up when the mucus is carried out of the trachea. The pulmonary surface is not ciliated and has a moist layer of cells covered by a material which prevent the collapse of the alveoli at the end of respiration. The inhaled pollutants, both gaseous and particulate thus have several opportunities to be removed before entering the alveoli. Gases can be absorbed along the pharynx or bronchioles and particulates can be deposited in the mucus layer and then removed through the action of the cilia.

The removal mechanisms or **clearance processes** are important in considering the effect of air pollution on the respiratory system. The material removed may be taken up by the blood stream, by the gastrointestinal tract via ciliary clearances of the bronchioles, or by the lymph system.

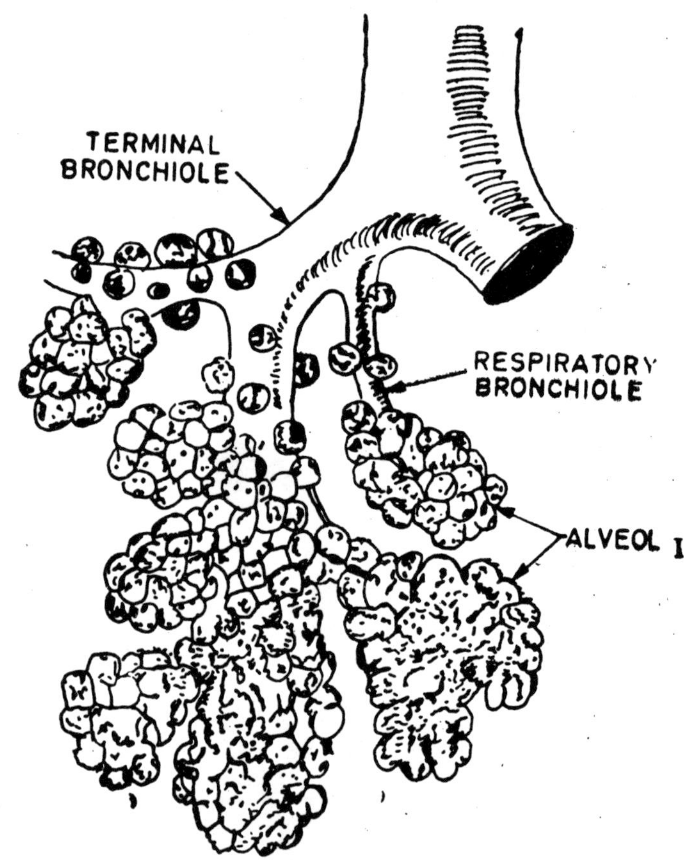

The terminal bronchial and alveolar structure of the human lung. (The diagram shows the pulmonary structure of the respiratory tract).

The cilia action is quite rapid. Experimental results indicate that clearance occurs in two phases, one of few hours and one of many hours. The respiratory system can clean itself to a large extent.

Obviously, it is the continuous exposure to high concentrations that overloads the clearance mechanisms and results in high equilibrium concentrations of deposited material in the respiratory system.

From this discussion it is clear that most of the effects of air pollutants on health are related to respiratory disease, *i.e.*, bronchitis, lung cancer, etc. Animal studies of individual pollutants have also shown that at high concentrations specific effects can occur; for example, the cilia may stop beating and thus cut off the main cleaning mechanisms of the system. In short, there are definite relationships between common air pollutants and physiological effects on the respiratory system.

Specific Diseases

Lung Cancer

According to Lave and Seskin (1970) there is definite evidence of an "urban" effect on lung cancer. Among smokers the likelihood of dying from lung cancer ranges from 26 per cent to 123 per cent greater if one lives in an urban area. For nonsmokers the range is wider and the urban effect appears to be even greater. The obvious question is : is this increased incidence of urban lung cancer caused by air pollution or is it at all related to air pollution?

Lave and Seskin (1970) argue that no reasonable man can object to the evidence of a substantial quantitative association between air pollution and several diseases, not just lung cancer. They further argue that the possible effects for city dwellers of greater smoking, occupational exposure, less exercise, additional tension and stress are not an explanation. It seems reasonable to conclude that while the above factors may have some effect, air pollution must also play an important role.

Bronchitis

The reviews by Goldsmith (1968), Lave and Seskin (1970), and the British Royal College of Physicians survey indicate that exposure to air pollution in the form of suspended particulates and sulphur oxides is a casual factor in promoting and aggravating bronchitis. Smoking habits and population density are also important variables found in epidemiological studies.

Other Cancers

There is evidence of an association between stomach cancer and air pollution. The cilia clearing action removes particulate from the bronchioles after which they are swallowed. Several chemicals which are **carcinogenic** (cancer-causing) have been found in the atmosphere. The chemical of most interest as a cancer-causing agent is **benzo (α) pyrene** which is one of the polynuclear aromatic hydrocarbons. This family of hydrocarbons may exist in the atmosphere as free molecules or can be adsorbed on carrier particles. The size of the carrier particulate affects its possibility of deposition in the lungs. In urban atmospheres there is an ample supply of particulate of about I micron size for deposition in the lung.

Some Specific Effects

Tolerance

Some pollutants seem to have the ability to initiate a tolerance response in laboratory animals. In particular, rats and mice exposed to ozone are found to develop a protection to second exposures. In fact, ozone exposure can provide protection to later exposure of NO_2, H_2O_2, H_2S, and other gases. Experiments with mice exposed to NO_2 also indicate reduced effects upon subsequent exposure.

Cilia Effects

Since the cilia are the primary mechanism by which the bronchiole tubes are cleared of particulate any effects of gaseous pollutants in reducing their effectiveness is of interest. At SO_2 concentrations of 12 ppm (far higher than ambient level) ciliary cessation has been shown to occur in about 4 to 6 minutes in rats. After exposure the cilia regained their mobility. Cigarette smoking can also stop the cilia action.

Increased Infection Susceptibility

Air pollutants are known to increase the susceptibility to bacterial infection. For example, NO_2 exposure at 0.5 ppm for several months caused increases in mortality of mice exposed to *Klebsiella pneumoniae* (Ehrlich and Henry, 1968). Other studies, at higher NO_2 levels, on squirrels, monkeys and hamsters showed similar effects. Ozone has also shown an ability to increase susceptibility to bacterial infection. Mice exposed to streptococcal aerosol after a 3-hour exposure to ozone from 0.08 ppm to 0.52 ppm showed an increase in mortality. Mice and hamsters exposed to higher ozone levels and then exposed to *K. penumoniae* also exhibited increased mortality. Thus it is certainly possible that air pollutants can increase human susceptibility to respiratory disease and mortality.

Table 5. Properties and Effects of Pollutants

Ozone	Colourless	
	Odour threshold	0.02-0.05 ppm
	Irritation of nose and throat	0.05 ppm
	Headache (in 30 minutes)	1 ppm (1962 $\mu g\ m^{-2}$)
	Industrial limit (WHO 8-h day)	100-200 $\mu g\ m^{-2}$
NO_2	Brown-red colour	
	Odour threshold	0.12-0.22 ppm
	Increased respiratory illness over 6-months test period	0.06-0.109 ppm
	U.S. industrial limit (8-h day)	0.5 ppm
SO_2	Colourless	
	Taste threshold	0.3 ppm
	Odour threshold	0.5 ppm
	Increased mortality	0.25 ppm (24 h measurement) with 750 $\mu g\ m^{-2}$ smoke
	Possible increased mortality	0.19 ppm (24 h measurement) with "low" particulate levels for "few" days.
	Increased respiratory illness in children	0.046 ppm with smoke at 100 $\mu g\ m^{-3}$ for long term exposure
	Throat irritaion	8-12 ppm
	Eye irritation	10 ppm
	Immediate coughing	20 ppm
	Industrial limit (WHO, 8-h day)	10-13 mg m^{-2}
CO	Colourless, tasteless odourless (faint odour at very high concentrations)	
	Central nervous system effect	15 ppm for 10 h.
	Industrial limit, (WHO 8-h day)	50 ppm

Source : Adapted from Stern, A.C. Air Pollution—Vol. I (2nd edn.) Academic Press 1968.

AIR POLLUTION AND PLANTS

Almost all air pollutants causing plant injury are gases, but some particulate matter or dusts may also affect vegetation, some gas contaminants such as ethylene, ammonia, chlorine, and sometimes mercury vapours, exert their injurious effects over limited areas. Most frequently they affect plants or plant products stored in poorly ventilated warehouses in which the pollutants are produced by the plants themselves (ethylene), or form leaks in the cooling system (ammonia).

More serious and widespread damage is caused to plants in the field by chemicals such as ozone, sulphur dioxide, hydrogen fluoride, nitrogen dioxide, peroxyacetyl nitrate (PAN), and particulates. In many localities the air pollutants spread into the area surrounding the source (s).

of pollution, become trapped and cause serious plant damage. More frequently, most air pollutants are transported downwind from the urban or industrial centers in which they are produced and may be carried by the wind to areas that are sometimes thousand of kilometres from the source.

How Air Pollutants Affect Plants

For the past 20 years, scientists have been growing plants under controlled conditions and exposing them to various concentrations of common air pollutants. Characteristic injuries to certain plants can be interpreted to show not only the presence, but also the relative concentration, of the air pollutant. Tobacco is an effective ozone indicator. Pinto bean plants are used to detect peroxyacetyl nitrate (PAN) and gladiolus to determine fluoride accumulations. Excellent detectors of SO_2 are dahlias, petunias, alfalfa, and cotton. High concentration of or long exposure to these chemicals cause visible and sometimes characteristic symptoms on the affected plants. The leaf is the primary indicator of injurious effects. More important economically, however, is the fact that even when plants are exposed to dosages less than those that cause acute damage, they can still suppress growth, cause dwarfing, early maturation and a decreased crop yield. Moreover, prolonged exposure to air pollutants seems to weaken plants and to predispose them to attack by insects and by some pathogens.

Damage to Leaves

Damage to leaves takes several forms. Plant damaging substances are termed **Phytotoxicants**. Some terms used to explain symptoms are given in Table 6.

Table 6. Some Important Terms

Abscission	Dropping of leaves
Chlorosis	Yellowing of normally green tissue due to chlorophyll destruction or failure of chlorophyll formation.
Epinasty	Downward curvature of the leaf due to higher rate of growth on upper surface
Mottle	An irregular pattern of indistinct light and dark areas
Necrotic	Dead and discoloured
Scalding	Large, irregular or diamond shaped blotches with dark brown or purple edges
Scorching	Burning of leaf margins
Senescence	Ageing

Three kinds of injury are commonly considered important.

1. Acute injury results from short-time exposure to relatively high concentrations, such as might occur under fumigation conditions. The effects are noted within a few hours to a few days and could, for example, be visible markings on the leaves due to a collapse and death of cells. This leads to necrotic patterns, that is, areas of dead tissue.

2. Chronic injury results from long-term low level exposure and usually causes chlorosis or leaf abscission.

3. The third form of injury is an effect on growth without visible markings. Usually a suppression of growth or yield occurs. This is more difficult to document since careful experiments must be done with and without the pollutants under carefully controlled conditions, such as in a greenhouse.

The full effects of chronic injury or growth suppression have been appreciated only recently as it is difficult to separate out the adverse affects of air pollution from other phenomena. Plant diseases may show symptoms which are very similar to those caused by air pollution. Poor plant care or high temperatures may also cause an appearance which is similar to that of a plant damaged by air pollution. Thus in diagnosis the effects of air pollution one must consider such variables as relevant plant disease, nutrition history, insect damage, weather damage, and the type of pollutants in the area.

The effect of concentration versus time is also an important variable in assessing damage. A high dose for a short time may cause an acute injury whereas the same total dose over a longer time may cause no visible effects at all. The concentration-time "law" may be expressed as

$$t(C—C_0)=K$$

where t = time in hours to produce a certain effect on a certain species

C = concentration of a specific gas in parts per million

C_0 = the threshold concentration of the gas, in parts per million, to cause injury.

K = an experimentally determined constant.

If we rewrite this equation as

$$C = \frac{K}{t} + C_0$$

then plotting C versus 1/t gives a straight line whose intercept C_0 is the threshold for injury. Specific pollutants and their effects are listed in Table 7 on p. 257.

The concentration at which each pollutant causes injury to a plant varies with the plant and even with the age of the plant or the plant part. As the duration that the plant is exposed to the pollutant is increased, damage can be caused by increasingly smaller concentrations of the pollutant until a minimum dose—injury threshold is reached. Plant injury by air pollutants generally increases with increased light intensity, increased soil moisture, and air relative humidity, increased temperature, and with the presence of other air pollutants.

Chlorine damage is often found near swimming pools and sewage disposal plants where chlorine is used as a disinfectant. Its damage to leaves in similar to that caused by SO_2. It is very difficult for the untrained eye to determine the cause of plant injury, since not only do the various air pollutants cause similar-appearing damage, but natural agents such as drought, frost, mineral deficiencies, and disease produce symptoms very similar to those caused by air pollutants.

Since different plant species differ greatly in their sensitivity to air pollutants, programmes for the breeding and dissemination of pollution resistant cultivates of crop plants and species of avenue trees are now under way. Unfortunately, these methods of control would result in loss of the use of several of our important and favourite plants.

(See Table on page 257.)

Table 7. Air Pollution Injury to Plants

Pollutant	*Source*	*Susceptible Plants*	*Symptoms*	*Remarks*
1	2	3	4	5
Ozone (O_3)	Automobile exhausts, Other internal combustion engines, (Released NO_2 combines with O_2 in sunlight —> O_3) From stratosphere. From lightning, from forests.	Expanding leaves of all plants, especially tobacco, bean, cereals, alfalfa, petunia, pine, citrus, corn.	Stippling mottling, and chlorosis of leaves, primarily on upper leaf surface. Spots are small to large bleached white to tan brown, or black. Premature defoliation and stunting occurs in plants such as citrus, grapes, and vines.	Enters through stomata. It is the most destructive air pollutant to plants. A major component of smog.
Peroxyacetyl nitrates (PAN)	Automobile exhausts or other internal combustion engines. (Gasoline vapours and incompletely burned gasoline $\pm$ O_3 or NO_2 →PAN).	Many kinds of plants, including spinach, petunia, tomato, lettuce, dahlia.	Causes "silver leaf" on plants, i.e., belached white to bronze spots on lower surface of leaves that may later spread throughout leaf thickness and resemble ezone injury.	Particularly severe near metropolitan areas with smog and inversion layers.
Sulphur dioxide (SO_2)	Stacks of factories, Automobile exhausts, and other internal combustion engines.	Many kinds of plants, including alfalfa, violet, conifers, pea, cotton, bean. Toxic at 0.3 to 0.5 ppm.	Low concentration cause general chlorosis, Higher concentrations cause bleaching of interveinal tissues of leaves.	Also combines with moisture and forms toxic acid droplets (acid rain)
Nitrogen dioxide (NO_2)	From oxygen and nitrogen in the air by hot combustion sources, *e.g.*, funaces, internal combustion engines.	Many kinds of plants including beans, tomatoes, Toxic at 2 to 3 ppm.	Causes bleaching and bronzing of plants similar to that caused by SO_2. At low concentration it also suppresses growth of plants	
Hydrogen fluoride (HF)	Stacks of factories processing ore or oil,	Many kinds of plants, including corn, peach, tulip. Actively growing, especially wet leaves, are most sensitive toxic at 0.1 to 0.2 ppb.	Leaf margins of dicots and leaf tips of monocots turn tan to dark brown, die and may fall from the leaf. Some parts tolerate HF up to 200 pm.	HF may evaporate or be washed out of plant and plant recovers slowly.

(Contd.)

Table 7 (Contd.)

1	2	3	4	5
Chlorine (Cl_2) and Hydrogen chloride (HCI)	Refineries, glass factories, incineration of plastics. Near swimming pools and sewage disposal plants.	Many kinds of plants, usually near the source. Toxic at 0.1 ppm.	Leaves show bleached necrotic are as between veins. Leaf margins often appear scorched. Leaves may drop prematurely. Damage resembles that caused by SO_2	
Ethylene (CH_2CH_2)	Automobile exhausts, Burning of gas, fuel oil, and coal. From ripening fruits in storage.	Many kinds of plants, Toxic at 0.05 ppm,.	Plants remain stunted, their leaves develop abnormally and senesce prematurely. Plants produce fewer blossoms and fruit. Fruit *e.g.*, apples, develop depressed, necrotic, dark areas (scald).	Ethylene is a plant hormone with numerous functions.
Particulate matter (dusts)	Dust from roads, cement factories. Burning of coal, etc.	All plants,	Forms dust or crusty layers on plant surfaces. Plants become chlorotic, grow poorly, and may die. Some dusts are toxic and burn leaf tissues directly or after dissolving in dew or rainwater.	
Carbon monoxide (CO)	Automobile exhausts Other internal combustion enginers.	All plants Visible effect at 100-10,000 ppm.	Abscission and Premature ageing of leaves. Initiation of roots on stems.	CO inhibit cellular respiration in plants by reacting with cytochrome-oxidase, which is vital to the utilization of sugars metabolic fuel.

Part VI

Wastes

- Solid Wastes
- Waste Water Treatment
- Sewage Treatment
- Human Waste
- Nuclear Wastes
- Nuclear Waste Disposal

SOLID WASTES

Introduction

Apart from Homo Sapiens, not one other living creature on Earth creates unmanageable waste. Living in a web of interdependency, from which we have foolishly cut loose, other life forms utilize every conceivable resource to advantage. Plants absorb energy synthesised from the sun. Graziers and browsers and a myriad other suckers and chewers consume the bounty offered by plants. Carnivores eat the plant eaters. What is left over is consumed by a pecking order of organisms ranging from termites, hyaenas, maggots, crowes, kites and vultures, to bacteria and fungi which help decay and break down nutrients to be returned to the soil.

However, litter along the roadsides, trash on the beaches and in the gardens, heaps of refuse in the streets, the pungent piles of rotting garbage are unpleasant reminders of untidy human habits. Thousands of backyards, vacant plots of land contain the remains of discarded food-waste, machine parts, parts of vehicles and so on. Open dumps scar the landscape. In some countries solid wastes are dumped into the ocean. The growing mass of solid waste produced annually the world over includes millions of tonnes of paper and paper products, plastics, billions of cans, bottles, tyres, junked machines, automobiles and millions of other articles and appliances of various sizes and kinds. Probably, the United States produce the largest solid waste. About 15 per cent of solid waste from every household in an affluent society consists of food waste.

Solid waste are a scourge in all Indian metropolitan cities. More than 4,000 tonnes of solid wastes are generated every day in Delhi alone. Unfortunately, in Delhi, the problem is growing from bad to worse. By MCD's own account, about 30 per cent of the garbage - about 1,500 tonnes - is left uncollected every day. Actual estimates of these 'left-overs' are much more. Most of these festering heaps lie in the heart of Delhi's congested areas.

The volume of garbage generated increases with increase in population as well as with increase in the standard of living. How much does this uncollected garbage add up to in a year? And will it lead to another potential Surat lurking around the corner? Frequent outbreaks of water-borne diseases in areas like Hauz Rani village in Delhi for instance, are stark indicators of coming events. Gastroenteritis, cholera, plague, dysentry, jaundice, food poisoning, malaria, fever etcetera are all inevitable consequences of such septic wastes.

The population of Delhi is fast increasing and by 2000 AD, it is likely to reach 12 million. Consequently solid wastes generation will rise steeply further accentuating an already acute problem. Delhi today is faced with a grave and grim challenge. Unless this challenge is met with decisively, much of Delhi will soon be buried under mountains of solid wastes that will continue to be generated.

However, Delhi is not alone is facing the challenge of Solid Waste Management. Other cities like Tokyo and New York are coping with this problem on a much larger scale. There is no shortage of required technology or expertise in our country. What is essential however, is an honest overhaul of the system.

SOURCES OF WASTE

Every human activity generates some waste. We create waste when we extract and process rock, minerals and fuels and when we grow and process food and fiber. When products wear out, become outmoded, or otherwise outlive their utility, we usually throw them out. Although a variety of materials constitute waste, for convenience of study, we classify waste by origin as municipal, mining, industrial and agricultural.

Components of municipal solid wastes are : bricks, tree branches, aluminium cans, newspaper, paints, sewage sludge, cinders, food, plastics and so on. The proportion of each constituent depends on the source of waste generation whether residential, or commercial. Wastes from residential areas are least hazardous. Commercial area wastes largely consist of paper, packaging, wood, and plastic and other recyclable matter. It is also high in calorific value and energy recovery is desirable. Table 1 shows the typical composition of refuse in a municipal landfill.

Table - 1

Composition of Refuse in a Typical Municipal Landfill.

Waste type	*Percent by Vol.*
Paper	50%
Organic	13
Plastic	10
Metal	6
Glass	1
Miscellaneous	20

The amount of municipal waste pales in comparison to the massive quantities of waste produced by mining, industrial and agricultural activities. Mining and mineral wastes include mill tailings, slag, and various mine wastes. Such wastes are generated at the rate approx. eleven times greater than the rate at which municipal solid wastes are produced. For example, for every metric tonne of copper produced by open pit mining more than 500 metric tonne of solid waste are generated. Air pollutants emitted in the process are not included in this total. However, mining and mineral processing usually take place in remote areas, so that these wastes are not visible as municipal wastes.

The industrial sector generates a wide variety of waste depending on the kinds of raw material used, energy source, and products. For example, coal burning electric power plants yield tremendous quantities of waste. Typically, about 10% of the mass of coal is noncombustible and accumulates as a residue known as **fly ash**. A 1000 megawatt plant must dispose of 250,000 metric tonnes of fly ash each year. If fly ash is dumped in a landfill, then during a plants life span (40 years), disposal of fly ash requires about 2.5 square kilometres of land.

Agricultural wastes are generated at a rate approx, fifteen times that of municipal wastes. About 75% of the agricultural wastes produced each year consist of animal manure, crop residue and various by-products of food production. However, most animal waste and crop residue goes into the soil enhancing its fertility and moisture retention capacity.

Hazardous wastes are produced in the manufacture of inorganic and organic chemicals, in the smelting and refining of ores, in electroplating, in petroleum refining, tannery, slaughter houses and hospitals.

PROPERTIES OF WASTE

Wastes persist in the environment if natural physical, chemical, and biological processes do not reduce them to harmless by-products. Heavy metals such as lead, mercury, and cadmium persist because these occur as elements and cannot be further broken down chemically. Our greatest concern is with persistent hazardous wastes that enters the food webs. These materials bioaccumulate from one trophic level to the other and thus are biologically magnified. Common practice describes the potential environmental persistence of a substance in terms of its biodegradability. **Biodegradable Waste** - is subjected to natural biological decay through the action of anaerobic or aerobic decomposers. **Non biodegradable waste** - resists such decay and thus persist in the environment unless degraded by physical or chemical processes. Most biodegradable waste is organic and includes paper, lawn clippings, and discarded wood. Agricultural wastes are mostly biodegradable. Urban refuse and industrial waste are complex mixtures of biodegradable and nonbiodegradable components.

In recent years, scientists have developed two new plastics that partially break down in the environment. In time, both biodegradable and photodegradable plastics fragment into small pieces. Biodegradable plastics contain embedded starch, which is decomposed by soil microorganisms. Photodegradable plastics lose their strength upon exposure to UV portion of solar radiation; the UV breaks the polymer chains composing such plastics.

While degradable plastics would appear to be an important innovation in view of today's widespread use of plastics, there are serious drawbacks to the two types developed so far:

1. While both biodegradable and photodegradable plastics fragment, the individual pieces do not further degrade.
2. The sunlight, O_2, and moisture necessary to break down such plastics are absent in landfills.
3. Both plastics have very limited use because they lack some of the important properties of other plastics, *e.g.* they lose strength upon contact with oil and solvents.

At this point, prospects for a completely degradable plastic with all of plastic's desirable properties are not promising. Hence, many solid waste experts feel that management efforts should focus on recycling plastics.

WASTE MANAGEMENT

Collection, transportation and final disposal of large volumes of wastes requires a high level of management and technical expertise. There is also a rising public awareness about the need for an environmentally acceptable management of solid waste.

The final disposal of solid wastes can be carried out by several methods—incineration, composting, landfilling and recycling. Certain hazardous wastes such as hospital wastes should be positively incinerated. Composting is suitable for wastes which have a high vegetable content. Surprisingly, composting or incineration are not practised by the MCD. Several incineration plants costing huge sums, are out of order. The composting plant at Okhla is lying idle. The only method of disposing of the solid wastes employed in Delhi is sanitary landfilling.

Recycling should be an integral part of waste management. It also makes economic sense. Producing aluminium and steel from recycled garbage requires 95 per cent less energy as compared to refining them from natural ores. Glass is 100 per cent recyclable and making jars or bottles from recycled glass also requires much less energy.

Thousands of tonnes of glass, steel, aluminium and other metals are thrown out of commercial and residential establishments every year. Similarly, paper and cardboards are dumped in landfills or burnt without energy recovery. Recycled paper is produced using much less energy, water and results in less pollution. Plastic production consumes a lot of oil. Recycled plastic, therefore, results in energy saving.

Littering and open Dumps

People often usually discard solid waste by throwing bottles, cans, fastfood containers, and other items on streets or out of car windows, creating visual pollution. People in rural areas in U.S. and even in urban areas in India dump their trash along roadsides rather than haul it to dumpster locations or country landfills. These unsightly open dumps support large populations of disease carrying rodents and insects and often contaminate ground water and surface water through leaching and run-off. The dumps also create air pollution when they catch fire from spontaneous combustion or when they are set on fire to reduce the volume of wastes. So open dumps are a nuisance to public health and should not be permitted.

Ocean Dumping

Until recently, ocean dumping was practiced by many coastal cities as a traditional method of solid waste disposal. Barges carrying the refuse used to travel some distance from the harbour and dump their loads into a natural trench or canyon on the ocean floor. In this way most of the trash is removed from sight, though not from the biosphere.

Ocean dumping upsets the ecological balance of the region of the sea. Many organisms are killed outright. Although certain plankton and fish survive in these areas, they are affected by the unusual environment. Biologists have found bandages and cigarette butts in fish stomachs. Concern about the dumping of wastes into the ocean focuses not only on toxicity to organisms, but also on the possibility of bioaccumulation in food webs. Also, because processes in the oceans are poorly understood, ocean dumping that seems safe enough at present may cause unanticipated problems in the future.

Sanitary Landfills

This basically involves placing the waste in compacted layers in a lined pit or a mound with appropriate leachate and landfill gas (mainly methane) control. Landfilling is not strictly a disposal method, but really one of containment or indefinite waste storage. Many landfills are used for disposal of both municipal and hazardous waste. The later are co-disposed with the municipal wastes utilizing the absorptive properties of the bulky municipal waste and the microbial degradation reactions which occur wherever putrescible material is buried. The advantages of landfilling are that it is relatively inexpensive; methane production can be exploited from municipal wastes and the wastes can be recycled or treated at a later date. The disadvantages are that there is a danger of leakage and pollution of ground water, a danger of explosion from methane and that there are fewer and fewer suitable sites available.

A sanitary landfill is not a dump but a well developed engineering technique designed to dispose of solid wastes on land without creating any nuisance or hazard to public health or any other environmental or aesthetic damage. Separation, sorting, recycling and energy recovery should be an integral part of landfill disposal. Pests, flies, insects, pathogens, as well as animals (cows, pigs and dogs) thrive on these landfills, spreading diseases and creating environmental problems. Periodic soil cover followed by compaction and levelling solves these problems effectively. In addition, no open burning is allowed at sanitary landfills. Furthermore, landfills can be designed to restore landscapes scarred by mining.

The foremost concern in designing a sanitary landfill is to safeguard groundwater quality. Where there is danger of groundwater contamination, the bottom of the landfill site is lined with a thick layer (2 metres or so) of impermeable clay (and in some cases, plastic sheets). In some areas, surface water that seeps through wastes (leachate) must be pumped out and then either piped or hauled to a sewage treatment plant. As an added precaution, special wells monitor ground-water around the perimeter of the landfill to determine whether it is being contaminated by unexpected leaks.

Municipalities practice three basic types of sanitary landfilling trench, area, and mounded depending upon the terrain and depth of the water table.

A trench landfill is most appropriate where the water table is deep and the subsurface soil and sediment are thick are relatively impermeable. Initially, a broad trench is excavated. Trucks dump their loads of refuse into the trench, and then special heavy equipment spreads and compacts the trash. Each day, the compacted waste is covered with a layer of clean dirt (obtained from the original trench excavation). This process is repeated until the trench is full. Meanwhile, another trench is dug and readied for new shipments or garbage.

An area landfill is developed in natural valleys or canyons as well as abandoned pits and quarries. Refuse is dumped on the bottom liner, compacted, and covered with soil each day until the site is filled. Area landfills are often more expensive to operate because of the need to haul in additional fill, but the cost of transporting waste is the key economic consideration. Hence, area landfilling may be chosen over trench landfilling if a suitable site is closer to the source of refuse.

In regions where the water table is close to the surface, **a mounded landfill** is often necessary to prevent groundwater contamination. Alternating layers of compacted refuse and dirt are mounded over a clay base on the surface to form a hill. A completed mounded landfill is contoured and vegetated to control erosion.

Besides the potential for groundwater contamination, another drawback of sanitary landfills is the possible migration of hazardous gases formed during the anaerobic decomposition of waste. Anaerobic decomposition produces methane (CH_4) and hydrogen sulphide (H_2S). These gases can seep through the sides of a landfill and follow layers of permeable soil and sediment into the basements of neighboring buildings. Since methane is explosive in confined spaces, such migration must be prevented. In addition, hydrogen sulphide can act as an asphyxiating agent. Therefore, if gas migration from a landfill is a potential problem, vent pipes and trenches must be installed to collect the gases so that they can be flared (burned), allowed to escape slowly into the open air, or used as an energy source.

Methane is the principal component of natural gas. The first methane-recovery landfill went into operation in 1975 at Rolling Hills Estates in California. Since then, many other large landfills have begun to recover methane by using a relatively simple technology. Perforated collecting pipes are inserted into vertical holes drilled into the landfill. A slight pressure reduction in the pipes draws methane from the landfill. Water vapour and other gases (such as hydrogen sulphide) are removed before the methane is piped to a nearby facility that uses it as a fuel. In 1992, methane was recovered at 114 landfills in the United States, supplying about 344 megawatts of power.

One popular myth about sanitary landfills is that the organic portion of the waste decomposes rapidly. Compacted garbage, however, is effectively sealed from oxygen and moisture, which, are necessary for rapid decay. In act, William Rathje, a University of Arizona archaeologist, and his students probed the contents of fourteen sanitary landfills in North America and found that only 20-50% of food and yard waste biodegrade within the first 15 years of burial. They also found that newspapers were still readable after more than a decade of burial.

Completed landfills cannot be used as building sites, but they are suitable for recreational areas, botanical gardens, or wildlife preserves. Even though landfills offer many advantages over open dumps for solid-waste disposal, they often encounter strong public opposition. Most people equate a sanitary landfill with a dump, and are also concerned about increased traffic and the noise or large garbage trucks. They simply refuse to accept a sanitary landfill in or near their neighborhood, even though it may eventually become a park that would benefit the entire community.

The Department of Civil Engineering at Jamia Millia Islamia, New Delhi recently carried out a survey of major sanitary landfill sites in Delhi. The survey was carried out recently as part of the M. Tech. Environmental Engineering programme financed by the UGC which noted that sorting, separation and recycling are not carried out at any of the sites. However, unauthorised ragpickers (estimated at more than 50,000) carry out this work, blissfully ignorant of the hazards to their health. They are both victims, as well as the vectors of diseases.

Levelling and compaction is fairly good at Mandawli and Ghazipur sites, but at all other sites this operation is poorly managed. This will only result in the settlement of structures when the sites are abandoned and put to other uses later on. Compaction should be carried out to achieve a level of shear strength and bulk density consistent with the intended land use in future.

Ground water protection measures are not undertaken at any of the sites. No distinction is made between clay and silt liners, even though silt may well be a million times more permeable than clay. Regular testing of underground water at landfill sties is not carried out. As a result, Delhities are bound to suffer with frequent outbreaks of water-borne diseases.

The solid waste collected from various parts of the city is dumped in the landfill sites at Ghazipur, Bhalaswa, Tughlakabad, Hasthal and Mandawli. Of these, the Ghazipur landfill site is the largest-spread over an area of 99 acres. It has been in use since 1984 and can be used for about a decade more. However, the Hasthal and Mandawli sites are almost full. Another at Tughlakabad will be filled in another year.

The Ghazipur land-fill site serves the entire East Delhi area. Waste from the Idgah slaughter house is also dumped here. Wastes from North Delhi central areas and parts of Rohini are dumped at the Bhalaswa site. Garbage from the rest of Rohini area and Narela goes to Mandawli, while Hasthal gets waste from Najafgarh and West Delhi areas. The Tughlakabad site covers South Delhi and New Delhi areas.

Table 2 : Existing Sanitary Landfill sites in Delhi

Landfill site	*Capacity (Tonnes/day)*	*Area (acres)*	*In Use since*	*Future Life (years)*
Ghazipur	1,200	99	1984	10
Bhalaswa	1,400	-	1992	-
Tughlakabad	1,200	7	1982	1
Hasthal	200	84	1986	Almost over
Mandawli	-	15	1991	Almost over

Table 3 : New Sanitary Landfill Sites

Landfill site	*Area (acres)*	*Future Life (years)*
Near Jaitpur village	42	13
Near Mandi village	282	72
Bhatti mines	-	100

INCINERATION

Many municipal areas burn their garbage than dump it. The process, as applied to waste disposal, is more complex than simply setting fire to a mass of garbage in the open dump. A municipal incinerator burns combustible solid waste and melts certain non-combustible materials. In a modern incinerator unit the trash is burnt in a carefully designed furnace.

Incinerator is better than landfilling from a health perspective, since the high temperature destroy pathogens and their vectors. The advantages of incineration are that it can dispose of 99.99% of organic wastes (including chlorinated organic xenobiotics) if properly carried out (high temperature 1200°C; adequate oxygen etc.); it reduces the volume of waste by about 80 to 90%. Hence disposing of the residue (ash) after incineration requires considerably less landfill space. Incineration has additional advantages over landfills of any kind because it does not directly endanger groundwater quality. Also, energy can be recovered from the process and utilised for electricity generation and combined heat and power.

However, there are drawbacks. Installation, maintenance and operational expenses of incinerators are high, so the average cost of incineration solid waste is generally greater than for a sanitary landfill. However, in regions where appropriate nearby sites for sanitary landfills are not available and where landprices are high, incinerators are often less expensive than landfills. This is usually the case in and around large urban areas, which produce large volume of commercial, household and industrial wastes. The solution in such situations may be a combination of incineration and landfills. Other disadvantages are that there is a danger of highly toxic pollutants being synthesized and emitted into the atmosphere if conditions are not controlled; and that there is an ash produced which needs careful disposal in sanitary landfills. For every 10 tonnes of waste fed into an incinerator, one tonne of ash is produced. This resulting ash is usually contaminated with toxic metals and dioxins.

Considerable progress was made in the 1970s and 1980s in incineration technology and in air-pollution control devices for incinerators. More than 50% of the municipal waste in Japan, Sweden and Switzerland is incinerated, and more than 20% incinerated in Belgium, France, Germany, Italy, Luxembourg and the Netherlands.

Composting

Composting is an age old process that simulates nature's method of recycling. It involves shredding and separating the putrescible fraction of municipal waste, often mixing it with other organic materials (including sewage sludge) and allowing decomposition reactions to take place by microbes, worms and other living organisms. This involves regular turning of the compost in order to promote aerobic decomposition processes and allowing the compost temperature to rise sufficiently to kill off pathogenic organisms and weed seeds and other undesirable constituents.

The finished compost is intended for use as a valuable fertilizer and soil conditioner and also as a growing medium for plants, especially as a substitute for peat which is a scarce resource from an ecosystem at risk of destruction with the accompanying problems of extinction of species.

Scientists at University of Florida's entomology and nematology department have found recently that by application of yard waste compost, not only soil's organic matter doubled (which lead to much higher yield), but it also helped keep chemicals out of ground water. The compost also helps control root knot, one of the most damaging pests to crops.

Almost any plant or animal matter, such as kitchen waste, paper, leaves and grass clippings, straw, sawdust will form a satisfactory base for a composting operation which can be decomposed in backyard compost heaps and used in gardens and flower beds to increase the humus content of soils. With food processing and other industries, large supply of organic waste can be collected and degraded in large composting plants. Composting, however, has some drawbacks, because most domestic sewage contains small amounts of toxic metals, and if sludge is used as fertilizer, these metals enter into our food chain. As a result, some food processors will not accept produce fertilized with sludge.

When seweage sludge and animal manure are composted, large quantities of methane gas are released. Methane is an excellent fuel. Some farmers have collected methane from cow manure, used the fuel to drive their tractors, and then recycled the compost as fertilizer. Although many such small-scale methane generators have been used throughout the world, few large-scale units are in operation.

Recycling

Recycling is the latest mantras in environmental circles. It has become a necessity, as many countries are unable to locate sites suitable for garbage disposal, since land is scarce. Further, dumping wastes in landfills causes pollution as when certain wastes decay, the emissions thus produced are harmful to living beings. The dwindling supply of raw materials also implies that we must extract the maximum from the existing resources. The financial element also attracts countries like ours to take to recycling, as we can save on valuable foreign exchange required to import commodities such as oil or raw materials.

Recycling involves separating materials, such as rubber, glass, paper, oils, and scrap metal, from refuse and reprocessing them for reuse. In the United States recycling (or **waste reclamation**) is a multibillion-dollar-a-year business, based primarily on the recycling of scrap metal from the nine million motor vehicles junked each year. Also, a significant fraction of the aluminium, lead, copper, and zinc used by industry is derived from recycling those metals. For example, industries that machine metal parts keep metal turnings and scraps separate from other plant refuse and recycle them. Automobile repair shops send wornout batteries to recyclers, who recover the lead they contain.

The primary step in recycling is garbage collection. The recyclable item has greater value if not mixed with the rest of the garbage. A tremendous amount of emphasis is therefore laid on "**Source separation**". That is, the trash is sorted into glass, plastic, paper, and the like by the consumers themselves. The recyclable articles can then be sold at "buy back" centres or at times it is collected from the individuals and then sold to recyclers. Source separation is advantageous as it involves the direct participation of society in quicker "**resource recovery**" which aids both the country's economy and environment.

In India, source separation is disliked by many. So persons are employed to separate garbage. People should be educated on the salient features of source separation. Buy back and collection of waste programmes which are already prevalent can be developed into a more organized industry.

In the United States paper makes up 18% of the total volume of compacted trash, newspapers are the largest single component of a landfill. Hence, public pressure is on for paper recycling. Nationwide, thousands of companies and government facilities sell waste paper (about 50% of which is high-grade) to paper recyclers. One-third of the nation's six hundred paper mills process waste paper exclusively, and about three hundred others utilize 10-30% waste paper that comes directly from collectors and some fifteen hundred waste-paper processing plants in the United States. In India, only 35% of waste paper is being recycled at present, with ample scope for greater use of old paper without resorting to virgin paper use.

Indian Rare Earths Factory in Alwaye, Kerala has taken up a recycling project. Wastes from the factory comprising of garnet sand, by a product of processing Kerala's beach sand to extract thorium, are being converted into synthetic granite at the Central Glass and Ceramic Research Institute in Calcutta. The technology involves mixing garnet sand with special additives, compressing it, sintering it under 1100°C and polishing the product to get a replica of natural granite.

Building industry can gainfully utilize **agro-waste** such as bagasse, banana leaves and stalks, saw mill waste, sisal fibres, rice husk, jute stalk etc. They can gainfully utilise these wastes for a variety of applications, and in many cases substitute the use of timber through products made from agro wastes processed along with suitable binders under pressure. Such treatment is used for making several kinds of insulation boards, panels, and roofing sheets. The products made from agro waste can be strong, lightweight and can also find easy aesthetic acceptance for use as exposed surfaces in residential and office interior construction. Thus large quantities of extremely precious timber can be conserved through increased use of substitutes made from such agro-waste.

Various paper and plastic wastes can be used as raw material in manufacture of fence posts, park benches, road furniture, pitch-fibre, pipes, electrical and drainage systems and asphaltic corrugated roofing sheets.

BP Chemicals and Research Laboratories at Sunbury on Thames, Surrey has carried out experiments on the plastic items that are thrown away. Researchers have succeeded in converting plastic into oil which can be used to synthesize virgin plastic. BP Chemicals hope to make it commercially viable towards the end of 1997.

Waste Reduction

A complement to landfilling, incineration, and resource recovery is waste reduction, that is, strategies that decrease the waste potential of consumer goods before they enter the waste stream. Waste-reduction strategies include cutting the amount of raw materials and packaging used for consumer goods, and designing products for durability.

Another waste-reduction strategy is the redesign of products so that each unit requires less raw material. That approach has a double advantage for automobiles. Cars that require less metal are smaller and lighter and hence more fuel efficient. By federal requirements to improve fuel economy, U.S. automobile manufacturers have trimmed the weight and size of most models now on the market. In fact, since 1975, the mass of a typical auto has been reduced by about 400 kilograms (880 pounds); about one-quarter of this was the result of substituting aluminium and plastic for steel. Plastic is more difficult to recycle than steel. In the future, U.S. Auto manufacturers will be designing cars whose components, including plastics, will be easier to recycle.

Packaging consumes large amounts of raw materials and creates tremendous amounts of waste. One way to reduce such waste is to develop packaging methods that conserve raw materials.

By the year 2000, The United States government hopes municipal solid waste can be reduced by 25%. While environmental scientists around the globe struggle to develop expensive, high-tech solutions to solve the planet's problems scientists at the University of Florida are finding Earth-friendly places for materials usually considered waste products. At the Ona Research and Education Center, scientists have been using leftover restaurant grease blamed for clogging human hearts and drains and mucking up landfills - as part of a liquid dietary supplement that is fed to beef cattle.

Individual Effort to reduce the volume of refuse

1. Evolve and participate in community recycling programmes.

2. If possible, use a mulching lawn mower, which returns grass clippings to the soil and reduces the need for fertilizer.

3. Compost yard and food waste.

4. As much as possible use cloth-bags for shopping Alternately reuse plastic bags.

5. Instead of disposing of old but still functional items such as clothing and appliances, donate them to charity.

6. Select durable rather than disposable goods (cloth rather than paper napkins, metal instead of plastic cutlery, ceramic coffee mugs instead of stylofoam cups).

7. Avoid overpackaged products (nonbiodegradable material, single-serving units).

8. Do not use disposable diapers, razors, flashlights, and so-on.

9. Save paper - save a tree ! Use both sides of a paper, be brief and reuse envelopes.

Florida House Learning Center - located on the grounds of the Sarasota County Technical Institute in Sarasota - is designed to illustrate innovative conservation techniques for builders, developers and homeowners. Like they show a driveway, made from crushed sea-shells, and the linoleum, made of cork, jute and linseed oil, carpet made from recycled milk jugs and tile made from recycled glass.

An enterprising arts teacher at Sanawar (H.P.) Mr. Attri, has made mats and bags out of dried maize and banana skins with Uncle Chips and Frooty covers as lining. Dyed in colourful hues they are environmentally and aesthetically far more appealing.

Tyres can be used for swings, crash guards, boat bumpers, can be shred and used for manufacture of new tyres, ground and used as additive in road construction. Of course in India, some cobblers recycle tyres to make shoe soles. Now Tyre Farms, California, has come up with another use : they can be embedded in fields or courtyards to trap rain water underground. The tyres are sliced in half and laid out on flag grounds, embedded them upto 15 inches underground. This technique could save on expensive irrigation for golf courses and patches of land that have to be kept green, but are not for serious cultivation.

If animal wastes such as fat, bones, feathers or blood are cooked (**rendered**), several valuable products can be produced. These include a fatty product called tallow, which is the raw material for soap, and a nonfatty product that is high in protein and can be used as an ingredient in animal feed. The raw material for a rendering plant contains wastes from farms, slaughterhouses, retail butcher shops, fish processing plants, poultry processors and canneries. If there were no rendering plants, these waste would impose a heavy burden on sewage treatment plants.

We have just started down a path that is going to take us a long way. As all people in the agriculture sector know, a weed is a plant out of place, and trash is a resource where you don't want it—until you can find a home for it, and that is what reducing waste is about.

PLASTICS : Reduce and Use with Caution

Plastics are everywhere : on the sole of your shoes, as the steering wheel of your car, your camera body and film, the midday coffee cup and may be, even inside you as a valve!

In the simplest of terms, plastics are synthetic substances - polymers - produced by chemical reactions, mostly made from petroleum.

In Indian metropolises, plastic waste is between 4 to 5% of the total waste generated, and of this, almost 90% is picked up by ragpickers and fed into the recycling trade. In Delhi, which is one of the biggest plastic trading centres of the country, almost 250 tonnes of plastic are dealt each day. Plastic waste management in India thus stands in sharp contrast to countries like the

US, where only 2% of plastic waste is recycled, or Germany, where under the Green Dot, about 7% of plastic waste is being recycled.

Table 4 : Types of Plastics

Name of common Plastics	*Characteristics*	*Common Objects of usages*
PVC - Poly Vinyl Chloride	Hardy, Chlorine compound, brittle unless treated	Sports goods, pipes, auto parts, buckets etc.
LDPE - Low Density Poly Ethylene	Moisture proof, inert	Trash bags carry bags
HDPE - High Density Poly Ethylene	Flexible, yet extremely strong	Milk pouches, cables, soft drink crates
PET - Poly Ethylene Tere-pthalate	Gas Permeation resistant, Shatter resistant	Soft drink bottles, storage jars
PP - Poly Propylene	Stiff, more resistant to heat and chemicals	Moulded Luggage, transparent plastic film
PUF Polyurethane Foam Thermocol and Foam Rubber	Extremely light brittle	Throw away plates, glasses and for packing

While the West has woken up to the hazards associated with recycling of plastic wastes, India finds it economical and employment oriented. Plastic is collected in the West only to be shipped to countries like India where it is reprocessed with minimal environmental and occupational health regulations.

Plastic waste exports to India are steadily on the increase. Shipments from the US had gone up from 3, 974 tonnes in first half of 1992 to over 7,841 tonnes in the consecutive period of year later. (97% increase in a single year!). Not only this, Australia had shipped 3,000 kg in 1990, 16,000 kg in 1992 and 74,000 kg between January and September 1993. Canada had brought in 42,275 kg. of plastic and polystyrene waste into India in 1992. It is believed that many other European countries, including Britain, have been exporting plastic wastes to India. The data for which is not available.

The fact that plastics have such a wide application is because they are flexible, can be easily adapted for a variety of uses, are not easily breakable, and because they are light and lightness of plastics has reduced transportation costs. But there has been little consideration for the impact these light and durable materials will have on the environment once they are discarded.

India had until recently, continued to use natural fibres like jute, coir, flannel cloth, paper, cotton etc. in the manufacturing of rugs, carpets, cushioning material, upholstery and even sacks. These are now being fast replaced by HDPE, PP, foam plastics, Nylon and PVC rexine.

But there are dangers inherent in this large scale conversion, which can endanger human health and life, and which are rarely ever taken into account while using plastics.

The increasing use of plastics in the consumer nondurable sector is also difficult to handle, and threatens to be our biggest waste problem in the next few years. Consider the example of the comparatively new 'soft squeeze tubes' for toothpaste, and shrink wraps for soap. While the earlier packaging of metal and paper/plastic were picked up for recycling, both these are now allowed to be disposed by the municipalities, which usually dump waste into crude, illdesigned landfills with dangers of leaching. Composite plastics, like tetrapacks, and laminates also follow a similar fate as they cannot be reprocessed.

Even if the plastic used can be recycled, its use in non-durables is not environmentally friendly either. For one, packaging has tended to increase and usually, this means an increase in the plastic output. If, as is the case in India, a vast majority of it is fed into the recycling trade, the problem of the process of recycling still remains a problem, for it is carried out in small shanties with no pollution control devices or worker protection. The worst culprit here is PVC, which accounts for 45 per cent of all the plastics recycled here. This material is particularly dangerous, for upon reprocessing it releases chemicals like dioxin, ethylene oxide, benzene, vinyl chloride and xylene. Apart from this, plastics contain lead, cadmium, barium etc. to give them their particular characteristics. Many of these chemicals are known carcinogens, others are flammable and explosive. Some of these are also known to be responsible for causing birth defects and damage to nervous system, blood, kidneys and immune system. Emissions can cause skin & respiratory problems. So recycling of such plastic is not only highly polluting, but exorbitant in developed countries.

What is not being understood is how a technology which is considered to be hazardous in America can be justified in India on grounds of economics and human health? How long will India continue to encourage highly polluting industries under one pretext or the other? When will we Indians learn a lesson from our neighbours in Southeast who have banned the impact of toxic wastes? In fact, two African countries, Nigeria and Cameroon, have gone a step ahead and imposed death penalty for those who import the toxic wastes. Recently, even China has passed a law to keep out useless solid waste and limit import of recyclable waste. Under the new law passed by the legislature on Oct. 30th 1995, Chinese Customs authorities will order violators to ship the waste back and pay fines

There have been suggestions about incinerating plastic in waste to energy conversions, but these have been decried by environmentalists worldwide for the pollution - both in the form of the heavy metal laden ash and gases - which are bound to escape. Yet, on the other hand few notice the incineration of plastics in the form of hospital waste with little or no environmental monitoring. In Delhi's prestigious Ram Manohar Lohia Hospital, for example, the only items to be incinerated are intravenous drip bottles, which are thrown into the chamber even as workers inhale the noxious fumes.

In the eighties, the claim of biodegradable plastics made from starch had been marketed as the ultimate 'be good-feel-good' material. The claims have already been contested, as tests have shown that even under ideal conditions, the plastic merely disintegrated into very small pieces on land, and did not dissolve in sea water, thereby making no difference to the environment.

In order to be recycled, a material needs to enter the chain of collection. When it does not do so, the harm it causes can be enormous. On the land, plastic carry bags choke the soil, and even compost made from improperly sorted waste can have similar effects. Villages near the Sultanpur Bird Sanctuary in Haryana last year found their fields choking with plastic bags that had entered as part of the manure they had applied. All operations were stopped as the farmers handpicked the bags and cleared the soil. On a large scale, the world's oceans and seas are being affected by plastic nets, bags, and similar trash to the extent that they kill sea birds who mistake them for food, while sea turtles mistake them for a meal of jellyfish, and then starving to death as the plastic gives a feeling of satiation.

Forests are being used as dumping grounds of polythene bags causing irreparable damage to the soil and vegetation. Presence of plastics in the soil make plant growth impossible as they secrete harmful chemicals, besides obstructing rain water from percolating to the sub-soil. Again, garbage when put in polythene bags becomes non biodegradable though it otherwise may be biodegradable.

It has been estimated that the world's merchant ships dump 639,000 plastic containers, including bags, into the seas each day. So distressing has this problem been that the Convention for Prevention of Pollution from Ships (MARPOL) was amended to tackle the problem of garbage disposal from ships.

The plastic carry bag leaves a trail of environmental pollution behind it. .Starting from its manufacture to its final disposal, it extracts a heavy environmental cost. It is made from used Low Density Polyethylene (LDPE) plastic which has been recycled at least a couple of time. Since the colour of the recycled raw material varies, bleaching and colouring chemicals are added to give the carry bags the familiar light green, blue or pink hue - colours which are toxic and not food grade, despite being commonly used for carrying food and vegetable. In fact, these colours contain metals like cadmium, which accumulates in vital body organs leading to kidney and respiratory disorders. Such raw materials which is not possible to tint otherwise owing to its poor quality, is dyed black to hide defects, and is used by meat shops all over.

The hazardous manner of the manufacture of these bags is a grave concern for health. Factories engaged in this activity are scattered all over Delhi, and many of them are unregistered/illegal operations. Toxic fumes from the process rent the air, and are freely breathed in by the workers who have no protection in the form of gloves, masks or any other protective devices. These workers and nearby residents (since many such factories are located in residential areas) have been known to regularly suffer from serious and sometimes fatal respiratory disorders. The machines also cause heavy vibrations and a disconcerting din which permeates the area.

Even after disposal, the menace of the carry bag does not end. In Delhi last year, sewer lines chocked by these low grade plastic bags, exploded owing to excessive methane levels. If burnt, irresponsibly they release highly toxic gases like phosgene, carbon monoxide, chlorine, sulphur dioxide, nitrogen dioxide, besides deadly dioxins. When dumped into land fills, they slowly leach heavy metals lead and cadmium, with rainwater which seeps into and contaminates groundwater. Cows and dogs have been known to have choked on them while feeding at garbage bins and recently a deer which gouged a carry bag died at the Delhi zoo.

So light and fragile is the bag, that it is not collected by ragpickers at garbage bins, as collecting them is a tedious and an economically unenviable affair. Hence most bags are dumped along with other garbage into landfills, where they occupy precious space and lie almost forever.

Recently, in Shimla the plastic carry bag has been banned. To encourage the use of newspaper and brown paper bags, the Himachal Government has also withdrawn sales tax on the paper used in making them. Other states and cities must follow suit. Or may be we should just carry our own bags instead.

The steps that Sri Lanka has taken to stop the use of plastic bags are worthly of emulation. In that small country, the people call them the 'plagues of shopping'. No customer accepts them. Instead, they are pleased to carry their shoppings in grey cotton bags made of cheap material. If in the schools even a single polythene cover is spotted in the fields, the authorities immediately swing into action. The parents of the child who brought it are called and admonished for that 'public nuisance'. In addition to this there are a number of voluntary social organizations there for educating people in various ways. The media too are concerned about it. The popular papers of Colombo are publishing regular theme pages on the subject.

WASTE WATER TREATMENT

The increased industrial activity in the last few decades has resulted in large scale pollution of rivers, lakes and seas so as to become a subject to major concern.

Sources and Types of Toxic Wastes

Domestic wastes contain ammoniacal nitrogen at levels up to 50-60 mg L^{-1} and may contain sulphide at levels up to 50 mg L^{-1}, both of which can cause damage to aquatic life, if discharged. Agricultural wastes may also contain even higher levels of both ammonia and sulphide plus a range of toxic pesticides and herbicides and the drainage from silage may be toxic due to low pH values. The majority of toxic wastes are associated with industry. Here the range of possibilities is too wide. Table 5, gives some indication of the main types of toxic industrial waste and the toxins that they contain.

Pre-Treatment of Toxic Wastes

Pre-treatment of industrial wastes prior to discharge to a sewer is widely practised. In some cases this is to comply with the restrictions imposed by sewer Authorities and Local Authorities on wastes entering a water. There are also a large number of specific toxic substances whose discharge to sewers is controlled.

The main advantages of treatment on site are the possibilities of recovering heat or specific substances in an uncontaminated condition and the economies which might result from treatment at higher temperatures and/or higher concentrations. Where the toxic materials are organic in nature there is often some difficulty in treatment due to inhibition of bacterial growth. It is often easier and cheaper to develop the necessary bacterial flora in an *on-site* treatment plant.

Table 5 : Sources of Some Common Water Pollutants in Industrial Waste Water

Pollutants	*Sources*
1	2
Acids — mainly inorganic but some organic causing pH<6	Acid Manufacture, Battery manu facture, Chemical Industry, Steel Industry
Alkalis—causing pH>9	Brewery wastes, Food Industries Chemical Industry, Textile Manufacture
Antibiotics	Pharmaceutical Industry
Ammoniacal nitrogen	Coke Manufacture, Fertiliser Manufacture, Rubber Industry
Chromium—mainly hexavalent but also less toxic trivalent form	Metal processing, Tanneries
Cyanide	Coke Production, Metal Plating
Detergents-mainly anionic but some cationic.	Detergent manufacture, Textile manufacture, Laundries, Food Industry

(Contd.)

Table 5 (Contd.)

1	2
Herbicides and Pesticides—mostly chlorinated hydrocarbons	Chemical Industry
Metals — mainly copper, cadmium cobalt, lead, nickel, mercury and zinc	Metal processing and plating, Chemical Industry
Phenols	Coke Production, Oil Refining Wood Preserving
Solvents — mostly benzene, acetone, carbon tetrachloride alcohols	Chemical Industry, Pharmaceuticals

This to some extent depends upon the concentration and toxicity of the substances concerned. In some cases dilution of the waste by admixture with sewage reduces the toxic inhibition making it preferably to treat the industrial waste and sewage together. Also many industrial wastes are deficient in some nutrient such as nitrogen or phosphorus. The desirable ratio of BOD : N : P is 100:5:1 and the ratio in domestic wastes is commonly 100:18:2.5 so that deficiencies in industrial wastes can be balanced.

Even where it is decided not to carry out pre-treatment of the toxic waste by chemical or biological methods it is often useful to install devices to improve the effluent quality by simple physical means. The includes some form of screening to reduce solids. Also some form of balancing to reduce variations in concentration, flow, pH etc. and some traps to prevent the escape of oil and grease and some grit arrestors.

The processes which are used may be classified as physical, chemical and biological. The physical processes are summarised in Table 6. Where the toxic wastes contain or are composed of organic materials it may also be necessary to provide some biological treatment especially if the effluent is to be discharged directly into a water way. Many different types of processes are used but the following are the most popular.

(a) High-rate filtration using plastic media with very high rates of recirculation.

(b) Activated sludge using contact stabilization.

Like all biological processes these can suffer from toxicity problems especially where the concentration of toxin is not constant. In general terms it is easier for bacteria and other micro-organisms to adapt to toxic substances than for organisms like worms, fly larvae etc. For this reason conventional percolating filters have not proved successful—the lack of grazing fauna has led to persistent ponding.

Due to a combination of high organic strength and inhibition from toxic substances it is unusual to obtain complete treatment of toxic industrial wastes by conventional primary and secondary treatment. The effluent from high-rate filters often has a BOD and COD similar to settled sewage and is either suitable for discharge to a sewer or for further biological treatment on site.

Table 6 : Physical Methods of Pre-Treatment

Process	*Aim*	*Examples*
Screening	Removal of coarse solids	Vegetable canneries, Paper Mills
Centrifuging	Concentration of solids	Sludge dewatering in chemical industry
Filtration	Concentration of fine solids	Final polishing and sludge dewatering in chemical and metal processing
Sedimentation	Removal of settleable solids	Separation of inorganic solids in ore extraction, coal and clay production
Flotation	Removal of low specific gravity solids and liquids	Separation of oil, grease and solids in chemical and food industry
Freezing	Concentration of liquids and sludges	Recovery of non-ferrous metals
Solvent Extration	Recovery of valuable materials	Coal-Carbonizing and Plastics manufacture
Ion Exchange	Separation and concentration	Metal processing
Reverse Osmosis	Separation of dissolved solids	Desalination of process and wash water
Adsorption	Concentration and removal of trace impurities.	Pesticide manufacture, dye stuffs removal

Chemical treatment of industrial wastes may be used in addition to and to some extent in place of biological treatment. The aims and somewhat different since biological treatment is mainly a way of oxidising organic matter or a way of converting it into a settleable form. Chemical treatment can also provide oxidation through chlorine, ozone, etc., but this is used only for oxidising particular compounds like cyanide since it is expensive and liable to lead to the production of undesirable chlorinated organics. It is mainly used for pH correction and improving the removal of solids. The commonest chemicals in use are shown in Table 7.

Table 7 : Chemicals used in Industrial Waste Treatment

Chemical	*Purpose*
1	2
Calcium hydroxide	pH adjustment, precipitation of metals and assisting sedimentation
Sodium hydroxide	Used mainly for pH adjustment in place of lime
Sodium carbonate	pH adjustment and precipitation of metals with soluble hydroxide

(Contd.)

Table 7 (Contd.)

1	2
Carbon dioxide	pH adjustment
Aluminium sulphate	Solids separation
Ferrous sulphate	Solids separation
Chlorine	Oxidation

Primary and Secondary Treatment

If the pre-treatment of toxic industrial wastes is successful, no difficulties should be encountered in subsequent treatment. However, no pre-treatment system is perfect and malfunction will occasionally occur mostly due to variations in the manufacturing process. As a result toxic material together with possible overload of organics and solids may be passed on to the subsequent treatment stages.

The effect of toxic materials on primary sedimentation is insignificant, since this is a purely physical process of sedimentation and flocculation. However, the effect of the primary sedimentation on toxic wastes can be very important. Toxic materials in suspension such as particulate metals are effectively removed. Also the flocculent material has a great capacity for adsorption and removes the majority of dissolved metals, pesticides and other toxic organics. In one respect this is beneficial since it renders the waste material less inhibitory for biological treatment but it selectively concentrates the toxins in the sludge and may give rise to problems in digestion and in sludge disposal.

The key to successful secondary treatment of wastes containing toxic materials is the adaptation of the microorganisms to the presence of the toxins. Bacteria and to a lesser extent protozoa show considerable ability to acclimates to the presence of toxic substances and a great adaptability in degrading new synthetic organics. Metazoa are rather less adaptable and for this reason activated sludge is generally better than percolating filters for treating toxic wastes. Experience with treating toxic industrial wastes on percolating filters has shown frequent ponding problems due to a lack of activity by the grazing fauna. High rate filters which utilise hydraulic scouring for film control have been used successfully and the high recirculation ratios help to dilute any incoming toxins. This helps to overcome the other disadvantage of filters which is due to the plug-flow nature of the process. Any shock loads of toxin are not as readily diluted as in a completely mixed reactor.

The activated sludge process is generally preferred for dealing with wastes containing an admixture of toxic materials. In particular the completely mixed version gives immediate dilution of any shock loads. The dangers with activated sludge are that :

(a) The toxin may reach concentration which inhibits enyzme activity.

(b) Some toxins also affect the bacterial surface and therefore affect settleability.

Nitrification is particularly sensitive to these problems. It is therefore important in treating a toxic waste by either biological process that a microbial population is developed which is acclimatised to the presence of the toxin and in the case of degradable toxins contains sufficient numbers of organisms which can metabolise the toxins. These twin aspects of acclimatisation require great care in the start up operation and may need a period of several months before successful operation is achieved.

Sludge Treatment and Disposal

Toxins which remain in solution during primary and secondary sedimentation do not appear in the sludges and thus cause to further difficulties. Toxicity in digestion may also occur due to soluble toxins in the treatment of industrial wastes by anaerobic methods. But the main problems are due to toxins which are in settleable form or are readily adsorbed, are selectively concentrated in the sludge and give rise to difficulties in digestion (and in subsequent disposal). The classes of toxins involved are mainly metals, chlorinated hydrocarbons, organic solvents and detergents.

SEWAGE TREATMENT

Sewage is the major source of water pollution. Many countries now treat sewage to remove pollutants and reduce the B.O.D., so that discharge into the streams and rivers has minimum impact.

The process of treating municipal sewage involves three steps **primary, secondary** and **tertiary**: The particular process used in a given situation depends on the volume to be treated, the location of the outfall the dilution factor, the potential hazard to users receiving the water, and in many cases, the cost of the project. Other countries still continue to discharge raw sewage into rivers and streams.

Primary Treatment

Primary treatment consists of removing floating and suspended solids by mechanical means (Fig. 1). More than one-half of the suspended solids can be removed by primary treatment. First the large solids are screened out and grease and scum are removed. This is followed by sedimentation in a basin to remove the remaining solids, called primary sludge. The screens, called trash racks, consist of steel bars about 2 to 4 inches apart. In some cases sand and other coarse material is removed by grit chambers to further protect pumps and other equipment from damage. Sometimes the waste water is then run through fine screens. Usually after screening and removal of grit, the waste water is run directly into setting tanks. The setting tanks may have skimming devices, or the removal of scum may be done separately.

The primary sludge is a burdensome problem because it is bulky and must be removed. It contains 94 to 99 per cent water. Usually the first step is to remove as much of the water as possible. In some cases the sludge is dried in beds with some of the water being removed by filtration. The residue is disposed of on land. Because the sludge itself can comprise a pollution problem, a better method is to bring about microbial decomposition in **sludge digestion tanks** before drying.

The other products of primary treatment are gases and the fluid or clarified waste water. The gas is mostly methane, which is usually burned as fuel to provide heat for the digesters and other equipment. The clarified waste water has highly objectionable properties and in most cases is put through a secondary treatment.

Secondary Treatment

Secondary treatment of waste involves the biological degradation of organic materials by microorganisms under controlled conditions (see fig. on p. 279) The usual method is to bring about the biological oxidation of the organic material under aerobic conditions, in which the waste is aerated to supply oxygen for the microorganisms. The degraded material settles out in secondary settling tanks and is therefore described as being removed by sedimentation. The sediment containing the microbial growths and their by-products is called secondary sludge or **activated sludge**. The clarified waste water is discharged in the outfall to rivers, lakes, bays, lagoons, or oceans. Some of the sludge is returned to aeration tanks where it is recycled with the incoming waste. Most of the sludge, however, must be removed, and this operation accounts for one-fourth to one-half of the operation.

Sludge

Both primary and secondary sludges are usually transferred to sludge digestion tanks where decomposition takes place under anaerobic conditions. Various methods for water removal are used. Because the removal of the sludge is a major part of sewage disposal, a great deal of effort has gone into drying and finding uses for the product, or locating disposal sites and transporting the sludge there.

A few of the treatment plants heat-dry the activated sludge, usually after some form of mechanical water removal, called *dewatering,* and sell the product as fertilizer. The process is costly. It is less expensive to incinerate the sludge in furnaces or to use it as landfill. Incineration sterilizes the sludge and reduces its volume. However, incineration has some disadvantages. It creates an air pollution problem and leaves and ash that must be disposed of. Still, disposing of a small amount of ash is easier than getting rid of a large amount of sludge.

Final Disposal

The methods used for disposal of sludges are usually those that are the least costly. The choice of disposal site whether on land or sea depends on the proximity of the treatment plant to suitable disposal locations. Dewatered sludges are commonly used as landfill. Liquid sludges are disposed of either on land or in bodies of water. Liquid sludge is used to fertilize or condition agricultural land. However, problems of odour, water pollution, and stimulation of insect and algal growth, as well as other aspects of public health and aesthetic values, must be considered.

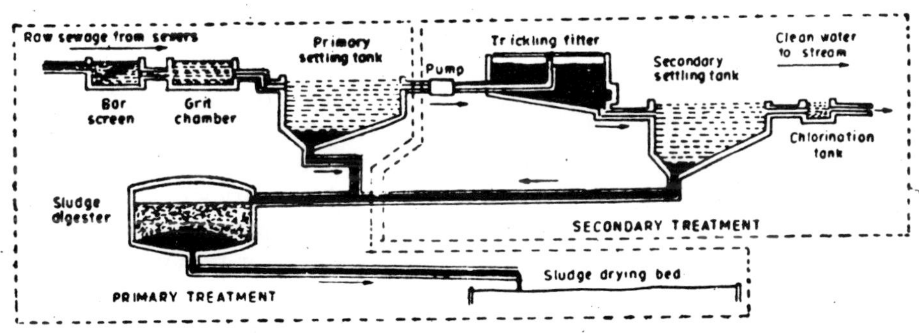

Sewage Plant Showing Facilities for Primary and Secondary Treatment.

Tertiary Treatment

Any treatment to further purify waste water beyond the methods commonly employed for secondary treatment may be called "advanced treatment" The term **tertiary treatment** is often used to refer to certain advanced treatment procedures for treating the effluent coming from secondary treatment. The purpose is the removal of the contaminants of waste water that remain after secondary treatment. These contaminants consist mainly of suspended solids, dissolved organic compounds, and the inorganic plant nutrients nitrogen and phosphorus. A thorough job of cleaning up sewage waste water is difficult and costly.

A number of alternative techniques depending upon the nature of pollutant are available.

The methods involve removal of nutrients by autotrophic plants, flocculation of colloidal particulate matter, use of adsorbants, use of oxidising agents like $KMnO_4$ or O_3, use of **reverse osmosis** by artificial pressure device across semipermeable membrane, or by specific chemical techniques. Chlorination is done to eliminate orthophosphates remaining in water after primary and secondary treatments which can be precipitated by Ca or some other metal ions. Aluminium sulphate, ferric chloride, ferrous (ic) sulphate and calcium hydroxide are good coagulants. Nitrates are removable with the help of *Nitrococcus denitrificans* and *Thiobacillus denitrificans* which denitrify and release nitrogen and water.

The hi-tech methods used for sewage treatment are highly expensive and beyond the reach of developing nations. In most of these countries sewage is discharged untreated into open drains and then on to rivers or open seas. Some countries including India and Thailand use a low-tech system of sewage treatment. The first step of this system is the same as the hi-tech system. The liquid waste is then left in an open pond for conversion of organic waste through photosynthesis. This low-tech system requires three times more space than the high-tech one.

HUMAN WASTE

Part of the living process is to get rid of unwanted material. Otherwise organisms die of the toxins in their own waste. The biological process for ridding the body of wastes is called **excretion** and takes place in some manner in all living creatures. In humans, the excreta of the alimentary canal are called faeces. They consist primarily of indigestible material, fats and bacteria, which comprise the most bulky portion of human waste. The urine, of which about 1.5 litres are excreted daily by a human adult, is mostly water with dissolved nitrogenous wastes and salts. The urine also carries away foreign substances such as drugs and other toxins.

When no more than 100,000 people populated the earth during the Ice Age, human excretion was not a problem. It was accomplished by walking to the woods or the rocks and leaving the waste for disposal by coprophagous organisms. The increased human population posed problems of disposal of human wastes. Civilization, with its concentration of humans living in cities, aggravates the problem of waste disposal beyond that which can be solved by natural means.

The people of the Harappan culture of the Indus Valley, circa 2500 B.C., were perhaps the first to become experts in sanitation, Household water supplies, bathing facilities, and drainage systems were widespread. Latrines with seats were built into some houses. There were sloping or stepped channels through the wall to either a pottery receptable or brick drain outside. At Mohenjo-Daro each house had a drain which ran into a central sewer system under the street which fed into cesspools. Manhole covers of brick were installed to allow for repair and cleaning. The ancient cities of Sumer and Babylon seem to have had similar hygienic devices.

The use of waste for productive purposes is an ancient solution to the disposal problem. Animal dung has been used as fertilizer for thousands of years. In the Orient including the Indian sub-continent, China and Japan, human dung has been used as fertilizer for centuries and is today a convenient answer to the dual problems of disposal and food production. The practice is not aesthetically or economically satisfactory in an industrial society, although dried sludge is produced by some modern sewage treatment plants as a by-product and sold as horticultural fertilizer.

It were perhaps aesthetic reasons, rather than hygienic reasons which stimulated the construction of sanitary facilities. **Cesspools** were dug everywhere and many of them were prolific breeders of contamination because of seepage into nearby wells. As a result, typhoid fever and cholera epidemics were rampant. With the passage of time the methods were improved. Cesspools are still being used in areas with smaller population, especially in hilly terrains. Such cesspools and sanitary fills pose special problems of groundwater pollution of wells especially when located at a higher level than the well. This danger can be partly tackled by locating pits at lower levels than wells and where this is not possible the well should not be in line with the pollution plume of the pit, which follows a straight path down a slope. London installed a sewer system in 1865 and many English cities followed the example. Modern sewer systems known as **combined sewers** (see fig. below) were designed to handle both human waste and surface drainage. These are now installed in all major cities the world over.

The composition of sewage is complex, and it differs depending upon the sources, the type of treatment or lack of it, and whether there is an admixture of storm drainage with industrial waste. A combined sewer receives surface runoff rain, sewage from street drains as well as household sewage, mainly containing the human wastes and used water. It is carried underground to a sewage treatment plant which removes solid wastes, brings about decomposition of organic wastes, kills micro-organisms and removes pollutants before discharging the treated sewage into a stream or a river.

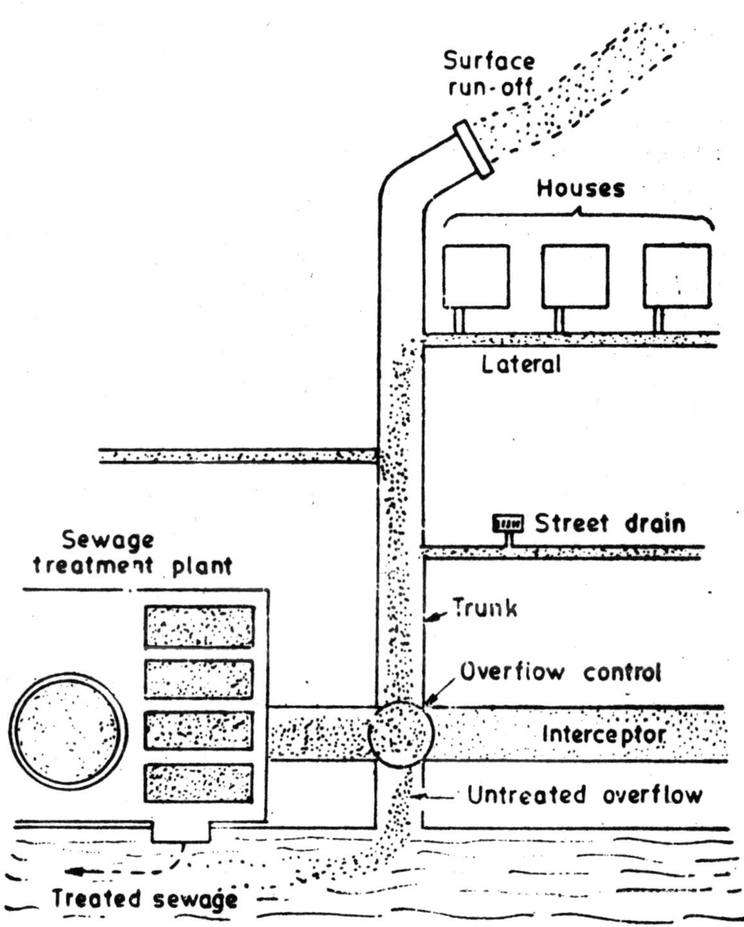

Combined Sewer System.

NUCLEAR WASTES

The fission products formed by the splitting of atoms in the nuclear fuel comprise the main source of radioactive wastes, and they are intensely radioactive. Must of the fission waste material is enclosed within the "cladding" or metallic material of the fuel rod tubes which forms the first containment barrier.

Atomic power plants stack both gases and suspended particulate matter. Particulate matter can be removed by high efficiency filters. In the pressurised water reactors, the gases are held in storage tanks for as long as 20 days or more, during which the radioactivity is reduced to low levels by natural decay of the nuclides. The gases are then released to the atmosphere via the plant stack. In the boiling water reactors, the large amount of gaseous wastes that is produced makes it difficult to hold them in storage for long periods. Only 30 minutes or so are allowed for decay before the waste gases are put through the plant stacks. Some of the radionuclides in the stack effluent have short half-lives, most of the radioactivity disappearing within minutes or hours: others remain radioactive for years. Some, such as strontium-90, are produced in large quantities than others. Some become distributed throughout the environment and are transmitted to humans via the food chain. When taken in by humans, certain radionuclides becomed concentrated in specific organs where they can be injurious to health. Consequently a larger amount of radioactivity is released to the environment. Materials that escape include radioactive caesium, hydrogen, and krypton.

Caesium-137 belongs to the family of elements called **alkali metals**, which also includes sodium and potassium. Caesium is the member most like potassium and is physiologically analogous to it. Most of the caesium taken into the body has a biological half-life of about 100 to 135 days, meaning that it takes that long for one-half of the caesium that is taken in to be eliminated from the body. During that time, damage to the body, especially the blood-forming tissues, can occur, Caesium - 137 has an environmental half-life of 33 years.

Hydrogen-3, or tritium, is produced in comparatively large quantities is pressurised water reactors. It is present in the reactor wastes as a component of water molecules and upon its release to the environment undergoes the same environmental and biological pathways as ordinary water.

Tritium emits beta radiation, as does radioactive carbon, strontium, and other radionuclides. The beta radiation from tritium is less energetic then any of the other important reactor waste products. It is less penetrating, and theoretically, less damaging to tissue. Moreover most of the water taken in is passed through the human system unchanged. Studies thus far show no evidence that tritium becomes concentrated in food chains. Radioactive hydrogen is important, however, because of the large amounts produced and the fact that hydrogen, both in water and as hydrogen nuclei, participates intimately in living processes, for example, photosynthesis in plants, global contamination of the hydrosphere could have injurious effects of undetermined magnitude on the food chain. The environmental half life of tritium is 12.3 years.

Another gaseous waste product, **krypton-85**, is a member of the family of elements called **noble gases**. Although krypton-85 is highly radioactive and has a moderately long half-life (10.76 years), the absence of any known participation in biological processes causes it to be an environmental contaminant of less concern than some of the other nuclear waste products. But because krypton is not chemically reactive and does not combine with other substances, it is long-lasting in the atmosphere; and, further, because of its long half-life, it accumulates in the environment. A radioactive element damages tissue by the emission of radiations; unlike non-radioactive poisons, it does not have to be involved in a chemical reaction to be injurious. It is estimated that the annual exposure from krypton will be about 0.02 millirem by the year 2000. Krypton-85 and tritium are released from nuclear reactors in much smaller amounts than from nuclear fuel reprocessing plants.

Iodine-131 comprises about 5 per cent of the fission products from a nuclear reactor. Radioactive iodine readily contaminates pastures and other plants, causing it to appear in milk and diary products (see fig. below). In the body, iodine becomes concentrated in the thyroid gland. Damage to the thyroid gland such as that observed to be caused by fallout from atomic testing can be serious, especially among children.

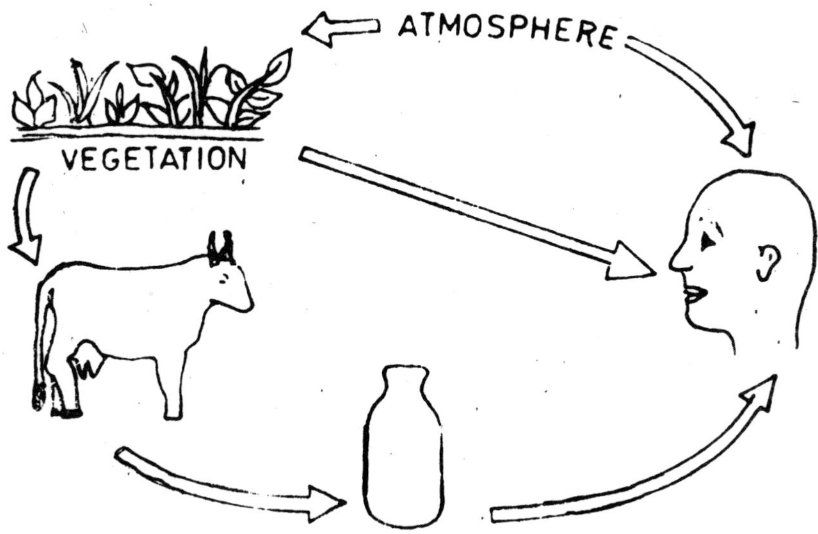

Iodine-131 pathways from fallout, Iodine-131 has a half-life of only 8 days but the radionuclide causes the greatest radiation exposure within a short time.

Strontium-90 finds its way into biological pathways because of its ready acceptance in biological systems as a substitute for calcium (See fig.) Contaminated food plants, meat, and dairy products appear to have normal nutritional properties for humans but cause radioactive strontium deposits in the bones and other tissues in place of calcium. Strontium, like radium, belongs to the alkaline earth family of elements, which also includes calcium. The physiological pathways of these nuclides are similar. About 99 per cent of the long-term radioactivity from either strontium or radium taken into the human body is found in the bones where damage can occur to bone cells and marrow—blood Oil-producing tissue. Because of the long radioactive life of strontium-90 (its half-life is 25 years), the potential for damage is very great.

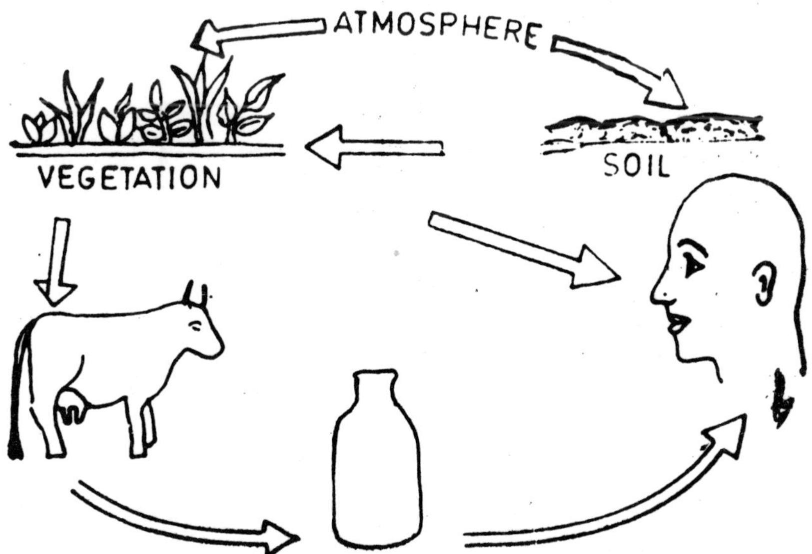

Strontium-90 pathways from fallout. Because strontium-90 behaves much like calcium in the body, it reaches man primarily through foods of plant origin and diary products. Strontium-90 has a half-life of 28 years; strontium-89, also produced by nuclear explosions, has a half-life of 51 days.

NUCLEAR WASTE DISPOSAL

In the operation of nuclear power stations, radioactive waste presents ever more pressing problems. **Mildly active** atomic waste consists of residues from filter and purification plants, from contaminated laboratory equipment and of sewage sludge from waste water separation. Its disposal has been effected by dumping in the sea, by burial in the ground, by depositing in salt mines, or in certain cases by discarding it in refuse dumps.

Moderately active nuclear waste consists of component parts of nuclear power stations rendered radioactive by neutron radiation, radioactive residues from purification processes (such as ion-exchange resins) and waste from nuclear research. Moderately active nuclear waste was formerly dumped in the sea and deposited in salt mines. As recently as 1972, 3800 tons of mildly to moderately radioactive waste packed in 7600 containers was sunk in the Atlantic by the European Atomic Energy Authority. As no containers will stand up to exposure to seawater for centuries, it must be expected that this and other radioactive waste deposited in the sea in large quantities will one day be released, be enriched in the food chain in the ocean, and sooner or later crop up in the foodstuffs of people.

The real problem, however, concerns **highly radioactive** nuclear waste which is left behind as a liquid upon separation of uranium and plutonium from burned fuel in reprocessing plants. No satisfactory solution for the disposal of highly active nuclear waste has yet been found, nuclear waste cannot be destroyed. In contrast to chemical waste, which can be transmuted and deprived

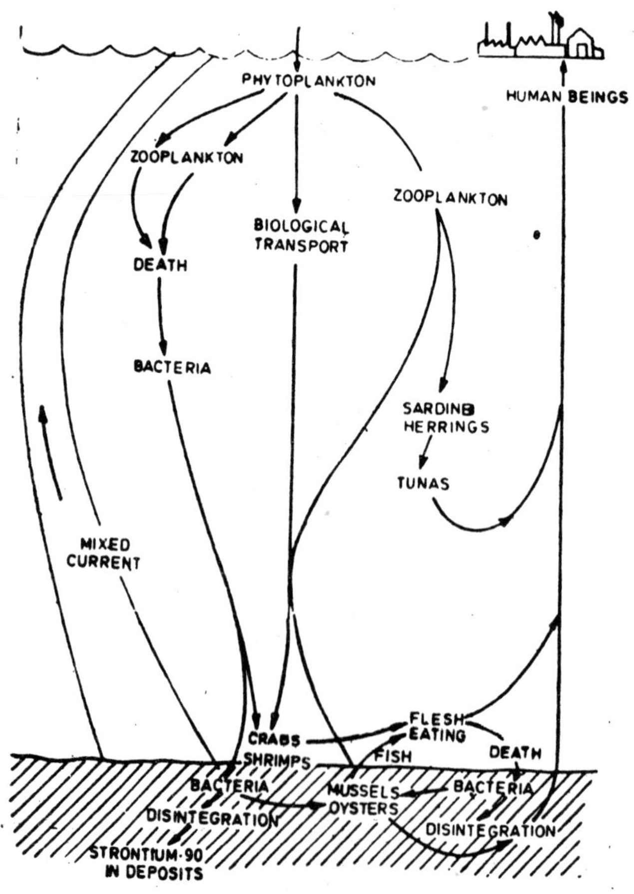

Biological consequences of Sinking Radioactive Substances.

of poison by chemical reaction, radioactive substance disintegrate according to a time scale (which cannot be varied) related to their half-life value. The principal problem thus concerns the disposal of the fission products which have a long half-life value, such as strontium-90 (with a half-life of 26 years), caesium-137 (30 years), samarium-151 (100 years), plutonium-239 (24,400 years), and iodine-129 (17,200,000 years),. The nuclear waste being produced today must therefore be stored away so securely, for periods hardly conceivable to man, that none of it can escape into the biosphere and so seep into the food chain of any living creature.

Some countries are exploring the possibility of sinking radioactive waste through the ice in the Antarctic. Existing international agreements concerning the Antarctic provide for special amendments if all the signatories agree. The heat released by radiation would make the nuclear waste containers sink gradually into the ice and melt their way through. In America it is proposed to ship the containers and transport them overland into the Eastern Antarctic, where the ice is particularly thick.

In view of the long-term nature of the problem of disposal of nuclear waste on the earth, the National Aeronautics and Space Administration (NASA) has carried out a study on behalf of the Nuclear Regulatory Commission (NRC) into the possibilities of extraplanetary nuclear waste disposal. Using the universe as a depository for atomic waste had already been put forward by various representatives of the Nuclear Regulatory Commission as the only safe method of atomic waste disposal.

Storage of Nuclear Waste in Aboveground Tanks

As a "solution" to the nuclear waste problem, many countries have considered storing highly active nuclear waste in tanks on the earth's surface. This method of disposal has been in use in the United State for over 20 years already. Each of more than 200 steel and concrete tanks contains over 3 million litres of highly radioactive liquid. As the radioactive waste constantly generates heat, the tanks have to be cooled all the time. This is achieved by piping the steam from the tanks into a condenser and mixing the contents with compressed air to ensure that no radioactive solid settle on the floor of the tank, which could result in serious local overheating.

The danger of radioactive contamination of the environment from storage tanks on the earth's surface takes several forms :

1. In conjunction with the strong radioactive radiation, the chemicals in the waste solution can cause corrosion of the tank container. An accident of this kind occurred in Hanford, Washington, which is the main deposit for atomic waste in the United States. As a result, 490,000 litres of radioactive waste seeped into the ground.

2. The tanks containing fuming nuclear waste must be constantly cooled. Approximately 9 kilowatts of energy are released by radioactive disintegration per cubic metre of highly active nuclear waste, or about 31 megawatts of energy per tank. Although each tank is equipped with two quite independent refrigeration systems, there is still a danger that the refrigeration may fail through power shortage, floods, earthquakes, sabotage or human negligence. In that event the contents of the tanks would heat up to over 1000°C. All the volatile radioactive fission products would then be released into the atmosphere. The results of such a mishap would be nothing short of catastrophic, both for the immediate vicinity and for remoter areas.

3. Owing to radioactive radiation, water molecules are split radiolytically into hydrogen and oxygen. The rate of this hydrogen production is so high that in the event of failure of the ventilation system the lower explosion limit of 4% for hydrogen in the air would be reached in a few hours. Through spontaneous heating in this way, an oxyhydrogen explosion could result, by which the tank would be destroyed together with the refrigeration system. Attempts have recently been made to provide more secure storage conditions for the radio-active residues

in liquid form. But the problem of the radiolytic disintegration of water and the risk of an oxyhdrogen explosion remains. The effects of a failure of the refrigeration system would be roughly the same.

The storage of nuclear waste in above ground tanks is thus accompanied by substantial risks, the effects of which cannot be finally assessed.

Part VII

Water and Water Pollution

- Water
- Hydrosphere
- Hydrologic Cycle
- Hydroponics
- Ground Water
- Ground Water Pollution
- Fluoridation
- Eutrophication
- Detergents
- Water Pollution
- Marine Pollution

WATER

A colourless, odourless, tasteless liquid that boils at 100° C and freezes at 0° C at atmospheric pressure. Water is essential to life, and no person can live more than a few days without it. Water is plentiful, but pure water is not.

Liquid of Life

Water is the essence of the living process. It is the most abundant and the most versatile of the chemicals of life. It is not surprising that in its physical and chemical properties alone, water is the most uniquely remarkable substance known. Its importance to life demands an understanding of its role in the process of living organisms and in human life.

Water forms most of our flesh and blood. Protoplasm is mostly water. The content varies in different tissues of the organism and in different plants and animals. In jellyfish, the protoplasm contains 95 per cent water. Human blood is 90 per cent water. Muscle and nerves contain 80 per cent or more water. The bones contain smaller amounts of water, and even the enamel of the teeth contains 2 per cent water. The total human body consists of 60 per cent to 90 per cent water. The variability being due to the fact that the tissues tend to become dehydrated with age.

With rare exceptions, animals and plants must have an abundant supply of moisture. Animals and plants that are fugitives from the sea and have managed to escape from the limitations of an aqueous existence cannot survive for long without access to moisture either in the air, the soil, or in food. Some organisms die within seconds if deprived of water. Others must have it within their systems but can survive for years in a desiccated environment. Some animals, for example certain rodents, can manufacture their own water. Human can withstand a loss of no more than 20 per cent of the water in the body without suffering an agonizing death. A healthy person can normally withstand no more than 7 to 10 days without water, but survival time depends largely on the amount of exertion and loss of water from heat and exposure.

The "Universal" Solvent

It is common knowledge that water easily dissolves sugar and salt as well as many other substances. Lakes, streams, rivers, springs, and seas are not pure water but contain a great deal of dissolved substances in solution. The ocean contains an average of about 3.5 per cent dissolved solids derived from the land through billions of years of erosion and run-off. The material in solution is mostly common salt (sodium chloride), but there is also an abundance of other mineral constituents, such as calcium, magnesium, sulphates, carbonates, and many metallic elements. Their concentrations are crucially important to the forms of plant and animal life that either select the water as their habitat or merely consume it for survival. The ability to take up and carry materials that are normally solid substances is perhaps the most important property of water in its role as the solvent. Water is the best solvent known. The solvent power of water enables it to carry the vital chemicals of life - minerals, salts, amino acids, and other organic substances-to the cells of the body. The return trip may be made with a load of waste products, and the water may be eliminated along with the refuse.

Terrestrial plants cannot take up mineral nutrients from the soil in the absence of water. Phytoplankton cannot absorb their mineral requirements unless the minerals are dissolved in the water of their environs. Food must be dissolved before it can enter the blood stream of animals. Even the oxygen and carbon dioxide needed by aquatic and marine organisms must be made available to most of them in solution. The products of metabolism are transported within plants in the aqueous solution called sap and within animals they are dissolved in the water of plasma. Many of the waste products of metabolism are carried away dissolved in water.

Transparency of Water

In the oceans, as on land, photosynthesis is the first stage in the nourishment of the food chain. Phytoplankton, like l and vegetation, thrive on sunlight, carbon dioxide, and mineral nutrients. Fortunately, water is transparent to light. Yet photosynthesis is by the producers in the ocean is limited by the relatively poor penetration of light through water compared to air. Some light penetrates clean ocean water to a depth of about 60 metres. However, most of the photosynthesis takes place in the upper few feet. In clear ocean water, the blue and green portions of sunlight penetrate the best, but these are the wavelengths of visible light that are the least efficient in photosynthesis. Red light is selectively absorbed and therefore is effective only in a narrow zone very near the surface.

Clear water, such as that found in much of the open seas and in many tropical waters, is relatively free of plankton, but there is the compensating factor that light penetration is also greater, resulting in photosynthesis taking place at greater depths. Tropical waters have a high rate of biological turnover because of the higher temperatures. But production of plankton is generally greater in cool waters than in tropical waters. A positive factor is the relatively high solubility of carbon dioxide and oxygen in water of low temperatures.

Ice

It is well known that ice floats on water. This is strange behaviour for a chemical substance but it is of great biological importance. Most materials that can be frozen and melted are heavier in their cold or solid state than in their warmer, liquid state. For example, a chunk of iron will sink in a pot of the molten metal.

The ability of ice to float is explained by the molecular behaviour of water. The molecules of water are constantly vibrating. When the water is cooled, the vibrations gradually slow down. As cooling proceeds, the molecules crowd together and form an increasingly dense pack. When a temperature of 4° C is reached, water is in its most dense condition. Below this point a sudden change takes place. The depressed molecular vibration combined with the strengthened attraction of the hydrogen bond causes the molecules to shift their positions to a geometric formation that forms a light, expansive latticework structure. As the water cools further, it continues to expand. When it freezes into ice, it is less dense than water in its liquid form and readily floats instead of sinking to the bottom. This can be explained as follows. As water is cooled, it becomes denser because the movement of the water molecules slow down as temperature drops. Water reaches its maximum density at 4°C But as the temperature drops further from 4° to 0°C, the attraction between the electropositive hydrogen atoms and the electronegative oxygen atoms-called hydrogen bonding—increases, and the water molecules become bonded together in a crystalline, lattice forming ice. Because of the greater space between the molecules, ice is less dense than water and occupies about 9 per cent more volume.

The fact that frozen water floats instead of sinking to the bottom is a profoundly favourable factor for life on the earth. If ice were heavier than liquid water, it would sink to the bottom of rivers, lakes, and frozen seas where it would not receive enough heat from the sun to melt it. Much of the water on earth would be solid ice. Large quantities of the earth's water would be entrapped in an unusable form. Evaporation and precipitation would be greatly reduced. Without moisture in the air, there would be little moderating effect on the sun's radiation and there would be extreme fluctuations in temperature. The world's climate would be drastically altered. Life would be difficult if not impossible. The biosphere as we know it would not exist.

Heating and Cooling

Where we heat water, its temperature increases until it boils. Continued heating causes no further increase in temperature. The average velocity of the molecules remains the same no matter how hot the vessel. All additional heat is absorbed, and the energy is put to work in breaking up the hydrogen bonds between the molecules, which must take place before the molecules can evaporate.

When the vapour returns to the liquid state it must give up its heat of evaporation. If 1 gram of steam condenses at 100°C, it gives off 540 calories (2.2 kilo Joule.) of heat, exactly what it absorbed when it evaporates. This principle accounts for the warmth imparted to objects when vapour condenses on them, and the cooling effect (withdrawal of heat) when moisture evaporates from a surface. This has an important moderating effect in biological systems such as the functioning of the sweat glands; when the temperature rises, the cooling effect of evaporation prevents an excessive increase in body temperature. Evaporation is also important for plants, and accounts in part for the cooling effect in areas of vegetation. It has a cooling effect on moist soil and prevents the surface from becoming as hot as it otherwise would from the direct rays of the sun. It has a similar homeostatic influence on the temperature of bodies of water and thus has an important moderating effect on land temperatures and world climate.

Specific Heat

The ability of water to store heat is a characteristic that accounts for much of its biological importance. Water can absorb great amounts of heat while the temperature increase is very little. If we walk bare feet over sand, rock, or pavement on a hot day, the heat may soon become unbearable and we will be relieved to step into a pool of water. Although the water has received the same amount of the sun's radiation, it remains refreshingly cool. On the other hand, in the evening the sand and pavement lose their heat but the water stays about the same temperature that it was during the heat of the day. Upon cooling, the temperature of sand drops five times faster than that of water. This great capacity of water to absorb heat, the slowness of water to warm up and cool off, and its ability to give up great quantities of stored heat is summed up in a property called its *specific heat.* The specific heat of a substance is the number of calories required to raise the temperature of 1 gram of the substance through 1°C.

The water in a lake or in the sea gives up 5 times as much heat as the same amount of soil, sand, or rock. That is why a large body of water has a moderating effect on the temperature of the surrounding area. During hot weather when the water receives large amounts of the sun's radiation, the water absorbs great quantities of heat. Equally large quantities of heat are given back to the air during cold weather. Much of the life on earth is dependent on the moderating effect produced by the three-fourths of the globe that is covered by water.

HYDROSPHERE

The total volume of water reserves available on land, oceans and the atmosphere of earth is known as hydrosphere. Water is the most useful natural resource on earth, economically, culturally, and biologically. Though water is seemingly abundant, the uneven distribution of usable water creates a serious conservation problem in many places where it is vitally needed. In such areas the purity of water becomes critical. Rivers and lakes in highly industrialized locations may carry an intolerable burden of chemical and human waste products to the point that aquatic life in its natural habitat is wiped out and human health is threatened. Even waters used for irrigation of crops may have excessively high concentrations of salts as the result of leaching or, along coastal areas, from underground seawater intrusion. Wells in such areas become nearly useless as sources of water for agricultural or domestic purposes.

About 300 B.C., Alexander the Great and his army of tough adventures thrust deeply into Persia and the unknown land beyond. They came to a mighty river with more than twice the flow of the Nile, and on its banks was an ancient civilization. A thousand years before Alexander, the

Aryans had invaded the continent and settled on the river. They did not give it a particular name but called it the Indus, the Aryan word for "river". The conquered land became India, the "land of the river." The Indus was destined to irrigate 23 million acres of land, the largest irrigated agricultural region on earth. Civilizations sprang up along the other waterways. In Egypt, the middle East, India and China, as well as in South and Central America, ancient civilizations had their origins in the development of the water supply for irrigation and the production of a dependable food supply.

Modern civilization is dependent on water for irrigation, industry, domestic needs, shipping and, of increasing importance for sanitation and waste disposal. Most of the areas of the world that are without developed water remain in the hunting and gathering or grazing stage. Civilization's further advance, and possibly the survival of our cultures will require intensive study of the development of water supplies and careful attention to the protection of water quantity and quality on a worldwide scale.

The Wet Planet

A man in space can look down on the earth and see that the surface of the planet is mainly an ocean. It is continuous except for interruptions by numerous islands. If Mount Everest, standing at 29,028 feet, were put into the deepest spot in the ocean, its peak would be more than a mile beneath the surface. The ocean occupies 70 per cent of the surface and contains 97 per cent of all the water on earth. Much of the remainder is frozen in the icecaps and glaciers. By comparison, the water in rivers and lakes is small. Less than 1 per cent is in the form of icefree fresh water in rivers, lakes, and aquifers. Yet this relatively negligible portion of the planet's water is crucially important to all forms of terrestrial and aquatic life. There is also a large underground supply of water. Much of it remains locked deep underground for long periods of time. But the soils near the surface also serve as reservoirs for enormous quantities of water, eventually lost through evaporation or by seepage into underground storage.

Some of the earth's water is in the atmosphere. The warmth of the sun and the air currents evaporate about 0.03 per cent of the water on the surface each year. The water vapour condenses and returns as rain and snow and eventually finds its way to the bodies of water to complete the cycle.

Much of the earth's water is in cold storage. Glaciers and the icecaps cover 11 per cent of the world's land area; icebergs and pack ice occupy 25 per cent of the ocean area. Permafrost—permanently frozen ground-holds another 10 per cent of the land area in its grip, while 30 to 40 per cent of the land is covered with snow at any given time. Three-fourths of all fresh water is locked up as ice, mostly in Antarctica and Greenland. The cold regions of the earth contain vast resources in minerals, petroleum, timber, and water. For this reason the rapidly increasing demand will surely bring more intensive development, and industrial activity in those areas. However, environmental problems of the cold regions differ greatly from those in the temperate and tropical parts of the world. Disturbances of the environment that may be mildly dusruptive in tropical and temperate areas could be excessively damaging in cold regions.

Small changes in world climate drastically affect the earth's distribution of water supplies. Glaciers are highly sensitive to climatic trends.

The distribution of water in the hydrosphere has not remained unchanged in various reserves throughout the course of history. World's climate has changed at least 4 times during the last 1 million years, experiencing 4 periods of glaciation when most northern portions of North America and Eurasia were covered by huge ice-sheets. The periods of glaciation alternated with interglaciation when most of ice-sheets melted to remain confined to polar region. If these ice-caps today melted completely, sea level would rise, by over 60 m inundating large parts of the continents. During periods of glaciation similarly the sea level is estimated to have fallen by upto 140 m. Such changes could have serious consequences.

HYDROLOGIC CYCLE

The oceans, the continents and the atmosphere the three great water reservoirs of the earth are not in any sense self-contained units but are in constant communication with one another. The relationships take the form of a circulation. Under the influence of solar radiation, water is constantly converted into water vapour and sent into the atmosphere. This process can occur in a number of ways. On the one hand, water evaporates directly from the surface of seas, lakes and rivers, from glaciers or straight out of the ground. Plants lose water in vapour form through *transpiration* as animals do through perspiration. A third process is the emission of water vapour by the combustion of organic materials such as wood, coal and oil, which are to be found all over the earth. The highest proportion of vaporized water, however, emanates directly from the oceans.

The water vapour content in the atmosphere is measured in terms of the relative humidity of the air. The rate at which the atmosphere receives vapour is largely determined by its relative humidity; the rate being faster at low relative humidity and slower at higher relative humidity. Whereas some portions of the vaporized water masses remain permanently suspended in the atmosphere in the form of moisture, others are condensed back by the cooling down of ascending airstreams and emerge in the form of clouds, fog, rain or dew. If the water cools still further, snow or hailstones may build up in the clouds. When the clouds get too heavy, *precipitation* begins and the moisture returns to the earth in the form of rain, snow or hail. If the precipitation falls directly into the sea or into lakes having no outlet, the water circulation will be complete. If it falls on land, some of it will accumulate by surface drainage in streams and rivers and so return to the ocean. Other forms of precipitation seep into the earth and replenish the groundwater, or if retained at an upper ground level will evaporate from there and is partly available to sustain plant life, which absorbs it through its roots and returns it to the atmosphere as water vapour by respiration and transpiration.

The hydrologic cycle is maintained by regular inflow of solar energy. Of the total amount of energy absorbed by the earth, about one-third is utilized in maintaining water circulation. The balance between the water received and water lost by a region, its water budget is of considerable importance in shaping its climatic and economic conditions. If the precipitation exceeds

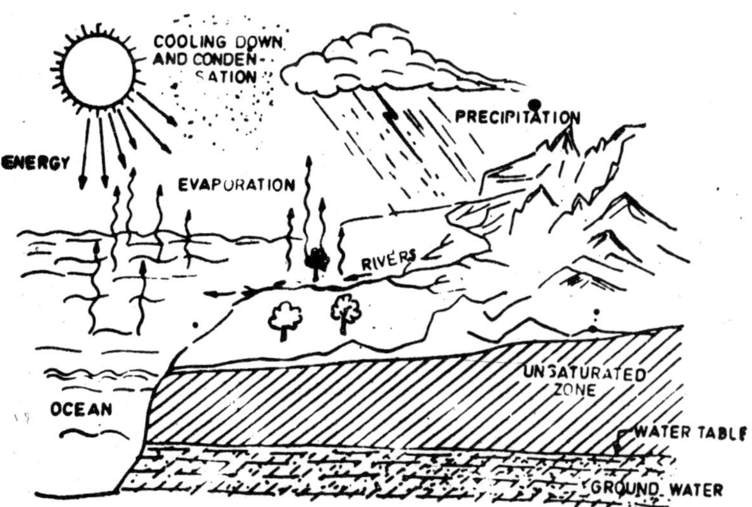

The Hydrologic Cycle.

the losses, the area has a favourable water balance for agriculture. If the reverse is true, the unfavourable water balance may severely affect the agriculture as in semi-arid and arid climates.

HYDROPONICS

Plants do not need solid to grow. All they need is air, mineral, nutrients and water. Growing plants this way is becoming increasingly common and the technique is referred to as *hydroponics* or *solution culture*. This technique was employed by John Woodward in the 17th century in an investigation of nutrition in mint (*Mentha* sp.) plants. Then in about 1860, three German plant physiologists—Julius Sachs, W. Knaps and W. Pfeffer—employed this technique extensively to determine the kinds and amounts of elements essential to the growth of plants.

The plants are usually grown in plastic sacs of water and mineral nutrients. Because their roots are in water and not in soil, plants grown hydroponically require support, either from their plastic containers or from frameworks over the containers.

To support the plant mechanically the roots are usually embedded in an inert material, such as vermiculite (hydrated magnesium—aluminium silicate), perlite, plastic or glass beads on to which nutrient solution is poured. The solution is either replaced at intervals as it becomes depleted or is allowed to flow continuously over roots. For optimal growth the nutrient solution must contain all the elements which the plant requires, in a suitable form and in approximately the right proportions. The pH value is adjusted to an optimal of 6.5.

The advantage of the hydroponic culture is that the nutrients supplied to the plants can be regulated precisely. The disadvantages are the need for root aeration. Another is the need to replace the solution every day or two for maximum growth; this is because the solution composition changes continuously as certain ions are absorbed more rapidly than others. This selective uptake not only depletes certain ions but also causes undesirable pH changes. Commercial greenhouse owners sometimes use the *nutrient film technique* in which the solutions are recirculated in a thin layer through troughs around the roots of valuable crops. Such solutions are pumped from tanks in which the pH and solution composition can be monitored and adjusted automatically. The pump forces the solution across the roots; then, when the pump is temporarily turned off, the solution is drained downhill, leaving a well-aerated film of nutrient solution on the root surfaces. Another disadvantage of hydroponic culture is its cost. It is much more expensive to buy land and hydroponic tanks for plants than merely to plant the crops straight in the ground. Large-scale hydroponic agriculture is, therefore, economically impossible, but hydroponics has an increasing role to play nevertheless. For instance, there are islands in the Pacific that are almost all rock and have little soil. Importing food to these islands is expensive, and it is cheaper to grow food on the islands using hydroponic techniques. Similarly small quantities of those plants, which are not commercially available, can be produced efficiently by this method.

GROUND WATER

A large proportion of the precipitated water (rain, snow, dew, etc.) falling on land sinks into the ground and supplements the underground water reserves. Two forms of subterranean water supplies can be distinguished: The capillary water in the upper *unsaturated zone* and the ground water below the water table in the *saturated zone*. Where the rainfall is light, the precipitation penetrates by seepage only into the upper strata and remains suspended in the highest levels. This is particularly evident when the ground consists of peat or other very clayey material. The highest stratum is then described as a water-retentive zone. If the precipitation is greater or the subsoil is very sandy or gritty, the percolating water fills up the whole pore space of the soil, but once the percipitation stops, the non-capillary pores lose their water by gravitational pull within a few hours. This water reaches the ground water; only capillary water remains back. In the lower-lying underground strata the seepage water spreads mostly over an impenetrable rock layer and so provides the groundwater. The water-carrying ground layers are described as groundwater layers.

Groundwater is in constant interchange with the rivers. As a general rule the ground water level is higher than the neighbouring river bed. The slope of water table from its point of elevation where it receives initial charge to its lowest point when it enters into a lake, stream or sea is called *hydraulic gradient.* In dry regions the groundwater lies considerably deeper than in wet regions. It may be near the surface in wet climates, especially during seasons of high precipitation or may be as deep as 100 m in arid climate during dry periods.

In the bedrock a distinction can be made between quarry water and groundwater. Quarry water seeps through small holes, in the bedrock, and serves gradually to enlarge them by physical and chemical processes. The chemical processes are particularly marked with a salt and gypsum subsoil or a limestone rock, owing to the high carbon dioxide content of the seepage water. Ground-stratum water is found primarily in sedimentary rock such as sandstone and loess.

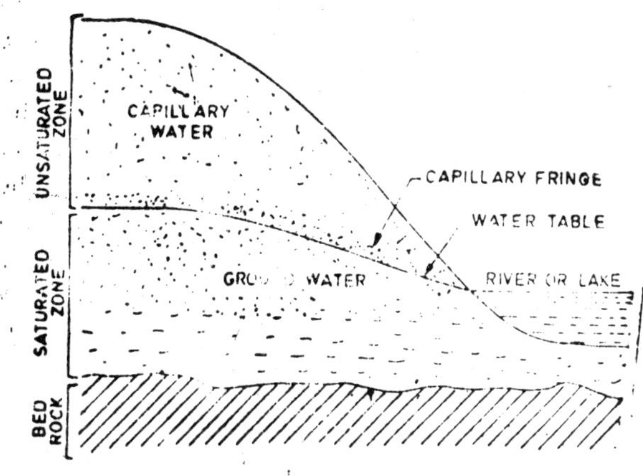

Sub-Surface Water Zones.

As a rule, groundwater is very rich in soluble matter. The concentration depends on the duration of the subterranean flow of water, the temperature, the solubility and composition of the rocks.

Groundwater is of immense importance in our biosphere. It helps to maintain the water levels of lakes and rivers. In addition the ground water is brought to the surface by digging wells and used for domestic and agricultural purposes. In great Australian Artesian basin the groundwater saturates the porous stratum of sandstone which dips down from high lands. This great reservoir of ground water (called *aquifer*) lies over a bed rock and beneath a cap shale both of which are impermeable when a well is dug through the shale, water rises up through hydrostatic pressure and no pumping operation is needed. Such wells are called *artesian wells.* Similar aquifers in the mountain areas on getting an opening form *springs.*

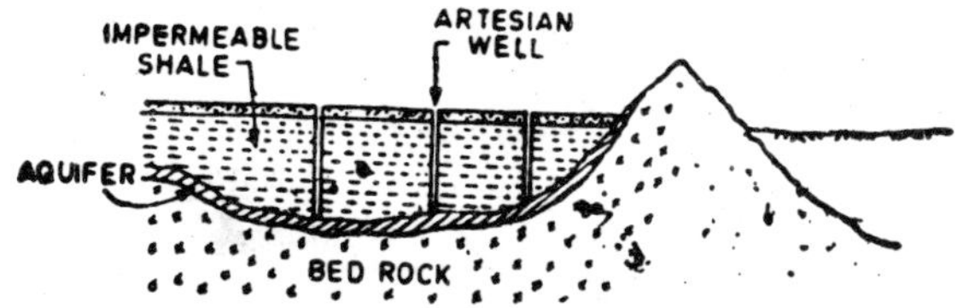

Artesian Well.

Although groundwater is commonly not available for direct use by the plants some of the ground water rises some distance through the profile, above the water table through capillary action to fill up the capillary porespace, whereas the soil above may be relatively dry. This zone of capillary rise above the water table is termed *capillary fringe.* The plants which are able to tap this capillary fringe are called *phreatophytes.* The water rising up the profile sometimes may reach the surface and on evaporation leave back the dissolved salts in ground water forming a crust of salts on the ground.

GROUNDWATER POLLUTION

The pollution of groundwater is a subject of a major concern firstly because of its increasing utilisation for human needs and secondly because of the ill effects of the increased industrial activity especially in the urban climate. Ground water gets polluted in a number of ways.

If raw sewage is dumped onto the soil, the liquid percolates into the ground. The soil being porous blocks large solid particles while allowing the liquid to pass through. Smaller particles and even molecules of contaminants, although not physically blocked, adhere to the soil particles, and they, too, are removed. After rain or irrigation, however, the percolating water dissolves mineral matter out of the soil and carries it to the ground water. In rare instances, such dissolved matter may be toxic, as for example when the water seeps through areas containing lead or arsenic minerals.

The pollution of groundwater from human sources presents special problems that are different from the pollution of surface water such as lakes or streams. There are two main reasons for this difference :

1. Most groundwater moves quite slowly through its zones a typical rate might be about 30 cm per day. Furthermore, the water does not mix as much during its motion through porous rock as it would, say in a river. Instead, the flow of water advances more like a column of marchers in a parade, who do not mingle much with the crowds on either side. Consequently, pollutants introduced into groundwater are not readily diluted. The pollutants commonly move along a *pollution plume* along the flow of ground water until discharged into a lake.

2. Groundwater does not have the access to air that is available to surface waters. Therefore, the oxidation that can purify or decontaminate surface water does not occur in deep aquifers.

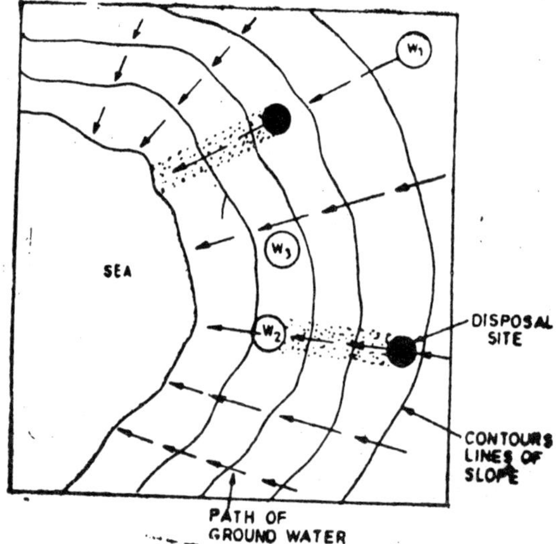

Wastewater is often stored in a basin, pitpond, or other such facility. The purpose of such storage is to hold the waste-water prior to treatment or simply to allow oxidation by air to decontaminate it. Many of them are unlined, and the soils beneath them are permeable, so the polluted water can seep down. The same problem is often associated with solid wastes, which are stored in landfills. These, too often, leak contaminants into the ground. The percolating rainwater dissolves and carries the leachate to ground water. Since the landfills usually form a mound, the water table beneath may also rise and the leachate reaching the water table may move more rapidly along the gradient to reach a stream or a depressed water table of a well.

Viruses, which can multiply in sewage and in landfills, are very difficult to remove from water. Viruses pass through ordinary filters, and are resistant to chlorination. Seepage from improperly designed landfills therefore imposes a danger of groundwater contamination by viruses.

It is thus important that landfills and sewage disposal sites should be located in depressions and at levels lower than well sites. On a slopy ground where the slope finally leads to a sea it is advisable that disposal sites lie downslope at lower level than the well, or when the same is not possible higher disposal sites should lie away from the flow path of the pollution plume. In both cases the pollutants will reach the sea without polluting the wells and readily disperse in wide expanse of sea water without much damage.

Bioresistant pesticides, which are sometimes heavily applied (or misapplied) to deal with large infestations of insects, can percolate down to groundwater.

Liquid chemical wastes that are injected into deep wells below the aquifers may sometimes leak or migrate up into groundwater sources. Pipes can burst under high pressure or can be corroded by acidic wastes.

Mine tailings can be very troublesome because they may contain toxic mineral matter. When these wastes are exposed to the outside, rainwater (which is slightly acidic naturally but more so if it is polluted) can dissolve some of these minerals and carry them down to an aquifer.

Ground water pollution can take alarming proportion in limestone regions where the groundwater dissolves away limestone to form underground interconnected cavities called *caverns* which often open to the surface by collapsing of their roof to form *sinkholes*. The sewage discharged into sinkholes can move freely through caverns to cause large-scale pollution of wells, ponds of lakes.

FLUORINE AND FLUORIDATION

Fluorine, belonging to the halogen family of elements, is the most corrosive chemical element that is known. It is a greenish-yellow gas, which, like the fluorine compounds, is highly toxic. The addition of fluorides to drinking water to prevent dental caries (tooth decay) had been opposed by some doctors and toxicologists but now has been endorsed by all international health organizations. The application of fluorides in toothpastes or for internal consumption hardens the enamel on teeth and so provides protection against destruction of the teeth by caries.

Water is necessary for life, whereas fluorine and its compounds can be toxic substances. Fluoride at high concentration produces a corrosive effect on glass, steel and a number of other metals. Drinking water is treated with 1 mg of fluoride per litre of water. At concentrations over 2 mg per littre people begin to react gradually with symptoms of mild dental fluorosis, as part of the daily intake of fluorine accumulates in the body. The fluorosis manifests itself as white flecks in the dental enamel. Pitting of the enamel can result from flouring concentrations of 3-4 mg per litre.

People do no absorb fluorine into their system only as an additive in drinking water, but fluorine is present in their diet in natural and artificial forms. This supply of fluorine in the daily diet varies in different countries and for different types of food, so that the total intake of fluorine in drinking water and foodstuffs cannot be assessed precisely. A further factor is the loading of the atmosphere with the fluorine waste gases from aluminum works and similar industrial plants, which can also result in an uncontrollable intake of fluorine.

Although fluorine has been shown active in a dental context only for children under 15 years of age, the entire water supply system is treated with fluorides. It is estimated that about 1000 litres of water would have to be treated with fluoride to enable a child whose teeth are still developing to drink 1 litre of treated water.

Caries is not a disease caused by fluorine deficiency but is the result of bacterial action aggravated by general nutrient deficiencies and lack of dental care. Fluorine medication can, at best, influence the symptom of caries, but not the disease itself.

EUTROPHICATION

Lakes, and ponds are constantly being enriched by organic matter from decomposing plant and animal remains, providing nutrition for algae and larger aquatic plants. This fertilization, whether from natural sources or man-made, is called *eutrophication*, from the Greek meaning "well nourished". This condition occurs often in nature and is not necessarily bad, for the organic material means food for fish and other aquatic life.

But too much mineral nutrient in the water can be more serious than too little. Eutrophication becomes excessive, however, when abnormally high amounts of nutrients from sewage, animal waste and detergents enter water causing excessive growth of micro-organism and aquatic vegetation. Many people have noticed the reddish or greenish colour of lakes, ponds or coastal areas caused by enormous numbers of algae and other phytoplankton, causing the death of fish and other aquatic life, sometimes on a massive scale. When the algae and other plankton die, decomposition of the superabundance of organic material causes the oxygen in the water to become suddenly depleted. Water cannot absorb replacement of oxygen fast enough to take care of the needs of the living organisms. Thus, they suffocate. Also, toxins are sometimes produced.

The "red tide" that is frequently seen along the coasts of California, the Gulf of Mexico, and India is caused by the buildup of microscopic *dinoflagellates*. They impart a red or dirty brown colouration to the water and at night create a spectacular display of luminescence in the froth of waves. Their most injurious effect is that production a lethal poison called *saxitoxin*, which accumulates in the bodies of clams and mussels. People who eat them are liable to be stricken with "paralytic shellfish poisoning,' a dangerous form of food poisoning having a high mortality rate. The red tide along the coast of India is thought to be caused by an over-abundance of mineral nutrients deposited by the heavily mineral-loaded run-off of flood water during the monsoon rains. Along the California coast, it is suspected that the nutrients from sewage outfalls contribute to excessive growth of the ogranisms when suitably high temperatures and other favourable conditions prevail.

Most of the open sea is a biological desert, an essentially barren area comprising 90 per cent of the ocean or nearly three-fourths of the earth's surface. Half the world's fish supply is produced in coastal waters and in a few offshore areas of comparably high fertility. The other half is produced in regions of upwelling water that total no more than 0.1 per cent of the ocean's surface. Thus, the most productive portions of the ocean—the coastal waters—are those which are the most polluted and where the marine life is the most susceptible to disturbances in the environment.

Since pollution of waters from sewage is a major danger to fish life, better sewage treatment is needed to prevent premature eutrophication of lakes and ponds. In addition to primary treatment (which removes large objects and suspended solids), and secondary treatment (which decomposes organic compounds and disinfects the water) a tertiary treatment (which removes nitrates, phosphates and other nutrients) is also necessary to prevent eutrophication of lakes ponds, etc. Excessive nutrient level which brings about eutrohpication may be hazardous when proliferation of weeds like water hyacinth, Eurasian milfoil and water chestnut occurs. These weeds often impair fishing, bathing and even navigation. Water hyacinth has become a major hazard in many parts of India to earn the name *'Bengal terror'.*

DETERGENTS

Any substance that reduces the surface tension of water specifically, a surface-active agent which concentrates at oil-water interfaces, exerts emulsifying action, and thus aids in removing soils. The older and still widely used types are the common sodium soaps of fatty acids which are relatively weak. Stronger synthetic detergents are classified as a cationic, anionic or nonionic, depending upon their mode of chemical action. The most widely used group comprises linear alkyl sulphonates (LAS) which are biodegradable. Detergents consist of *surfactants. Surfactants* are surface active agents, substances having properties similar to soap. Mixed with conditioning and water-softening agents, mainly polyphosphates. The most commonly used surfactants now are linear alkyl sulphonates, *e.g.*, sodium dodecylbenzenesulphonate. These compounds are rapidly degraded by bacteria. Materials that are decomposed by micro-organisms are referred to as *biodegradable*, and do not normally constitute a serious environmental hazard, unlike the branched chain alkyl compound, *e.g.*, tetrapropylene benzene sulphonate, which they replaced and which were resistant to bacterial degradation. However, all detergents have appreciable toxicity and, if released in large amounts, may reach concentrations which kill many organisms. This is most likely to occur when they are being used as dispersants to break up oil slicks on beaches or in natural waters. Even if the detergents do not kill the ogranisms directly, they may damage cell membranes sufficiently to allow other toxicants to act. Thus, the benefits of rapid oil dispersal by detergents must be weighed very carefully against the risks involved before their use is approved.

Detergents used for cleaning purposes contain compounds called *builders.* The most common builders used are polyphosphates. Sodium tripolyphosphate, the most widely used phosphate builder in detergents, in one of a family of phosphates used extensively as water softeners.

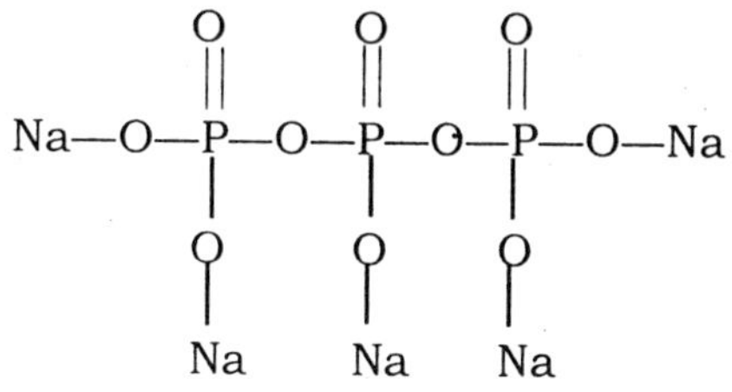

```
           O       O      O
           ||      ||     ||
Na—O—P—O—P—O—P—O—Na
           |       |      |
           O       O      O
           |       |      |
          Na      Na     Na
```

sodium tripolyphosphate

The value of the polyphosphates is in their ability to form soluble complexes with alkaline earth metal ions, mainly calcium (Ca^{2+}) and magnesium (Mg^{2+}) found in hard waters. Compounds having this action are called *sequestering agents.* The polyphosphates sequester divalent ions such as calcium and magnesium so that they do not interfere with the cleaning action of the surfactant. Builders also make wash water alkaline, thus improving dirt removal.

The polyphosphates hydrolyse to phosphate when released into the natural environment, and this may contribute to excessive eutrophication, enrichment of the water with nutrients, accelerated growth of algae, oxygen depletion of the surface water, and impaired fish production. The ecological balance is destroyed.

However, substitutes for polyphosphates have all got drawbacks and it seems likely that their use will continue, but with care being taken to remove them from waste water before it is released. This can be done quite readily by precipitation with aluminium sulphate.

WATER POLLUTION

The addition to water of an excess of material (or heat) that is harmful to humans, animals, or desirable aquatic life, or otherwise causes significant departures from the normal activities of various living communities in or near water bodies.

Types of Impurities in Water

Chemically pure water is a collection of H_2O molecules—nothing else. Such a substance is not found in nature—not in wild streams or lakes, not in clouds or rain, not in falling snow, nor in the polar ice caps. Very pure water can be prepared in the laboratory but only with considerable difficulty.

Toxic Chemicals

Water is an usually good solvent. It is especially good at dissolving mineral salts, which typically consist of negative and positive ions. The positive ions are those of metals, including many that are poisonous, such as the ions of copper, cadmium, mercury, and lead. Water is also a good solvent for many organic compounds that contain oxygen, such as alcohols, sugars, and organic acids. Furthermore, many materials that are normally considered insoluble in water are in fact very slightly soluble. Hydrocarbons are said to be insoluble in water, yet benzene, for example, dissolves in water to the extent of about 0.1 per cent.

Insoluble particles, if they are small enough, may settle so slowly that for all practical purposes they remain in water indefinitely.

Nutrient matter is metabolized by living organisms in water, and the resulting waste products may be pollutants. A soluble substance may react with an insoluble contaminant and bring it into solution. For example, acids in water react with many minerals and thus dissolve them. Finally, a contaminant may pollute water simply by floating on it. Water is denser than almost all hydrocarbons; therefore, petroleum floats on water. A floating oil spill is a water pollutant. (See individual entries for Acid Rain, Arsenic, cadmium. Lead, Mercury, Oil spills, Pesticides etc.)

Disease-Causing Agents

To humans the most serious problems of the pollution of drinking water are water borne diseases, particularly typhus, salmonellosis, dysentry and cholera. In many regions of the world, water purification systems are either inadequate or non-existent. Millions of infants die at an early age from drinking polluted water, and even adults suffer from stomach disorders. Disease-causing bacteria and viruses generally enter the water supply from human or animal fecal matter.

Classification of Impurities

Impurities in water may be divided into three classes-suspended, colloidal and dissolved.

Suspended Particles, which have diameters of more than about 1 micrometre, are the largest. They are large enough to settle out of water reasonably quickly and to be retained by many common filters. They are also large enough to absorb light thus making water containing them look cloudy or mucky.

Table 1 : Impurities in Natural Water

Source	*Particle Class*				
	Suspended	*Colloidal*		*Dissolved*	
1	*2*	*3*	*4*	*5*	*6*
Atmosphere			*Molecules*	*Positive ions*	*Negative ions*
	← Dusts →		Carbon dioxide, CO_2 Sulphur dioxide, SO_2 Oxygen, O_2 Nitrogen, N_2	Hydrogen, H^+	Bicarbonate, HCO_3^- Sulphate, SO_4^{2-}
Mineral soil and rock	← Sand silt → ← Clays → ← mineral soil particles →		Carbon dioxide, CO_2	Sodium, Na^+ Potassium, K^+ Calcium, Ca^{2+} Magnesium, Mg^{2+} Iron, Fe^{2+} Manganese, Mn^{2+}	Chloride, Cl^- Fluoride, F^- Sulphate, SO_4^{2-} Carbonate, CO_3^{2-} Bicarbonate, HCO_3^- Nitrate, NO_3^- Various phosphates
Living organisms and their decomposition products	Algae Diatoms Bacteria ← humus → Fish and other organisms	Viruses Organic colouring matter	carbon dioxide, CO_2 oxygen, O2 nitrogen, N_2 hydrogen sulfide, H_2S methane, CH_4 various organic wastes, some of which produce odour and colour	Hydrogen, H^+ Sodium, Na^+ Ammonium, NH_4^+	Chloride, Cl^- Bicarbonate, HCO_3^- Nitrate, NO_3^-

Colloidal Particles are so small that their settling rate is insignificant, and they pass through the holes of most filter media; therefore, they cannot be removed from water by settling or by ordinary filtration. Water that contains colloidal particles appears cloudy when observed at right angles to a beam of light. The colours of natural waters, such as the blues, greens, and reds of lakes or seas, are caused largely by colloidal particles.

Dissolved Matter does not settle out, is not retained on filters, and does not make water cloudy. The particles of which such matter consists are not larger than about one-thousandth micrometre in diameter. Natural waters contain substances in all three categories as shown in Table 1.

Sources of Water Pollutants

The most serious destruction of water quality comes from municipal wastes and industrial wastes. The runoff from feedlots; the drainage of acids from mines; the erosion of soil from farms, roads and construction sites; the spillage of oil from tankers and pipelines; the disposal of solid and liquid wastes by community sewers and water treatment plants; the leaching of salts from irrigated lands; and the release of heat by way of water cooled generators etc., are important causes of water pollution.

MARINE POLLUTION

Nearly three-fourths of the Earth is covered by oceans, in a continuous ecosystem broken only by intermittent land masses. Oceans account for 97 per cent of water on the earth. The oceans perform an important role in moderating the earth's climate: they are also the source of an enormous range of plant and animal life.

Phytoplankton flourish in a thin layer of water stretching to 100 metres in depth, exploiting the energy from the sun and nutrients in the water to expand and multiply. They form the food supply for zooplankton. These two types of plankton are so abundant that, they are estimated to generate, respectively, 16 billion and 1.6 billion tonnes of carbon (the basic material of living tissue) each year. They form the food source for a variety of larger creatures, which in turn have an annual productivity of around 160 million tonnes. Plankton-eating species, particularly the herring family, supply humans with about two-fifths of their fish.

Areas of greatest plankton abundance are found where the seas are rich in nutrients (i.e. useful food). The most productive places are along the coasts of the land masses: these receive a number of nutrients washed from the land to the sea, which feed a range of life forms in the warmer, shallower waters around the coast.

Less than 1 per cent of the seas are occupied by coastal ecosystems (the vast bulk of ocean being the less productive open seas, or the slightly more productive continental shelf). The four main types of coastal ecosystem are : estuaries; tidal wetlands of temperate zones (*i.e.*, *saltmarshes*); their equivalents in tropical zones (*i.e.*, mangroves); and coral reefs. All are facing particular problems in relation to human activity, connected with development or pollution. Pollutants can be transported into the marine environment along rivers or can arrive via the air. Substantial quantities of pollutants are deposited deliberately in the seas via pipelines or from ships.

The various sources of marine pollution are oil, sewage, persistent pollutants and radioactive materials.

Marine pollution through oil attracts world attention when huge oil tankers meet with a disaster, spilling huge quantities of oil on the ocean surface. The oil spreads fast forming a layer at the surface, cutting off gas exchange with atmosphere and suffocating large quantity of marine life to death. The Torrey Canyan, a huge oil tanker which sank off the coast of Cornwael, England spilled 100,000 tonnes of oil killing much of the coastal marine organisms in the region, other major episodes include wrecking of super tanker Amoco Cadir in 1978 along French Coast line, Kowloon Bridge grounding on west coast of Ireland in 1986 and grounding of Exxon Valdez in 1989 in Alaska. These accidental spills, however, contribute no more than one-tenth of the total oil discharged into the oceans from flushing of tanks offshore wells, refineries, rivers and drains. The total discharge into the ocean is estimated to be between 2-5 million tonnes annually. One of the recent event of oil pollution was the enormous volume of crude oil which discharged into the gulf waters during the beginning of 1991 from Kuwaiti coastal oil facilities during the major war in the region. This was perhaps the biggest single discharge of oil which threatened many coastal water treatment plants, as far away as the coast of Saudi Arabia, and resulted in the death of yet unaccounted marine life.

Sewage is another major source of marine pollution in areas close to major sewage outlets. The dangers from this type of pollution are limited to areas within a few kilometres from the point of discharge, but they have caused health problems and economic losses. Even in relatively developed regions, sewage is pumped into the sea untreated: in the Mediterranean, an estimated 90 per cent of the sewage generated around the coast is still dumped raw into the water.

Many countries now use secondary sewage treatment before discharging effluent into the sea. In Canada, the U.S. and Japan, secondary treatment is carried out in 75 per cent of treatment plants; in Sweden, tertiary treatment is carried out in 80 per cent of plants.

Finally, even treatment of sewage can cause marine pollution when the waste products are dumped at sea. After the liquid and solid sewage is treated, a sludge remains which can contain up to 60 per cent of the original contaminants. The UK is the only nation to continue dumping this sludge in the North Sea—some 9.4 million tonnes a year, at 13 sites round the coast, with the majority being dumped in the Thames Estuary. Dumping increases the input of nutrients such as nitrogen and phosphorus into the marine environment, increasing the risk of eutrophication. Incidences of fish disease in the US have been associated with sewage sludge dumping.

The two best-known persistent pollutants in the marine environment are the pesticide DDT and the polychlorinated biphenyls. Organochlorine pesticides generally are persistent in the environment : they are part of a number of synthetic organic compounds called 'xenobiotics', which break down extremely slowly and so circulate in ecosystems for many years. Xenobiotics build up in the food chain (they are eaten by small fish, which are eaten by larger fish, etc., with the quantity of xenobiotic becoming larger as it is absorbed by larger animals or people) and thus represent an increasing threat to people who unwittingly eat polluted fish.

Another group of persistent pollutants are the heavy metals, such as mercury, cadmium and lead, which are released in enormous quantities from industry.

Metal pollution first came to prominence in the *Minamata Bay* episode in Japan. Since the 1930s, the Chisso chemical plant had been discharging methyl mercury into the bay which local villagers used as a source of shellfish. This mercury was known to be toxic, but it was heavily diluted in the waters of the bay, and was believed to be dispersed.

In the 1950s, the villagers began to exhibit what came to be known as '*Minamata disease*'. The symptoms of methyl mercury poisoning include: severe mental retardation, seizures, loss of co-ordination, partial or total loss or vision of hearing. It also affects embroys in the womb, so that apparently healthy mothers were giving birth to severely affected children. All in all, 1,385 villagers suffered mercury poisoning and 649 died.

Generally, when mercury levels in edible fish exceed one part per million (1 milligram per kilogram), in Europe and North America, people are banned from eating it. This has frequently occurred in the North Sea over the past few decades, and sporadic bans have been applied.

Radioactivity

Radiation pollutes the world's oceans from four separate sources: atmosphere fallout, dumping of radioactive waste, discharges from nuclear installations and discharges from nuclear-powered vessels. The latter is usually accidental.

The marine—and terrestrial—environment suffer continuing fallout of radiation from the airborne testing of nuclear weapons in the 1950s and 1960s. The dumping of radioactive waste in the sea, carried out by the UK, Belgium, The Netherlands and Switzerland between 1967 and 1982 in the Atlantic Ocean off the coast of Spain, was halted by a decision of the London Dumping Convention in 1983.

Greenpeace points out that at least 250 kg of plutonium, theoretically enough to kill 250 million people, have been discharged into the Irish Sea, making it the most radioactive sea in the world.

Radioactive leaks from nuclear-powered vessels and craft carrying nuclear weapons are responsible for a small amount of radioactive contamination. Though small in scale, these episodes could have significant local effects, which would increase in the event of a large-scale nuclear accident.

There are other marine pollutants, including: *Plastics*, which are not easily broken down by natural processes, and thus tend to accumulate in the oceans. Birds and mammals often eat plastic, mistaking it for food: *World Resources 1987* estimates that up to 1 million seabirds and 100,000 cetaceans (*i.e.*, whales, dolphins, porpoises, etc.) die every year because of plastic pollution.

The North Sea countries—Belgium, Denmark, France. The Netherlands, Norway, Sweden and West Germany—are keen on international action to clean up the area. At a meeting of the International Conference on the Protection of the North Sea in November 1987, the states agreed to end the dumping of harmful wastes into the North Sea by end 1989, improve controls over input of dangerous substances from rivers and prohibit the dumping of garbage from ships. Britain alone insisted on continuing to dump sewage sludge.

However the problems in the North Sea continue; in 1988 a 'killer algae', *Chrysochromulina polylepsis*, started growing in waters off the Swedish coast, and multiplied into a massive algal bloom. This jelled into a floating mass many kilometres wide, which at one time was doubling in size every 20 hours. As the 'killer algae' spread, it suffocated fish over large areas of sea, and other marine life such as mussels, starfish and crabs.

In Sweden, at least 80 tons of fish were lost: Norwegian fish farm losses included 500 tons of dead salmon. Long-term effects on the industry are expected to be serious. To multiply like this, algae need high concentrations of nitrogen and phosphorus and warm water. The former are supplied by pollution: the latter may be a consequence of global warming.

A killer algae growth in the Adriatic, off the Italian coast was reported in 1987. Piles of rotting fish washed up on to Italy's north-eastern coast, with local scientists noting the disappearance of most fish life up to four miles off the shore, in some places stretching up to 15 miles. Life had disappeared in a marine area of 400 square miles.

The Po, Italy's biggest river, is the source to two-thirds of the pollution threatening the Adriatic. It carries raw sewage from Milan and Turin, the effluent from thousands of factories, pesticides and huge amounts of fertilizer residues. Scientists working for Kronos 1991, a national environment group, analysed the Po water and calculated that every year it dumps into the Adriatic at least 60,000 tons of nitrates, 13,000 tons of ammoniacs, 7,000 tons of phosphates, 5,400 tons of detergents, 9,500 tons of crude oil, 1,551 tons of lead, 950 tons of zinc, 850 tons of chrome, 243 tons of arsenic and 65 tons of mercury. Itlay's government was unable to confirm or deny this analysis: the last official survey of the Po was in 1976.

On the positive side, the United Nations Environment Programme has sponsored a multi-billion dollar scheme committing coastal states to protect the sea. Italian government's decision to set aside $3.5 billion to clean up the Po signalled growing determination to deal with the problem.

Overfishing in sea waters has been another reason of deterioration of the marine environment, disturbing the vital ecosystem.

In 1987 and 1988, Norwegian fishermen witnessed the invasion of hundreds of thousands of harp seals. These were literally starving to death and becoming entangled in fishing nets in their attemptts to get food. The phenomenon was eventually attributed to the collapse of stocks of capelin (a kind of sardine) which had been overfished. A similar type of incident occurred in Scotland, 1988, when the sand eel fishery collapsed.

Part VIII

Energy and Environment

- Energy
- Energy Reserves
- Hydroelectric Energy
- Geothermal Energy
- Solar Energy
- Tidal Power
- Wind Power
- Energy from Waste
- Energy from Non-conventional Sources
- Energy Conservation
- Radioactivity
- Nuclear Fission
- Nuclear Fusion
- Nuclear Energy
- Nuclear Energy and the Environment
- Nuclear Power Plants
- Background Radiations
- Chernobyl
- China Syndrome
- Three Mile Island
- Power Generation in India

ENERGY

Like space, time and information, energy is one of the fundamental quantities. It is present in a number of forms which differ basically from one another and together constitute the physical reality of our universe. All physical processes can be construed as a transition from one form of energy to another.

The most important forms of energy are the following :

1. Mechanical energy. In order to raise a body of mass m by height h on the earth (acceleration due to gravity $g = 9.8$ m per sec^2), work (= energy) $E=mgh$ must be spent. This energy consumed in raising the body is then imparted in the form of potential energy (energy of position) to the body. If the body drops the same distance, the energy can be released again. In order to raise the speed of a body of mass m to the level v, energy must be imparted to the body so that $E= \frac{1}{2}mv^2$. If this energy is imparted to the moving body and the body makes impact with an obstacle, it can be released again (kinetic energy, energy of movement).

2. Thermal Energy. In order to heat a body of mass m and of a specific heat c, (which depends on the material of the body) by a fixed temperature gradient ΔT (in ^{0}C), a particular amount of energy $E=mc\Delta T$ is required. This thermal energy can be given off again by the body into the atmosphere, for example in the form of heat radiation. The supply of thermal energy need not necessarily result in a raising of the temperature of a body; but as the melting of boiling point of the body is approached, it can be used for a change in form from "solid" to "liquid" or from "liquid" to "vapour".

3. Electric Energy. If a current of strength l (measured in amperes) and of force V (measured in volts) flow for a time t, the amount of energy E produced by the current = Vlt. With the aid of this energy from a flowing electric current, thermal energy can be released (for electric heating) or a machine can be driven. In order to create an electric field (or to separate two electric charges), energy is needed which will be released again when the field collapses (energy from the electric field). The energy of a charged condenser with capacity C and force V between the plates of the condenser = $\frac{1}{2}CV^2$. If the condenser is short-circuited, this field energy stored in a condenser can be reconverted into the energy of an electric current. Energy can also be stored in a magnetic field corresponding to the energy of an electric field .

4. Chemical Energy. In chemical compounds, energy is stored in the linkage of atoms to molecules. This can vary greatly in amount. Compounds are accordingly described as "poor in energy "or "rich in energy." An energy-rich compound is fuel oil, in the combustion of which (that is, in its chemical reaction with oxygen), part of the chemical bond energy is converted into and released as thermal energy. In the reverse direction, there are chemical reactions in which energy must be stored so that energy-rich compounds are created. One example of this is the photosynthesis that occurs in plants.

5. Energy in Matter. Matter consists of concentrated energy. The connection between energy and mass of matter is given by the formula $E=mc^2$ (in which c=velocity of light). If, for example, a negatively and a positively charged electron collide, they convert their mass completely into radiant energy. Part of this energy can be released by nuclear fission or nuclear fusion (atomic energy).

The Language of Energy

Abbreviation	*Item measured*	*Explantion*
1	*2*	*3*
Mtcc	National energy	=million tonnes of coal equivalent. A million tonnes of coal theoretically

(Contd.)

1	2	3
		produces a certain amount of heat. Other forms of energy are measured in coal equivalent, or sometimes oil equivalent. 1 tonne of coal is equivalent to 0.59 tonnes of oil, 700 cubic metres of gas.
J	energy	=Joule. The joule is an internationally recognized, and precisely defined unit of energy—though rather a small one. There are 26.6 million billion joules in a million tonnes of coal. This would usually be written as 26.6 PJ.
PJ	energy	=petajoule=a million billion joules.
W	rate at which energy is supplied	=watt. 'Power' is the rate at which energy is used or delivered. A watt is a joule per second.
kW	rate at which energy is supplied	=kilowatt=a thousand watts.
MW	rate at which energy is supplied	=megawatt=a million watts. A large power station can supply 500 to 600 MW from one turbine. Three such plants grouped together means the station has a capability of 1,500 to 1,800 MW.
GW	rate at which energy is supplied	=gigawatt=a thousand megawatts.
kWh	energy use over 1 hour	=kilowatt-hour. If the electric fire burns for one hour, it supplies 1 kWh of electrical energy. kWh is usually used to indicate the cost of system for producing electricity.
GWh	energy use over 1 hour	=gigawatt-hour= a million kilowatt-hours.
TWh	energy use over 1 hour	=terawatt-hour=thousand million kilowatt-hours

Uses of Energy

All living organisms have to transform energy to enable them to maintain their vital processes. They obtain his biological energy from their food intake, in which the solar energy obtained from the photosynthesis of plants is converted into chemical bond energy and stored.

For hundreds of thousands of years, every human being has been using up about 2000 calories of energy a day, as biological energy to maintain their metabolism. Today the inhabitants of highly industrialized countries consume some 200,000 calories per person each day, which (besides the biological energy they need) includes the energy required for heating, transport, television, refrigeration, industrial production, consumption and luxuries. For each calorie of energy needed to sustain life biologically, we thus consume over 100 calories for other purposes. It is now estimated that if energy growth continues at the present rate, in 10 years time about 400,000 calories, and in 20 years' time about 800,000 calories, will be produced and consumed in the industrial countries per capita of the population every day. By way of contrast, many residents of developing countries do not obtain even the 2000 calories a day needed to sustain their metabolism.

Energy in the form of oil, coal, gas and electricity is used predominantly in three sectors: industry, households and transport. In the transport sector, some 92 per cent of the energy is used in the form of gasoline and diesel oil, and the remainder in the form of electrical energy for public transport. Household consumption consists 60 per cent of oil, 22 per cent of coal and gas, and 18 per cent of electricity. Industry, on the other hand, uses about 70 per cent oil, 10 per cent solid fuels and gas, and 20 per cent electric power. In transport, the energy is used almost exclusively as the motive power for vehicles: and a small portion is used for heating the interior of vehicles. In the domestic sector, energy is used for space cooling heating and for household appliances, television, lighting etc. In the industrial sector, a large part of the energy consumed is swallowed up by the metal-producing industry (coal for blast furnaces and electric power for aluminium production) and the chemical industry. The energy is used primarily to provide heat to sustain chemical reactions. A large part of the energy consumed in the chemical industry goes into artificial fertilizer production.

Such energy consumption makes possible the increase in automation and mechanization of industrial production. In highly industrialised countries there is a systematic compulsion to produce as economically and profitably as possible so that each individual can conduct his affairs in accordance with his own particular standard of living. What it previously took ten workmen a week to produce can today be produced with ease by one worker with a machine that only needs fuel to run it. Human energy and labour are being increasingly replaced by technical energy and machinery.

ENERGY RESERVES

There are two fundamentally different forms of energy reserves, energy **raw materials,** which are available in only limited quantities on the earth and are consumed in power stations for energy production (oil, coal and natural gas) or are split for the same purpose (uranium and thorium), and **regenerative natural reserves** which can be used by man for energy production (sun, wind tides, the heat of the earth, differences in temperature in the oceans, etc.). These latter sources of energy have the great advantage that they are never exhausted, as no raw materials are consumed. When energy raw materials are consumed waste materials are produced that contaminate the environment (atomic waste, carbon dioxide, sulphur dioxide). When recourse is had to regenerative natural reserves no waste substances are produced.

Despite these enormous advantages, energy research and energy consumption have been concentrated in the past almost entirely on the utilization of fossil and nuclear fuels. This onesidedness is a primary cause of the existing worldwide shortage of energy.

Supplies of oil and gas are running out. UNEP estimates that we are using up available reserves of oil at a rate of 3.2 per cent per year, and natural gas at a rate of 1.7 per cent per year. This means that the world will run out of exploitable oil in 31 years, and of gas in 58.

World Resources 1986 reckons that oil production could continue at its present level until some time between 2025 and 2050, but would then drop off sharply; and natural gas should be able to maintain its current output for several decades.

The Worldwatch Institute calculate that ultimate depletion of world oil resources is between 50 and 88 years away.

There is a possibility that large oil and gas fields are yet to be discovered, although in the case of oil, more than three quarters of the world's sedimentary areas have been explored. Although the exact date of oil and gas depletion remains the subject of discussion, they are going to run out, probably in the first half of the next century.

Of fossil fuel reserves 89 per cent are in the form of coal. By far the largest proportion of available fuel reserves is in the form of coal. The established and estimated reserves of pit coal and brown coal on the earth together amount to some 8.8 trillion tons. On the assumption that world energy requirements will remain at 10 billion tons of pit-coal units, which corresponds to the established world energy requirement for 1975, coal alone, disregarding, all other sources of energy, should be able to satisfy world energy requirements for the next 880 years.

Though the world does have abundant supplies of coal, it may not be wise to burn them, because of the increasing carbon dioxide build-up in the atmosphere. Carbon dioxide release is unavoidable when coal, oil, or gas is burnt and the quantities emitted so far, together with pollution from other sources, are already large enough to cause an increase in the world's average temperature (see Greenhouse Effect).

This means that there are serious problems with all three principal sources of primary energy. Oil and gas are in relatively short supply, and the use of coal may be discouraged around the world for climatic reasons.

Hydroelectric power has been successful in smaller installations, and has been used in countries such as Iceland where there are a number of fast-flowing rivers. The environmental degradation, and inefficiency in operation, of mega-schemes (such as the Sardar Sarovar and Indira sagar projects in Gujrat and Madhya Pradesh in India) funded in the Third World by the large financial agencies has made multilateral agencies more wary of hydro power.

Nuclear power has been discredited in all but a few countries as too expensive and too dangerous (see Nuclear Energy).

Thus, within the lifetime of the current generation, there is a likelihood that the sources of energy commonplace today will have run out or become unusable. The nature of world energy production will have to change. Certain countries have already recognized this, and are investing in energy technologies such as energy conservation, solar power, wind energy and fuel cells.

HYDROELECTRIC ENERGY

Many early settlers in North America used the power of falling water to drive their mills and factories. The energy of water dropping downward through a dam is used to produce electricity. Energy produced in this way is called hydroelectric power. The principle here is simple. The heart of a hydroelectric system is a fan-shaped device called a **turbine**. Water falling through a pipe in the dam flows past the blades of the turbine. The blades are forced to rotate, thereby driving an electric generator. No fossil fuels are used, and the power supply is renewable continuously as water evaporates from the ocean and falls to the Earth to collect in the mountains.

Today, about 5 per cent of the total world consumption of energy is supplied by hydroelectric generators. However, there are large regional differences in the use of this resource. In the United States, 4 per cent of the total energy is supplied by falling water, whereas this total is 26 per cent in India, 50 per cent in Norway and around 75 per cent in parts of China. According to experts only a small percentage of the total worldwide potential for hydropower has been exploited. On the one hand, engineers are starting to plan huge projects while on the other hand, environmentalists have raised many serious questions about large-scale hydroelectric projects. Some of the *pros* and *cons* of the use of this energy source are listed below :

Benefits of Large-Scale Hydroelectric Plants

1. Cheap, pollution-free, renewable energy is made available.
2. Water stored behind dams can be used for irrigation, thus increasing productivity of local agriculture.
3. Cheap electricity provided by dams can be used to manufacture fertilizer.
4. Dams establish lakes that provide a habitat for fish, which in turn can be used for a food supply.
5. Lakes produced by dams are prime recreational sites.
6. In many areas a dependable water supply created by dams has increased public hygiene.
7. Hydroelectric energy reduces the need for fossil fuel, which produces carbon dioxide, which may disrupt world climate. In this way, world climate is protected.

Problems of Large Scale Hydroelectric Plants

1. Silt from soil erosion upriver can fill in the lakes behind large dams, reducing their utility within a few decades. Then the expensive dam projects become useless.
2. Dams flood valuable farmland, reducing food production in many areas.
3. In a free-flowing river, soil nutrients released from regions upstream are carried to the valleys below. During flood time, some of these nutrients are spread out on lowland farms, renewing the fertility of the soil. When the river cycle is disrupted by the dam, the flow of nutrients is cut off, reducing soil fertility near the river mouth.
4. The traditional flow of nutrients to the sea also fertilizes ocean estauries, providing food for salt water fish. When this flow is reduced, populations of ocean fish have been disturbed. For example, the sardine catch on the mouth of the Nile River declined by 18,000 tonnes annually when the Aswan dam was built. In some cases, dams disrupt migration of spawning fish such as salmon. Also, water taken from considerable depths below the surface of the dammed lake is often much colder than surface water and therefore changes the biota of the stream below the dam.
5. Great aesthetic loss occurs when beautiful natural canyons are obliterated.
6. In many areas disease organisms breed in dam-produced lakes or irrigation canals, leading to sickness and death.
7. There is a fear that major dams planned for the future might disrupt the flow of warm water into the Arctic, thereby disrupting world climate.

In view of the environmental problems outlined here, many people believe that a better answer might be construction of many small-scale hydroelectric facilities rather than a few large dams. One of most controversial cases of this kind is the Tehri Hydro Development Project in Garhwal Hill in Uttar Pradesh, India.

GEOTHERMAL ENERGY

Geothermal energy is the heat of the earth's interior. It is generated within the earth in such enormous quantities as to hold potential for future development on a global scale. The heat is produced by radioactive decay of uranium, thorium and potassium. The rate of decay is so slow and the quantity of these elements within the earth is so large that this internal heat production scarcely diminishes with time and the energy source can be regarded as practically inexhaustible.

The temperature of the earth at a depth of 100 km in outer layers of the mantle is about 1100°C, close to the melting point of rocks. The temperature shows a gradual falls upwards and in the uppermost 3 km of the earth's crust the temperature fall towards the surface (increase downwards) or thermal gradient averages 3°C per 100 m. The heat transfer is thus so slow that it is of little significance as compared to solar energy. At certain points, however, the molten magma of the mantle, makes its way towards the earth's surface to eject out in the form of volcano or settles below the surface to slowly from the intrustive igneous rock. The latter form of magma provides the source of thermal energy, especially when a rock still hot comes in contact with vast stores of ground water. The resultant heated water may escape to the surface in the form of **hot springs** with water near the boiling point. In other places jet like emissions of steam and hot water occur at intervals. These are called **geysers**. At other places the vent may emit superheated steam upto a temperature of 320°C. This is termed **fumarole**. Although some other gases may also be present in the fumarole, usually 99 per cent is steam. Fumaroles are common in the region of current and recent volcanic activity where the heated ground water fails to reach the surface, the energy cab be tapped through bore wells, termed **geothermal wells**, to draw hot water and steam to the surface.

In regions where ground water is not adjacent to or recently formed igneous rock, but rock is still very hot, the hot dry-rock energy can be tapped by drilling into the hot rock and then shattering the rock by hydrofracture forcing down water under pressure. The surface water is then pumped down one well and heated water pumped up another well. Such a well is called **'dry well'**.

Another source of geothermal energy is the **geopressurised energy** in the wedge of gulf coast region, where water under high pressure contains dissolved natural gas. These are avoided in drilling for petroleum as the high pressure makes it difficult to control the well, but has potentials for generating hot water for generating electricity. It is estimated that in gulf area alone upto 115,000 megawatts of power can be generated from geopressurised energy.

Hot springs have been used for bathing ever since the ancient Romans built their luxurious thermal baths at Vicus Calidus (now Vichy) in what is now France, and Adquae Solis (now Bath) in Britain. The Maoris made practical use of the geysers in New Zealand for cooking their food. The first geothermal well was obtained in Hungary in 1867 is an attempt to bring mineral water to the surface for drinking. In the 1930s, natural hot water was piped into a section of Budapest to heat houses, and in Reykjavik, Iceland, wells have been drilled more than a mile deep to reach hot water, which is then used to circulate through the walls of concrete houses and for heating greenhouses. Similar use is made of geothermal waters in other parts of the world. But heating is only the simplest application. The most important future for geothermal energy is to convert it into electric power.

The first geothermal electricity was generated in 1904 at Lardarello, Italy. The Italians first tried lighting electric bulbs by a small generator driven by a steam turbine. The next year they built the world's first geothermal power plant, which still furnishes electricity to help propel Italy's electric trains.

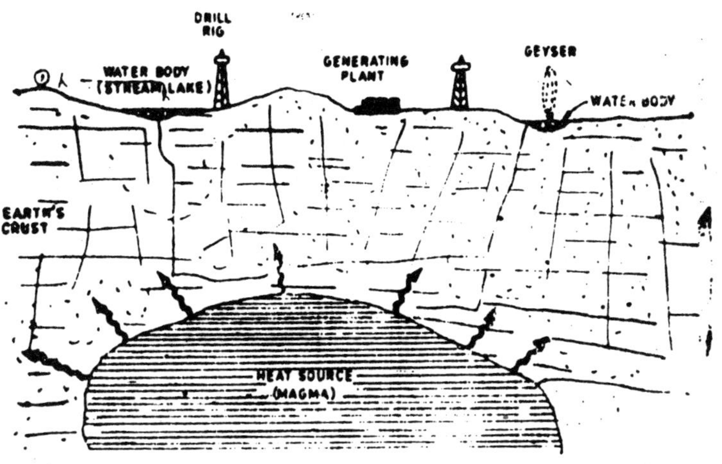

Geothermal Energy : Heat from magma converts ground water into streams which rises up through the fractures or is reached by drilling.

Geothermal electric power is still considered a potential resource that is largely untapped. Geothermal power plants are in operation in Italy, Iceland, Mexico, New Zealand, Japan, El Salvador, Russia, and the United States, but they produce only an extremely small fraction of the world's electricity. The production of pollution-free energy from natural thermal sources is not always as simple as it might seem. The hot water and steam are in some locations contaminated with mud, corroding salts, and poisonous gases. Hydrogen sulphide, a highly toxic gas, and radon, a natural radioactive material that can cause lung cancer when inhaled, have been found in underground steam sources used for power plants. Drilling wells that draw off large amounts of steam under pressure might cause underground disturbances and, as in the area of the Salton Sea, might cause the intrusion of salt water into the subterranean basin.

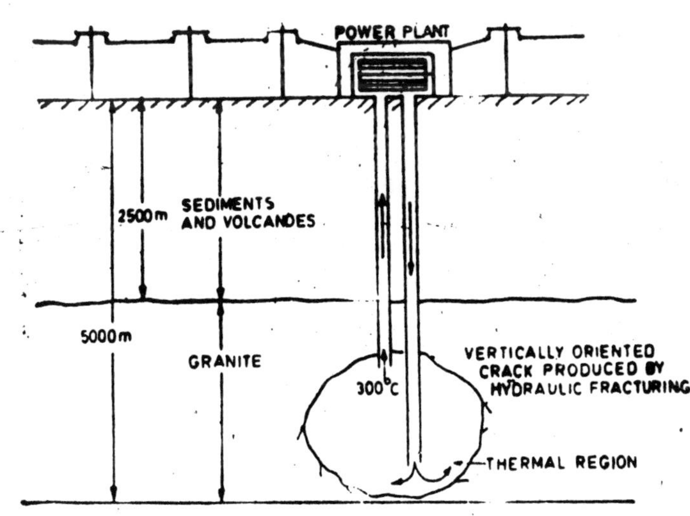

Schematic diagram showing proposed water flow in a "dry well: Geothermal Power Plant.

Research and development activities relating to cold storage unit and 5 kw power plant, both based on geothermal energy at Mani Karan (Himachal Pradesh) in India are in full progress. Pilot cold storage plant in Mani Karan using geothermal energy has shown positive results.

SOLAR ENERGY

History

Solar energy has actually been put to work by humans for thousands of years. The earliest applications were probably developed to dry fruit the crops, distill water, and heat dwellings. Many mountain and cliff dwellings employed large natural overhangs to block the high summer sun, shading the dwellings and keeping them cool, while exposing them to the low winter Sun, warming living space during the cold months of the year.

These early applications of solar energy that employ no mechanical equipment or apparatus and utilize easily obtainable, inexpensive, local materials are called passive systems.

Mechanical systems employing solar energy were being developed in the mid 1800s. At the 1878 Paris World Exposition, a large solar steam generator was displayed. It consisted of a large parabolic mirror that reflected incoming solar radiation to a focal point which produced temperatures high enough to generate steam. The steam was used to power a pump and other types of mechanical equipment. Similar applications were built in the southwestern United States in the early 1900s, also producing steam to run mechanical equipment. One of the simplest and commonly used method of solar heating is the use of glass panes to admit Sunlight into the room and capture longwave reradiation, a greenhouse principle.

The use of the sun's energy is not new, but we now have new materials, new knowledge and experience, and a broader understanding of the earth's ecosystem and the worldwide need for a safe, clean, and widely available energy source.

Nature of Solar Energy

Energy from the sun is created by the fusion of hydrogen into helium. The earth's atmosphere receives solar energy at the rate of 1.5 X 10^{18} kilowatt hours per year. About 30 per cent of the radiation reaching the atmosphere is immediately reflected back into space. Approximately 50 per cent is absorbed by the atmosphere, land, and ocean masses to contribute to the temperature of the environment. About 20 per cent is used in the evaporation, convection, and precipitation processes of the hydrologic cycle. A small fraction powers the movement of air (wind) and the circulation of oceans. An even smaller amount is converted into plant energy in the chlorophyll of green leaves. It is this tiny fraction that has produced all the fossil fuels on the earth, along with our food and other vegetation.

Every day, the sun's energy reaching the earth is several thousand times as much as we can possibly use. In less than 3 days the solar energy reaching the earth is greater than the estimated total of all known reserves of fossil fuel. The problem is not that there is too little energy available. The energy reaching us from the sun is actually much more than we need. The problem is how to harness and store this energy for use on demand. There are many different types of collectors of solar energy.

Flat plate collectors are usually stationary and can utilize heat from diffuse solar radiation as well as direct, making it possible to collect that on bright as well as cloudy days. Focusing collectors must be moved during the day to track the sun remaining normal to the sun's rays for they utilize only direct radiation, but can produce much higher temperatures.

Flat Plate Collectors

There are many variations in design and materials of flat plate collectors. All, however, are built on the same principle. Solar radiation is absorbed by a blackened metal surface, usually copper, aluminium, or steel, heating it to temperatures as high as 149°C depending on materials used and climatic conditions. This heat is transferred to a fluid, most often water or air, flowing over the metal surface. If water is used, it is usually circulated through tubes welded or moulded into the metal plate, or it flows down channels to grooves in the metal surface. In solar air heaters, the air flows over a baffled or corrugated surface. The baffles provide more surface area for the transfer of heat to the moving air than a flat metal surface.

The collector is insulated on the back side (away from the sun) to prevent unwanted heat loss, and is covered on the front side (facing the sun) with one or two panes of glass. The glass traps the absorbed heat, creating a "greenhouse effect", much like a car with its windows rolled up on a hot, sunny day. The glass transmits the short-wave solar radiation but reflects the long-wave radiation (heat) emitted by the black collector surface.

Focussing Collectors

Most focusing collectors are nearly spherical or cylindrical in shape and usually parabolic or circular in cross section. The focussing may be done with glass lenses on curved mirrors or metal surfaces. Parabolic mirrors reflect the incoming paralled rays of direct solar radiation onto a focal point (or line in the case of cylindrical collectors), achieving very high temperatures. This type of collector can achieve temperatures approaching 3000°C or less at the focal point, depending on the area of the collector.

Small focussing collectors (1.2 or 1.5 metres in diameter) made of polished steel, aluminium, or chrome, or many small mirrors are often used in nonindustrialized countries to boil water and cook foods. Larger collectors are common in many African countries and the Australian continent for industrial purposes. The high temperatures generated usually produce steam to operate steam-powered equipment. Perhaps the largest focusing collector is located in France, where a solar laboratory operates a focusing collector that is 10 stories in height.

Major Applications

The solar energy has been put to a number of uses, and a number of new methods are being developed. Solar heating of dwellings by putting glasspanes and making use of greenhouse effect is a common practice. Houses, hotels and establishments are now being warmed by circulating heated water through a system of pipes and using collectors for trapping solar radiations. One encouraging use of solar energy is solar-fired furnaces that save consumption of fossil-fuel and production of exhaust gases. The world's largest solar furnace is located in pyrenees region of southern France. An array of wall-sized mirrors are mounted each on a rotatable pedestal to reflect solar rays on a large fixed mirror, 40 m high and 54 m wide. The latter reflects rays to a focus within the furnace chamber.

Major Research and development in tapping solar energy is towards generation of electricity. Silicon solar cells originally used in space programmes to power space vehicles are being used commonly for power generation. Solar cells can also be made using cadmium sulphide or gallium arsenide. Solar cells are also called photovoltaic cells and have promising prospects in arid climates where sky is generally clear, especially at higher altitudes, for example the Ladakh region in NW Himalayas.

Among the more ambitious plans is to construct *Satellite solar power stations* (SSPS), comprising a huge space vehicle bearing large array of solar cells mounted on large panels, and having a synchronous fixed equator orbital position above the earth. The power generated by space station would be sent to earth in the form of microwave beams to a earth receiving station on ground.

The Indian Scene

Systematic and ceaseless efforts are being made to tap scientifically the abundant solar energy available in the country. A simple and common mode in solar energy utilisation is solar thermal conversion. A large number of applications in areas particularly those where low grade thermal energy is needed, have already become commercial. These include cooking, water heating, water desalination, space heating, crop drying, etc. Efforts are also afoot to develop economically viable solar collectors for high temperature applications. Installation of solar thermal systems and devices like solar water-heating system both for domestic and industrial application, desalination system, timber kiln, air conditioning, etc., covering an area of 1.23 lakh square metre of collector area, has helped save/generate energy to the extent of 350 lakh kwh per annum. Over 1.40 lakh solar cookers were in use in 1989-90.

A 50 kw solar power plant has been established at Gwal Pahari in Haryana to generate electricity and energise 10-tonne capacity cold storage for preservation of milk or green vegetables and fruits, etc. A number of feasibility reports have been prepared to establish 30 mw capacity power plants in Rajasthan, Punjab and Gujarat. Such power plants may prove economical and effective to supply pollution-fee electricity in rural-remote areas which are quite away from coalpit heads.

Solar Photovoltaics

Solar Photovoltaics offer instantaneous conversion of solar energy into electricity without any pollution. Under solar photovoltaic demonstration programme, over 23,000 solar photovolatic street light, 546 community lighting/TV systems, 1,104, solar photovoltaic water pumping systems, 1,443 domestic lighting units and 1,294 battery charging units have been installed throughout the country. So far 6,228 villages have been provided with photovoltaic street lights. It is proposed to install 2,300 street lighting systems, 50 water pumping systems, 1,000 domestic street lighting systems and 100 kwp capacity in power plants and various other applications which also includes installation of collector area of 30,000 square metre and promoting the use of over 31,000 additional solar cookers.

Importance of this system is very high as it provides electricity at an economic cost in areas away from grid and which require small amounts of electricity for various uses. Small power packs of 1 to 25 kw are doing great service in remote and far-flung areas. A capacity of 25 tonnes for production of solar silicon (basic raw material) has been established in the country. Besides production of silicon, wafers and modules are being manufactured by a number of units. Research and development efforts are in full swing for amorphous silicon. The Amorphous Silicon Pilot Plant set up at Solar Energy. Centre with an annual capacity of 500 kwp on a single shift basis, is expected to be commissioned shortly. This programme of solar electrification in remote villages was undertaken in consultation with Rural Electrification Corporation and state electricity boards. In addition to Bharat Heavy Electricals Ltd., Central Electronics Ltd., and one more public sector company, namely, Rajasthan Electronic Instrument Ltd., have entered the field of solar photovoltaic production.

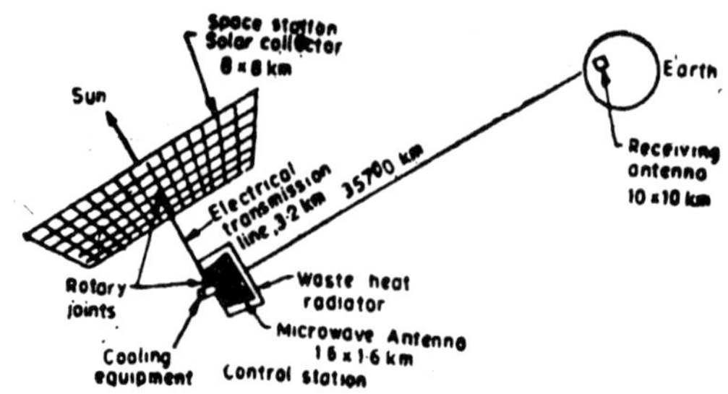

Proposed satellite solar power station designed to provide 10 million kilowatts of electricity.

TIDAL POWER

Ocean waves, tides, and currents are global in extent and extremely powerful. To-date, the most significant attempts at harnessing the energy of the ocean have been related to tidal power. Tides result from the gravitational attraction of moon. Moon has 2.25 times stronger gravitational pull as compared to the sun.

In many coastal regions, the flow of the tide naturally funnels through narrow entrances into bays and estuaries, and strong currents are established. Twice a day the water flows inland, with **high tide** or **flood tide** and twice a day it rushes outward with the **ebb tide** or **low tide**. The highest tidal ranges are exhibited in March and September during equinoxes when the sun, moon and the earth are in line. These are called **spring tides**. Lowest range is in June and December during solstices, when moon and sun are aligned at right angles to the earth. These are called **neap tides**. Some of the energy from this movement of currents can be harnessed if a tidal dam is built and a turbine is installed.

Devices to harness tidal energy go back to at least eleventh century. Tidal mills were common along northern coasts of Europe and North America even in early twentieth century. The advent of less expensive electric power rendered these incapable of economical competition, and were mostly discarded. Over the last few years, with fossil fuel resources dewindling at faster rate, and their accompanied pollution has forced a second look at the vast reserves of practically non-exhaustible tidal energy. The power generation from tidal energy, however, is a costly affair in comparison to hydrogenerators. They have to handle water flowing in both directions, one way during high tide and other way during low tide. Since the power availability depends on activity of moon rather than desires of human beings, some means of energy storage are essential for a useful output. The generators must be built to withstand saltwater corrosion. In cold regions, ice must be kept out and in warm waters, fouling by marine life presents problems.

The two major tidal power plants in operation are located in France and in USSR. La Rance plant on the Brittany coast of France was commissioned in 1966. The plant has 24 conduits to power as many generating units. Each unit has adjustable blades to permit water generation from water flowing in either direction. The units are constructed of high quality concrete, stainless steel and aluminium bronze to prevent salt water corrosion. The plant has a generating capacity of 240 megawatts.

Although there are a number of likely locations around the shoreline of the world, but in view of the expenditure involved, their utilization in near future is not contemplated.

WIND POWER

History

Wind power was put to work centuries ago for pumping, irrigation, and other types of mechanical work. Large wind mills and small, multivaned wind pumps found on farms throughout the midwestern United States are a familiar site. It was not until this century that extensive experiments with wind generators began—converting wind energy to electrical energy by transferring the rotational motion of the blades to a generator or alternator. In 1950, there were an estimated 50,000 wind generators in use in the midwestern. United States alone, but rural electrification programmes made them (temporarily) obsolete. At present, there are a number of commercially manufactured wind generators of various sizes (200 to 6000 watts) available for residential use.

Principles

Basically, a wind generator converts wind energy to electrical power by transferring the rotational motion of the blades to an alternator or generator which produces electricity. At present, a three-bladed propeller design is believed to be the most efficient and therefore most often used, although two-bladed models are not uncommon. The entire blade and hub assembly, called the rotor, rotates in a direction perpendicular to the wind. The speed of this rotation varies with the blade design (sails, airfoils, etc.), with the fastest speed obtained by a propeller—airfoil design. The rotor is usually connected to the alternator by means of a gear or belt-drive assembly. Gears are used to increase the relative rotational velocity of the alternator. A few wind generators are direct drive, meaning that there is no gearing between the rotor and the generator. These designs, however, use specially made low-speed generators which are both very heavy and expensive to build. Most alternators and generators are designed to work only at high rpms, and therefore make the gearing necessary. The alternator generates current by rotating a metallic coil (usually copper) through an electromagnetic field. The induced current is then distributed to a control panel, which measures the amount of power and current generated. The electrical power is stored in a batter storage bank where 5 days of electrical energy may be stored for use during calm spells. Batteries store electric power in the form of direct current. The power can be used directly in this form or may be converted to alternating current by means of an inverter. Most appliances have both direct-and alternating-current ratings, but a few require just alternating current. A gas-fired alternator is usually included in the system to provide electric power for unusually long periods of calm weather.

An important conceptual utilization of wind power for power generation in plains, where high winds are common is to install wind trains consisting of oval train track about 8 km long and 1.5 km wide loaded with flat cars. Air foils are mounted on each car. The air foils are computer controlled. Computer simulation at the university of Montana have provided encouraging results.

India has a wind energy potential of about 20,000 mw for power generation in coastal and windy sites. Efforts are being made to develop and install wind generators, windmills and battery charging systems, such as 200 kw or higher capacity wind generators with increasing indigenous content. A total capacity of 23.55 mw out of an aggregate capacity of 31.4 mw, has already been established at 12 wind farms. Projects worth a capacity of 35 to 55 mw are under implementation. About 2,524 wind pumps have been installed for drinking water and irrigation.

ENERGY FROM WASTE

Energy from Urban Waste

Many societies in the developed world are incredibly wasteful. Half of the household trash in the United States and Canada is paper. Huge piles of bark, wood scraps, and logging wastes rot slowly near many sawmills. If people collected these wastes and used them as fuel, considerable quantities of energy could be salvaged. In France, there are 20 generators that burn garbage for use as domestic energy sources. A large facility in Paris produces electric power for 130,000 people and 20 per cent of the total steam used for heating in the entire city. In North America a few facilities burn trash for fuel, but in most regions garbage is simply buried in a landfill.

A pilot plant to utilise solid municipal waste for conversion of energy into electricity of 3.75 mw capacity has been established at Timarpur, Delhi. Besides energy, the plant provides for disposal of huge urban waste and environment upgradation.

Biogas

Fuel can be produced from other waste products as well. When organic waste such as sewage garbage, manure, or crop residues decompose in the absence of air, methane gas is released. Methane produced in this manner is called **biogas**. Despite the difference in names, however, methane produced from wastes is identical to methane extracted from an underground gas deposit. In an ideal sequence, a farmer would collect cow manure and deposit it in a specially designed underground concrete pit with a steel cap to hold the gas released. The manure would be allowed to rot naturally, and the gas would be collected and used. The decomposed manure remaining in the concrete tank would then be removed and spread on the fields as a high-quality fertilizer. Studies in China have shown that a small biogas digester using human wastes and manure from a few cows can supply enough fuel to cook three meals and boil 15 litres of water a day for one family. Construction costs are returned in one to three years in the form of fuel savings. Many thousands of these units have been built throughout Asia.

Under National Programme on Biogas Development 12.40 lakh biogas plants have been installed in India producing gas equivalent to the saving of over 43.8 lakh tonnes of fuel wood valued at Rs. 175.6 crore annually. Besides they are generating manure to the extent of above 211 lakh tonnes worth over Rs. 174.5 crore annually. Similarly, over 500 institutional/community, biogas plants have been installed for providing energy for cooking, lighting, etc., for village community as a whole. Significant progress has been made to utilise alternate feedstocks such as human excreta, agricultural residues, cotton willow dust and water hyacinth. Construction of latrine-linked biogas plants is gaining popularity which besides providing cooking energy is improving sanitation and environment. Additionally it dispenses with age-long practice of manually handling human excreta. About 71,000 latrine-linked biogas plants had been installed by December 1989.

Energy from Agricultural Waste

Department of Non-Conventional Energy Sources in collaboration with Punjab is setting up a 10 mw agro-thermal power plant based on rice husk at Jalkheri near Patiala.

The Department is cooperating with Punjab Agro Industrial Development Corporation (PAIDC) for a joint sector project of Oswal Agro Furane being set up in Punjab for power generation component of the plant. The project envisages de-husking of large quantities of rice and utilisation of husk in the first stage to extract furfural and then to fire husk in a fludised boiler linked with co-generation plant to generate 12 mw of electricity. This cent per cent export-oriented plant will have about 5 mw of surplus electricity to be connected to the grid. This electricity can be used, *inter alia*, for irrigation purposes.

It has been estimated that sugar industry in India can generate almost 2,000 mw as surplus electricity during crushing season. For instance a 2,000 TCD sugar mill using energy conservation principles with high steam condition boilers and back pressure turbine can generate 10 mw of electricity and sufficient process steam. After meeting factory's requirements, almost 6 mw can be surplus to be fed to the grid. Since sugar factories are normally in the rural areas, this surplus electricity could be used for irrigating sugarcane fields or industrialisation of rural areas.

ENERGY FROM NON-CONVENTIONAL SOURCES

Other than the most important sources of energy for the present and immediate future—oil, coal, hydropower, and nuclear power—there are several promising alternate sources. They vary in the amount of potential energy they can supply and in the amount of research and engineering needed before their full use can be realized. Those under study and exploration include : solar energy, geothermal energy, nuclear fusion, solid waste, wind, tidal and wave action, hydrogen fuel, thermionics and fuel cells, uranium and thorium is magmatic rocks, deuterium and uranium in sea water, and fission of boron-11.

Indian Renewable Energy Development Agency Limited

Indian Renewable Energy Development Agency (IREDA) is registered under the Companies

Act with authorised capital of Rs. 10 crore as an undertaking of Department of Non Conventional Energy Sources. IREDA aims to promote projects in non-conventional energy sources by providing soft term loans to manufacturers, traders and users. It acts as a financial intermediary with financial institutions on behalf of new and renewable sources of energy (NRSE) industries etc. It has also received grant-in-aid from the Netherlands in respect of project financing. IREDA received a total of 133 projects till 31 December, 1989 demanding loan assistance worth Rs. 62.85 crore. It had sanctioned 63 projects for Rs. 14.78 crore till the said date. Further, 35 projects have been sanctioned by IREDA and another 62 are in the process of appraisal. Projects sanctioned are spread over 11 states in the areas of biomass, solar, thermal, biogas, industrial affluents and battery powered vehicles, etc.

ENERGY CONSERVATION

There are abundant sources of energy all about us in the sun and its manifestations in the wind, water, and oceans. Each of these are free, renewable, non-pollutant, decentralized sources of energy which can be utilized and integrated into a whole, self-sufficient system for survival, dependent only to a small degree on fossil-fuel sources.

Although energy from the sun and wind is plentiful, there are still problems to be overcome in utilizing and storing energy from these sources in some areas of the world. To make solar energy and wind power immediately practical and feasible requires a policy of energy conservation to minimize the energy demands of a building or dwelling. We must begin to design with "energy-conscious" techniques to reduce our need and consumption of energy. The use of proper materials and design techniques responding to the sun and wind can significantly reduce the energy needs to heat and cool a residence.

There are many other "low-energy techniques" to significantly reduce the energy demands. Techniques, which respond to the forces of nature and local climatic conditions rather than trying to conquer, and which use local, natural building materials, are perhaps the most important steps towards an effective energy-conservation approach which makes solar heating systems and wind-powered electric systems feasible and practical. There are no new technological developments necessary to begin designing and building with these techniques. The only new development would be the integration of all these techniques and alternative energy sources into one living unit. It can be done; all that is needed now is the commitment to do it.

RADIOACTIVITY

An atom can be visualized as having a central *nucleus*, consisting of a number of particles clinging tightly to one another, surrounded by ***electrons*** which circle round it rather like planets round a sun.

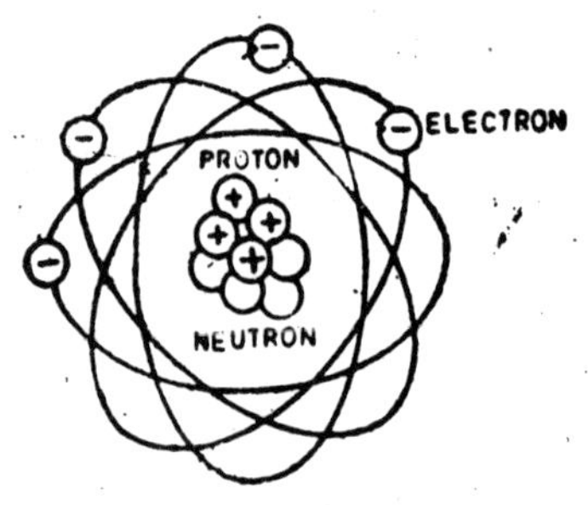

The Atom

The electrons carry a negative electric charge, which is balanced by the positive charge of some of the particles (**protons**) in the nucleus. The number of protons and electrons in an atom are the same.

The other particles in the nucleus are called *neutrons*, and have no electrical charge. The atom itself is neutral since the number of protons and electrons is equal. Similar atoms bond together to form a chemical element.

Each different element has its own particular number of protons in the nucleus. If an atom loses or gains a proton, then the new formation becomes an atom belonging to another element.

On the other hand, an element can occur in a variety of forms, with varying numbers of neutrons. These different forms are called **isotopes**; uranium-238 has 92 protons and 146 neutrons (92+146 = 238) whereas uranium-235 has 92 protons and 143 neutrons (92+143 = 235). The atoms of isotopes are called **nuclides**.

A number of nuclides are unstable; because of their atomic formation, they are trying to change to something else and so become stable. This is done by releasing a chunk of the atom; if a proton is lost, then the element changes to another element *e.g.* :

* Uranium-238 releases two protons and two neutrons, thus becoming thorium-234 (new atomic structure : 90 + 144 = 234).
* Thorium-234 is also unstale. A proton in the nucleus turns into a neutron, so a negatively-charged electron is released and escapes. With one proton less, the thorium 234 turns into protactinium -234 (new atomic structure : 89+145=234).
* The process continues until lead-206 is formed. Lead-206 is a stable element, and the process of metamorphosis stops.

This process of transforming is called *radioactive decay*, and the unstable nuclides are **radionuclides**.

The French physicist Henri Becquerel discovered in 1896 that uranium minerals spontaneously emit energy in the form of radiation. Thus, if a photographic film is held near a uranium mineral in the dark, the film becomes exposed, just as it would be if it were held near a light. This emission of radiation from an element was called **radioactivity**. A series of careful, tedious separations carried out by Marie and Pierre Curie resulted in the discovery of new radioactive elements, the most important of which was **radium**.

Some elements exist both as nonradioactive (stable) isotopes and as radioactive isotopes (**radioisotopes**). Radioactive elements undergo spontaneous **radioactive decay** in which they decompose giving off particles and radiation and leaving another isotope behind. The products of radioactive decomposition may be radioactive in their turn or they may be stable.

Nuclear Reactions

Radioactivity and other nuclear changes can be represented by simple equations. For example :

$^{226}_{88}Ra$	—>	$^{4}_{2}He$	+	$^{232}_{86}Rn$
radium		helium nucleus		radon

This equation tells us that a $^{226}_{88}Ra$ nucleus decomposes to give one $^{4}_{2}He$ and one $^{232}_{86}Rn$

nucleus. ($^{4}_{2}He$ nucleus is called an alpha particle) Note that both the mass numbers (226=4+222) and the atomic numbers (88-2+86) are balanced.

Nuclear reactions may also involve neutrons. The symbol $^{1}_{0}n$ is used to represent the neutron. (Its charge is 0 and its mass number is 1).

The first artificial nuclear reaction was carried out by Ernest Rutherford and his co-workers in 1919, when they bombarded ordinary nitrogen with helium nuclei to produce oxygen-17, which is nonradioactive :

$$^{14}_{7}N \quad + \quad ^{4}_{2}He \quad \rightarrow \quad ^{17}_{8}O \quad + \quad ^{1}_{1}H$$

Fifteen years later, in 1934, Irene and Frederic Joliot-Curie (Mme, Curie's daughter and son-in-law) converted boron to nitrogen-13, which is radioactive. This was the first artificially produced radioisotope.

$$^{10}_{5}B \quad + \quad ^{4}_{2}He \quad \rightarrow \quad ^{13}_{7}N \quad + \quad ^{1}_{0}$$

Radioactive Radiations

There are three kinds of radiation in radioactive decay :

1. 'alpha' radiation, which is the emission of a chunk of the nucleus containing two protons and two neutrons;
2. 'beta' radiation, which is the emission of an electron;
3. 'gamma' rays, which are pure energy emitted from very excited nuclides.

Radiation from radioactive decay can be extremely dangerous to human health. Alpha and beta radiation, and high-energy electromagnetic radiation such as gamma rays or X-rays, produce electrically-charged particles (*ions*) in material that they strike. This process (**ionizing radiation**) frequently results in changes to living tissue, which can injure it. Non-ionizing radiation, such as radio or ultra-violet waves, does not normally have this effect.

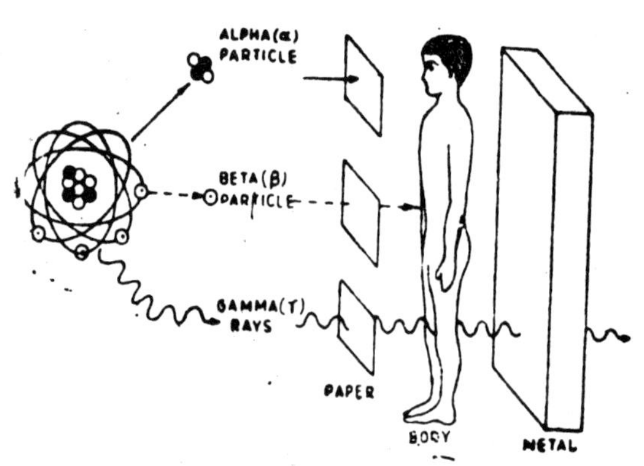

Penetrating Capacity

The ionizing radiations have different abilities to travel. Alpha particles, are heavy particles and (two protons and two neutrons) can be blocked by a sheet of paper and will not penetrate the skin. Alpha-emitting nuclides pose a serious health risk if they get into the body through an open wound, or are breathed in or eaten. Once inside the body they continue to be radioactive and are particularly damaging.

Beta particles are more penetrating, but can be stopped by relatively thin layers of water, glass or metal. Beta emitters can also be hazardous if taken into the body.

Gamma rays travel at the speed of light and will go through almost anything before being absorbed. The rays can be stopped by a thick layer of lead or concrete. X-rays have similar penetrating abilities.

Half life

The half-life is the time it takes for half the nuclei in a sample to decompose. The half-life of radium-226 is 1600 years. Therefore, if 1g of radium-226 were placed in a container in 1991, there would be only 1/2 g left after 1600 years in the year 3591 and only 1/4g after another 1600 years (in the year 5191) and so on. Each radioisotope has its own characteristic half-life.

The concept of half-life does not imply that after 1600 years half of the radium will suddenly decompose. The half-life is an averaged value for all the radium nuclei. This means that there is a chance for some decompositions to occur in *any interval of time.* Since there are many atoms in a sample of radium, some may be decomposing every second.

The rate at which the radiation is emitted from a sample of radium-226 depends on its quantity. Since each atom of radium-226 that decomposes is converted into another element (radon-222), the quantity of radium-226 in any sample constantly decreases. However, the radon it produces is also radioactive, and it, too, decomposes to produce another radioisotope. This series of radioactive disintegrations goes on through a number of other radioisotopes until finally a stable isotope, lead-206, is produced. These radioisotopes have half-lives ranging from fractions of a second to about 20 years. Therefore, the total radioactivity produced by a sample of radium together with its radioactive waste products is more than that produced by the radium alone.

A pertinent question arises; if nuclei of radioactive elements are unstable, why are there any left on Earth? One conceivable answer is that these survivors are all descendents of radioisotopes with very long half-lives. The half-life of natural uranium-238, for example, is 45×10^8 years. The radiations from such materials plus the effect of radiation that comes to the Earth from outer space is called the **background radiation.**

Radioactive decay

Type of nucleide and radiation	*Half -time*
Uranium-238	45×10^8 years
$\alpha \downarrow$	
Thorium 234	24.1 days
$\beta \downarrow$	
Protactinium-234	1.17 minutes
$\beta \downarrow$	

(Contd.)

Uranium-234	24.5×10^4 years
α ↓	
Thorium 230	8,000 years
α ↓	
Radium-226	1,600 years
α ↓	
Radon 222	3.823 days
α ↓	
Polonium 218	3.05 minutes
α ↓	
Lead-214	26.8 minutes
β ↓	
Bismuth-214	19.7 minutes
β ↓	
Polonium-214	0.000164 seconds
α ↓	
Lead-210	22.3 years
β ↓	
Bismuth-210	5.01 days
β ↓	
Polonium-210	138.4 days
α ↓	
Lead-206	stable

NUCLEAR FISSION

The splitting of a nucleus is called **atomic** or **nuclear fission**, Fission releases energy because the uranium nuclei have more energy than their products.

In 1939, Otto Hahn, Fritz Strassman, and Lise Meitner, realized that neutrons caused the splitting, or fission, of uranium nuclei. Further studies showed that extra neutrons were also released in the fission reaction. If the fission reaction is started by neutrons and then also releases neutrons, a new possibility arises—the opportunity for a **chain reaction**. This discovery changed nuclear science from a study of purely theoretical interest to an issue of utmost importance.

A chain reaction is a series of steps in a process that occur one after the other, in sequence, each step being added to the preceding step like the links in a chain. If the lengthening of one chain proceeds at a given rate, the production of 10 branches means that 10 reactions are going on at the same time, so that the rate has increased tenfold. The condition under which a chain reaction just continues at a steady rate, neither accelerating nor slowing down, is called the **critical condition**.

The production of the atomic (fission) bomb and of nuclear reactors depends on branching nuclear chain reactions. The process is initiated when a neutron strikes a uranium-235 nucleus and can proceed in any of several different ways (see fig. on p. 327). For example, uranium-235 plus a neutron forms barium-142 and krypton-91, while releasing three neutrons.

$$^{235}_{92}U + ^{1}_{0}N \rightarrow ^{142}_{56}BA + ^{91}_{36}Kr + 3\,^{1}_{0}n$$

These neutrons can initiate three new reactions, which in turn produce more neutrons, and so forth. This is, therefore, a branching chain reaction.

The amounts of energy involved in nuclear fission are very large compared with those of chemical reactions. If the branching chain reaction continues very rapidly, it produces an atomic explosion. If the chain branching is carefully controlled, energy can be released slowly and used to drive turbines to generate electricity. Fission reactions produce radioactive wastes, barium-142 and krypton-91, which are both radioactive. The reaction represented by this equation is only one out of many that occur in atomic fission. Many different radioisotopes are produced by atomic fission. Also, the fission products, as a group, are much more radioactive than the uranium from which they are produced. A compound with a short half-life decomposes quickly, but in its early stages it emits radiation at dangerously high levels.

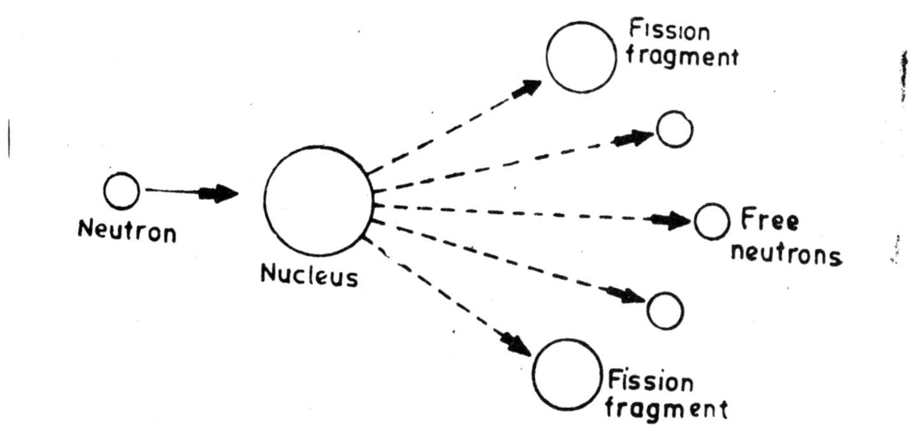

A typical fission reaction. An atom such as that of uranium-235 can undergo fission when a free neutron strikes its heavy central nucleus, splitting it into two pieces (fission fragments) that fly apart at high speed, and releasing two or three additional free neutrons. The released neutrons cause a chain reaction when they strike the nuclei of other uranium-235 atoms; and when the flying fission fragments collide with other atoms, their kinetic energy is converted to usable heat.

NUCLEAR FUSION

Combination or fusion, of nuclei of certain light elements, particularly certain isotopes of hydrogen.

The great advantage of nuclear fusion is that it promises abundant energy with much less environmental danger than is faced from fission reactors. However, no useful fusion reactor has yet been developed.

Any fusion reactor would utilize hydrogen nuclei. There are three isotopes of hydrogen : **protium**, **deuterium**, and **tritium** (Table 1) Fusion reactions can occur between any two hydrogen isotopes. The heavier isotopes fuse more readily, but the lighter ones are more abundant.

Table 1 : Isotopes of Hydrogen

Symbols	*Natural Abundance (%)*	*Remarks*
${}^{1}_{1}H$	99.985	"Ordinary" hydrogen Protium
${}^{2}_{1}H$, or ${}^{2}_{1}D$	0.015	"Heavy" hydrogen ; Deuterium
${}^{3}_{1}H$ or ${}^{3}_{1}T$	Almost none	Tritium; Radioactive half life : 12- year

Unlike the fission reaction, however, fusion cannot be triggered by neutrons. What is required instead is very high temperatures. The resulting fusion is therefore called a **thermonuclear reaction**, the type of reaction by which the Sun produces energy. If a large mass of hydrogen isotopes fuses in a very short time, the reaction goes out of control; this is the explosion of the "hydrogen bomb". One the other hand, useful energy could be extracted from fusion if it were possible to devise a controlled thermonuclear reaction.

Two deuterium nuclei can fuse to form a helium-3 nucleus plus a neutron.

$${}^{2}_{1}H + {}^{2}_{1}H \rightarrow {}^{3}_{2}H + {}^{1}_{0}N$$

To start this reaction, the temperature would have to be raised to about 400 million degrees Celsius. The problems imposed by this requirement are so severe that it is not even being attempted. Instead, all efforts to control fusion are being directed to a cooler reaction (40 million degrees Celsius)—the fusion of deuterium with tritium to give helium-4 plus a neutron:

$${}^{2}_{1}H + {}^{3}_{1}H \rightarrow {}^{4}_{2}H + {}^{1}_{0}N$$

This process requires a source of tritium, which is not naturally available on Earth.

Tritium is produced by the reaction between neutrons and the light isotope of the metal lithium, ${}^{6}_{3}Li$. This isotope constitutes 7.6 per cent of natural lithium and is about 10 times as abundant in the Earth's crust as uranium-235.

$${}^{6}_{3}Li + {}^{1}_{0}n \rightarrow {}^{2}_{1}H + {}^{4}_{2}He$$

Thermonuclear reactions are extremely difficult to control because no materials can withstand the required temperatures. Instead, all substances decompose into their free atoms, and the atoms themselves decompose into a mixture of positive nuclei and free electrons that is called a **plasma**. No rigid container exists that can survive long enough to confine a plasma for the useful production of thermonuclear energy.

NUCLEAR ENERGY

A general term for any energy released by radioactive decay, by nuclear fission or fusion. Common synonyms are atomic energy, nuclear power. The principal uses of energy are civilian and military. Nuclear energy for civilian, commercial purposes is today made available only by the method of nuclear fission.

Atomic, or nuclear energy is not a direct source of energy for consumer use; it is a means of producing electricity. An enormous amount of electric power can be produced by a small amount of atomic fuel. On a weight for weight bases, uranium contains 3 million times as much potential energy as coal. One gram of fissionable material releases 23,000 kilowatt hours of heat. A tonne of uranium would provide as much energy as 3 million tonnes of coal or 12 million barrels of oil. In practice, however, only a small fraction of the potential energy in atomic fuel is utilized during one cycle in nuclear reactor. Much of the energy is lost in the form of heat and waste products. Also, after each cycle, the fuel must be purified before it can used further.

How Nuclear Power Works

Oil-and coal-burning power stations heat water to produce steam, which drives a turbine and thus generates electricity. Nuclear power stations generate electricity in exactly the same way.

Nuclear power technology depends on the inherent instability of uranium, a substance in which the protons and neutrons in the nucleus of the atom are bound together less tightly than in other elements. If this nucleus is hit by a stray neutron from another atom, it will split into two parts, at the same time releasing two or three neutrons of its own. This 'splitting' is called **nuclear fission**.

If two of these neutrons in turn hit other uranium nuclei, then four to six neutrons will be released; and the process accelerates. If there is enough fissionable material in close proximity, the result is a chain of events in which nuclei split and their neutrons then split other nuclei. This is called a **chain reaction**.

Do this with maximum speed, and the result is an atomic explosion; control and regulate the reaction, however, and the result is merely enormous quantities of heat. This boils water to power the electricity generators; and the result is called nuclear energy.

Historical Development

Nuclear energy and nuclear weapons are two of the most controversial environmental issues of the late twentieth century. Born together during intensive scientific research in the 1930s and the Second World War, it was the more aggressive twin that appeared first, in explosions at Hiroshima and Nagasaki in 1945. Nuclear energy began to be developed for civil uses only during the late 1950s. Since then, it has expanded rapidly: achieving an impressive rate of development.

At present about 17 per cent of the world's electricity is generated through nuclear sources. For example, Japan has as many as 38 nuclear power plants generating 30,000 MWe (megawatts-electric). USSR, even after the 1986 **Chernobyl** accident, continues to operate 56 units producing 33,000 MWe, And USA, despite its 1979 mishap at **Three Mile Island** has 110 units in operation generating 1,00,000 MWe.

The advocates of nuclear power maintain that mishaps like Chernobyl and Three Mile Island are very rare occurrences. They claim that the technology is clean, safe and cheap and the nuclear power industry has an excellent track record of safety. The opponents say it is environmentally destructive, dangerous and expensive. Debates about the future of the nuclear industry, however continue.

NUCLEAR ENERGY AND THE ENVIRONMENT

In assessing the radiological effects of nuclear technology, the individual power plant is only "the tip of the iceberg." To enable it to operate, a nuclear power plant has to be linked with a fuel cycle that emits radioactivity at many points and so contaminates the environment.

The first stage in the nuclear industry is the uranium mine. The ore is pulverized and subjected to a leaching process to separate the uranium. The waste from this separating process, known as tailings, contains the subsidiary products of the natural radioactive series and is deposited on dumps. Rainwater washes the soluble nuclides into the ground, into groundwater and into surface waters, where it is absorbed by organisms in the soil and water. By the extraction of uranium from the mines, the natural radioactive substances that normally lie deep down in the earth's crust are brought to the surface and released into the biosphere.

The uranium extracted from the mines then passes into the factories for preparing fuel elements, where the fissionable uranium-235 isotope recovered from nature in small quantities is enriched. In this highly energy consuming enrichment process further radioactive waste if formed and released into the environment. Once the uranium has been sufficiently enriched it is fed, in the form of small tablets, into fuel elements for nuclear power plants. In the neuclear power plant the process of nuclear fission takes place, whereby a large number of mostly radioactive fission products are formed from the uranium-235. These partly gaseous radioisotopes escape to some extent in the nuclear power plant and are released into the air and water below the legally permitted limits for radioactivity.

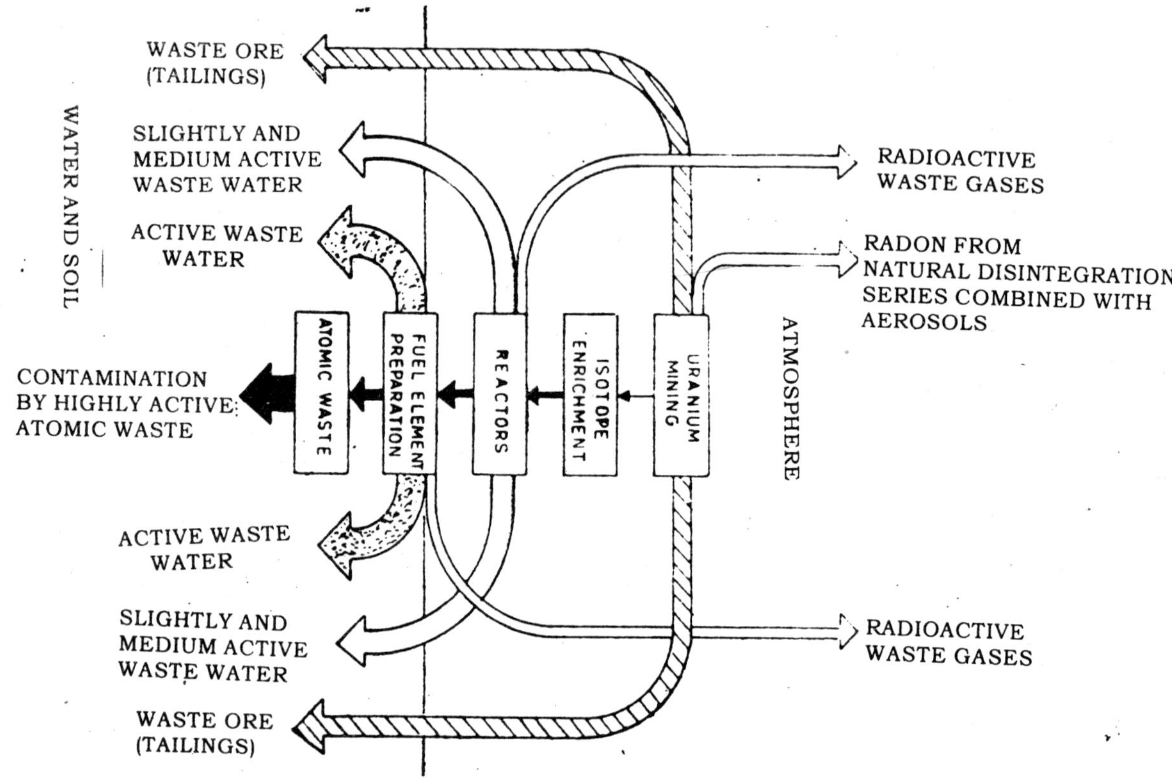

Environmental Contamination by the Nuclear Fuel Cycle.

After about a year the fuel elements are exchanged and, after a period of intermediate storage in the nuclear power plant to permit decay of the short-period isotopes, are transported to a reprocessing plant. For reconditioning purposes the fuel cells are split by remote control, the individual rods are cut into short sections, and the uranium oxide is washed out of the sheaths with nitric acid. By this disintegration of the fuel elements the radioactive inert gases formed by nuclear fission are released. As inert gases cannot be combined by a chemical process, they are emitted from the reprocessing plant through a chimney into the atmosphere. Other radionuclides released in gaseous and liquid form in this process are kept out of the waste air and water of the plant by means of filters. But as these filters are not 100 per cent absorbent, a certain amount of the radioisotopes escapes into the environment.

The uranyl nitrate solution produced in the dissociation of the fuel elements is subjected to a multistage extraction process, by means of which the uranium, the plutonium and the fission products are separated from one another. The refined end products, uranium and plutonium, are processed again by manufacturers to provide new fuel elements.

Special problems arise in reprocessing plants in connection with the production of tritium and plutonium.

The fission product solutions remaining after the separation of uranium and plutonium cannot be reutilized and must be disposed off as nuclear waste.

NUCLEAR POWER PLANT

The essential difference between a nuclear power station and a conventional power station consists in the way in which energy is provided. In both cases energy is used to heat water to produce steam to drive a generator by means of a turbine. This produces electrical energy. In a nuclear power plant the energy is obtained by fission of uranium nuclei.

Construction

Uranium is piled up in **fuel rods** in the form of uranium dioxide tablets. Some tens of thousands of these fuel rods hang in the **core** of the reactor. The nuclear fission takes place inside the fuel rods. The thermal energy released spreads outward and is absorbed on the surface of the fuel rods by the cooling water which is propelled through the core by pumps. The water in the primary cooling circuit gives off its energy through a heat exchanger to a secondary circuit. It is the steam from the secondary circuit that drives the turbines. The secondary circuit is a safeguard to ensure that any radioactive contamination caused by a defect in the primary circuit cannot reach the turbines.

Besides the fuel rods, control rods are installed in the core. These consist of neutron-absorbent material, by means of which the chain reaction can be controlled. If the reactor has to be shut off suddenly, the control rods are injected into the core where they absorb a large amount of the neutrons and reduce the rate of nuclear fission.

The reactor pressure vessel that surrounds the core is enclosed in a shell, a concrete casing with walls about 2 metres thick. The entire primary cooling circuit is enclosed in a spherical safety casing made of steel about 50 metres in diameter. The safety tank also contains a fuel element storage basin, a loading and unloading device for fuel elements, a waste water tank, and various safety appliances such as an emergency cooling system and a pressure reservoir.

As small leakages occur regularly in the large number of fuel rods, the water in the primary cooling circuit is always subject to mild radioactive contamination. This radioactivity is partly removed by purification of the primary cooling agent, and the rest escapes by leakage of steam from the primary cooling circuit into the safety tank. In order to prevent leakage of this radioactively contaminated air into the environment, a partial vacuum is maintained in the safety tank. Part of the air is removed by constant suction, fed into various filter systems and (due regard being given to the legally prescribed limits for radioactivity) discharged into the air through

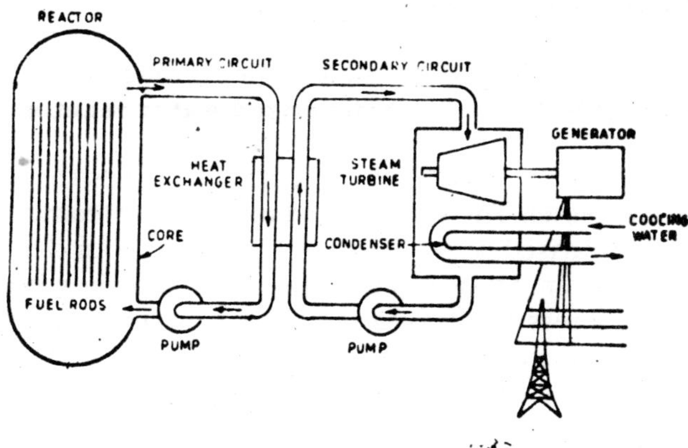

Pressurised-Water Reactor.

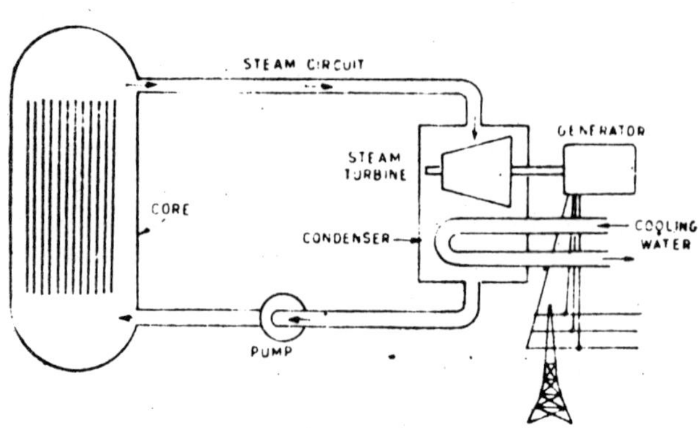

Boiling-Water Reactor

a chimney. The radioactive waste water resulting from the decontamination of the primary cooling water is retained for some time and then, after being decontaminated down to the prescribed limits, is mixed with the cooling water and allowed to run away.

Owing to the strong neutron radiation, in time the components inside the reactor become radioactive. A nuclear power plant can be expected to last for 20 to 40 years. At the expiration of that period, the nuclear plant must be either demolished or bricked up.

Apart from the type of light water-cooled, pressurised-water reactor described here, there are a number of other types of nuclear power plant. The boiling-water reactor operates with only one cooling circuit. The pressure inside the reactor is relatively low. The steam produced inside the core is fed into turbines directly., High-temperature reactors are gas-cooled. Their fuel elements are contained in globes cased in graphite. High-temperature reactors have a high operating temperature and consequently a high efficiency coefficient.

Types of Reactors

Nuclear reactors are classified according to the fuel, coolant and moderator used to support the nuclear chain reaction. There are mainly five types of reactors :

1. **Pressurised Water Reactors (PWR)**
2. **Boiling Water Reactors (BWR)**
3. **Pressurised Heavy Water Reactors (PHWR)**
4. **Gas-Cooled Reactors (GCR), and**
5. **Liquid Metal Cooled Breeder Reactors (LMBR)**

The two most commonly used types of reactors, depending on the operating temperature are the **boiling water reactor (BWR)** and the **pressurized water reactor (PWR)**.

In the boiling water reactor the heat from the fuel rods causes the water to boil, producing steam at the top of the reactor vessel. The steam is fed directly to steam turbines which turn the electric generators. In the pressurised water reactor, the water is under sufficiently high pressure to prevent boiling even at a temperature above the normal boiling point of water. The high-temperature water, still under pressure, leaves the reactor vessel and enters a device called a *heat exchanger*, which contains a separate, secondary water system. The lower pressure in the secondary water system causes the water to boil, producing steam which is delivered to the steam turbines.

There are 434 units in operation in 26 countries with a capacity of 316,000 MWe and 100 units under construction with a capacity of 80,000 MWe the type wise distribution is as follows :

Units in Operation (434)*

Type of Reactor	*Number*
PWR	238
BWR	87
GCR	30
LWGR	27
PHWR	26
AGR***	14
Others	12

Units Under Construction (100)*

Type of Reactor	*Number*
PWR	66
PHWR	18
BWR	9
LWGR	5
FBR**	2

* Data as of June 30, 1989

** Fast Breeder Reactor

*** Advanced Gas-Cooled Reactor

Source : Nuclear Power Corporation (A Govt. of India Enterprise), Bombay-400005

Safety of Nuclear Power Plants

Concern about the risks of operating nuclear power plants followed a reactor accident in 1979 at the **Three Mile Island** power plant near Harrisburg. Pennsylvania (USA). A series of equipment failures, misleading instrument readings and human error resulted in an overheated reactor core and prevented routine emergency systems from operating properly. Small amounts of radioactive gases were released from the plant during that crisis.

The crisis reversed earlier confidence that major reactor accidents are unlikely. The overheated reactor core might have lead to further overheating and to a *meltdown*, the term for an uncontrolled nuclear reaction that causes the core to melt through the containment vessel into the earth and release radioactive gases.

Sabotage or an earthquake, as well as human error or equipment failure, could cause nuclear plant crises—such as an accident in which a main pipe in the primary cooling circuit fractures. In that event, the control rods would switch off the reactor, immediately eliminating the nuclear fission process. However, the radioactive disintegration of the fission products cannot be arrested. In a nuclear power plant of 650 megawatts, the heat formation of radioactive disintegration amounts to about 200 Mw three seconds after the reactor is switched off, to 30 Mw after one hour, and to 12 Mw after 24 hours. In fact, the radioactive disintegration continues at a substantial level for several months.

Under normal operating conditions for the reactor, the external surface of the fuel casing has a temperature of about 350°C, while the interior of the fuel rods is very much hotter about 2200°C. This temperature is approaching the melting point of the material. If the cooling liquid is lost, the outer surface of the rods would begin to heat up rapidly, owing both to the high temperatures inside and to continuous heating from the fission products. Within 10 to 15 seconds the fuel casing would begin to break down: and within a minute the casing would melt and the fuel rods themselves would begin to melt. Unless the emergency cooling of the core comes into operation in these first few minutes, the whole of the reactor core, the fuel and the supporting structure would all begin to melt away and to collapse on the floor of the innermost fuel tank.

To meet an accident such as this (loss of the cooling agent). The nuclear power plant is equipped with several emergency cooling systems quite independent of one another. These are based on purely theoretical calculations. The first experimental test made on them in the United States proved a complete failure: the emergency coolant could not cool down the reactor core sufficiently because it was able to escape through the leak in the cooling circuit. Owing to a layer of steam forming between the hot surface of the fuel rods and the emergency coolant, the rest of the emergency coolant was not able to carry away the heat imparted by the fuel rods. In 1978, a successful loss-of-coolant test was conducted in the USA. In that test loss of pressure in the coolant system triggered the reactor to shut down and the emergency system to flood the reactor core. The temperature rose to only 488°C—meltdown would begin at 1650°C—and the test was over in two minutes. However, just a few months later at Three Mile Island, the emergency system did not operate properly.

If upon the failure of the cooling system the reactor core melts, emergency coolant added at this point would only make the situation worse. The melted metals in the fuel would react violently with the emergency cooling water to produce a great volume of heat: steam and hydrogen would be released in such quantities and at such pressure that the pressure tank would be brought to bursting point. If the containing tanks did not explode, the melted mass of fuel and accompanying mounting would get melted down further under the extra heat produced by the radioactivity of the fission products. At this point there would be no other technical measure whereby the melting process could be arrested. It would be beyond control. How deep the molten reactor core would sink into the ground and what form the material would finally assume is not precisely known. But it is reasonably certain that practically all the gaseous fission products and some of the volatile and nonvolatile products would be given off into the atmosphere. Inside a nuclear reactor of the present-day size (1000 Mw), the long life radioactive fission products accumulated after one year would approximate the amount that would be released by about 1000 atom bombs of the Hiroshima variety.

BACKGROUND RADIATION

Natural radioactive radiation emitted by the trace concentration of radioactive materials in the rocks and soils of the Earth as well as the cosmic rays coming from the Sun and outer space. 'Natural background radiation' comes from the environment, rather than being artificially produced. A number of radioactive materials occur naturally and can be found all over the earth. Uranium, for example, is a thousand time as plentiful as gold, and nearly as common as tin, nickel and zinc. It combines easily with other elements, forms many compounds and is distributed in minute quantities throughout the earths' lands and seas.

Naturally radioactive materials are so widespread that they have become assimilated in our water supplies and food sources, our building materials and even the air we breathe. Some areas contain rocks which are quite significantly radioactive. There is a lot of natural background radiation around.

Natural background radiation varies from place to place for a number of reasons : the higher an area is above sea level, the less atmospheric protection there is against cosmic rays, so greater is the exposure to radiation. For the same reason, air travel involves a greater dose of cosmic radiation than staying on the ground.

The main radioactive materials in rocks are potassium-40, rubidium-87, and two series of elements arising from the radioactive decay of uranium-238 and thorium-232, long-lived radionuclides that have been on earth since its origin. Studies in France, Germany, the USA, Italy and Japan suggest that about 95 per cent of people live in areas where the average dose rate varies from 0.3 to 0.5 millisieverts a year.

This radiation exposure comes from outside the body. On average, two-thirds of the radiation exposure received from natural sources is internal—it comes from breathing, eating and drinking radioactive substances, especially radon.

Fish and seafood accumulate lead-210 and polonium-210, so people eating large quantities of these can expect to receive high doses. Reindeer meat contains substantial amounts of radioactive materials, because the animals graze on lichen (*Cladonia rangifera*) which accumulate radioactive substances. People living in Western Australia, where there are large amounts of uranium, receive up to 75 times the normal dose from the sheep and kangaroo meat they consume.

CHERNOBYL

The accident at the Chernobyl nuclear power station in the Ukraine, USSR, has been described as the worst-ever disaster in the world. A sequence of operational errors resulted in an explosion that blew the roof off reactor No. 4, hurling radioactive debris into the air, and starting an enormous fire. Radioactive clouds trailed from the stricken plant, polluting more than 20 countries up to 2,000 km away. Estimates of the number of deaths from the accident vary from a few thousand to hundreds of thousands of people. What can be said with certainty is that Chernobyl-related deaths will continue to occur for the next 50 years.

Reactor Design

The Chernobyl reactor which blew up was an RBMK design, cooled by water, with a graphite moderator to control the speed of the chain reaction (*see* nuclear reaction). The Chernobyl plant was one of the best in the country. It contained safety features not found in certain Western installations, including substantial steel and concrete containment walls round the reactor; modern control equipment, including sensors and measuring equipment, duplicate and well-protected power control cables; and a honeycomb of primary cooling circuits.

The Accident

According to the detailed analysis presented by the USSR to the International Atomic Energy Authority in 1986, human error was largely responsible for the Chernobyl accident. The plant operators were conducting an experiment, with the reactor 'ticking over' at very low output-7 per cent of its normal operating capability. The purpose of the experiment has not been stated. However it is assumed that it was concerned with the need to keep a supply of coolant water to the nuclear core in the event of an unplanned reactor shutdown.

Coolant in the RBMK is usually pumped in using power from the reactor itself, emergency pumping could be supplied by a standby diesel generator, but this takes 50 seconds—too long for continuance of safe operation—to reach full power. The engineers were trying to find out if a freewheeling turbine, previously connected to the working reactor, could keep the pumps going for a short period before the diesel generators cut in.

To simulate this event, the operators reduced the output of the reactor but failed to maintain power at the minimum safe operating level. They also turned off an automatic shutdown safety device, and the emergency core cooling system, so that the experiment would not be interrupted.

Reactors running at low output need to be carefully monitored to ensure a uniformity of neutron density. If this does not occur, the reaction can occur more quickly in parts of the core, leading if uncorrected to a **runaway chain reaction.** At low levels the reactor can thus become unstable and difficult to control.

Closing down the safety mechanisms was explicitly against written safety procedures. The operators compounded their errors by trying to increase the flow of coolant through the reactor. The extra water reduced the amount of steam being generated, which in turn reduced the overall pressure in the coolant circuit (this should have activated the emergency shutdown systems which had been turned off). Eventually, the pressure dropped so low that air bubbles formed in the water near the pump intake, the coolant circuit overheated, and the water flashed to steam.

Steam absorbs neutrons much less efficiently than water, so in parts of the reactor core the reaction went 'prompt critical' as the excess neutrons started a runaway chain reaction. Almost instantly, the energy generated in this area increased by a factor of more than 100, the fuel exploded, and other explosions were set in motion. The energy released was sufficient to lift the 1,000 tonne plate of the reactor core and shatter the building housing the reactor. The radioactive core was now exposed to the air.

At 1.23 A.M. on 26 April 1986 reactor No. 4 at Chernobyl was in a state of silent running. A few minutes later, it was an inferno.

There were immediate casualties. On the evening of the accident, 129 patients were sent to Moscow Hospitals, followed next day by 170 more. American surgeon Robert Gale, called in to give medical assistance, reported that doctors had to use the 'triage' system first adopted in the First World War patients were divided into those who were going to die, those who might live with treatment, and those who did not need immediate care. The middle group received priority attention, but surgeons needed to operate cautiously, as the affected people had absorbed so much radiation that internal organs, blood and urine had become radioactive.

Thirty-six hours after the explosion, began the evacuation of people within a 30 km zone round the power station. Eventually 130,000 people were taken out of the area and relocated, in an exercise which took about ten days.

At the plant itself, the first task was to stop the fire spreading to reactor No. 3 and the other two operational units. This was achieved by extraordinary heroism on the part of Soviet firefighters, working in areas which they knew to be radioactive. Many died as a to result. More will perish from the long-term effects of radiation exposure.

The scientists did not know how to contain the radioactive holocaust in unit 4, as the problem had never before been encountered on such a large scale. They dumped 5,000 tonnes of inert material on the reactor by helicopter—boron carbide, dolomite, clay, sand and lead. The mixture was designed to quench the fire, absorb neutrons and yet be porous enough to allow the core to be cooled by a continuous flow of gas.

After five days, the temperature inside the core started to increase again, with a resulting increase in radioactive emissions. Nine days after the accident, cold nitrogen was fed into the reactor space, successfully stabilizing the core. On the tenth day, the radioactive release was contained.

Thirty-five people died in the attempt to contain the inferno at Chernobyl. According to assessments, 24,000 Soviet citizens will die from Chernobyl caused-cancer related deaths during the next 50 years.

The disaster could have been even worse had not a number of coincidences occurred :

The accident occurred at night, when a few hundred people were working at the plant rather than the thousands on the day shift. These would all have been exposed to a large radiation dose. The time of the event also meant that people around the plant were in their houses asleep: if they had been outside, they would have received ten times the radiation doses that they were actually exposed to.

The fierceness of the fire sent a *radioactive plume* high into the air, thus minimizing fallout in the immediate vicinity; and the wind was from the south, so the radiation was blown to the less-populated areas north of the plant.

The most fortunate occurrence was the absence of rain. If it had been raining, enormous quantities of radioactive materials would have been washed down to affect the local population.

In different weather conditions, many of the local residents would have been evacuated, not to new accommodation, but to hospitals and mortuaries. Thousands would have died in the year after the accident. As it was, the immediate death toll was surprisingly small.

The Implications

According to Dr. Robert Gale, the American scientist who assisted in treating casualties after Chernobyl, another Chernobyl-type disaster is unavoidable while present technology is operated by humans. He rates "the likelihood of another major accident somewhere in the world at not less than 25 per cent, and in the United States the probability of a core meltdown within the next 20 years at about 50 per cent. Dr. Gale predicted that during the next 50 years there would be up to 60,000 extra cancer-related deaths, 1,000 birth defects and 5,000 cases of genetic abnormality as a result of the Chernobyl accident, with about 40 per cent of the deaths occurring in the Soviet Union. Other contaminated European countries would also suffer.

The incident has slowed down, but not stopped, the onward march of nuclear power. After the disaster, plans to build nuclear power plants were abandoned in China, Austria, Finland and the Philippines, and put on hold in Belgium, Yugoslavia, Switzerland, Italy, Spain and Greece. In Eastern Europe, nuclear plant construction is officially continuing as before, although it will be some time before this can be confirmed or denied. All RBMK reactors in the USSR received major refits.

In the West, the nuclear industry has devoted much effort to distancing itself from the disaster focusing on technical differences between Russian, American and Canadian designs to draw the conclusion that "it couldn't happen here".

Every reactor design has advantages and disadvantages compared to the others. The lesson of Chernobyl is not that one form of nuclear technology is preferable; it is that all nuclear reactors contain within themselves the capacity for enormous environmental disaster.

CHINA SYNDROME

In a nuclear reactor if the radioactive core gets overheated the reactor vessel could melt, and the molten mixture of fuel and fission products might spread out and possibly escape to the environment. Such a disaster would be the result of a series of accidents, each having its own probability of occurring.

Imagine, first that a pipe carrying cooling water to the reactor bursts. Since the water is also the neutron moderator, the chain reaction in the core would be terminated. However, there is another source of heat in the reactor that cannot be turned off—the energy released by the radioactivity of the accumulated fission products. It has been predicted that parts of the reactor would heat up to 1480°C in about 45 seconds, at which temperature water would react with the metal surrounding the fuel to release hydrogen gas, which can explode in air.

To prevent such an occurrence, an emergency core cooling system automatically introduces another stream of water. Even if this works as expected, large quantities of water and steam carrying radioactivity will still spray out of the reactor at high pressure. The containment structure is designed to prevent any of this material from escaping to the outside. However, if the emergency cooling system should fail to operate—or if it should fail to hold the reactor below the 1480°C danger point—the fuel rods may buckle or even break and thus block the passage of water completely. In such an event, the temperature would continue to rise until, in about half an hour, the fuel would melt. Within a few hours the molten fuel would break through the reactor vessel, and tons of white-hot radioactive material would melt its way into the ground. This scenario is sometimes called the **China Syndrome**, meaning that the molten mass is moving toward the other side of the globe. Large quantities of radioactive matter could be released from the ground to the air, to subsurface water or, if the containment structure itself failed, directly from the reactor core to the atmosphere.

THREE MILE ISLAND

The nuclear power station accident at Three Mile Island, Pennsylvania on 28 March 1979 was the first acknowledged large-scale nuclear disaster. Three Mile Island occurred over a period of three days, and was reported as it happened around the world. Until Chernobyl, TMI was regarded as the most significant example of what could go wrong with nuclear power.

The Accident

Mark Stephens (1980) gives a detailed account of what happened to the TMI Pressurized Water Reactor known as Unit 2. In 1979, the plant was new, having been licensed in February 1978. The nuclear industry was confident about nuclear power in general and this type of plant in particular.

The accident, was caused by a number of factors. At 4.00 A.M. two feedwater pumps failed in the secondary cooling system of Unit 2 : there was no longer any water circulating to remove heat from the reactor, and the temperature and pressure of water inside the reactor increased. Emergency systems swung into operation: within nine seconds, 69 baron and silver control rods dropped into place among the 38,816 zirconium fuel rods. These rods absorbed neutrons, and stopped a chain reaction. After this, things began to go wrong. A relief valve had opened to release some of the super-pressurized cooling water; as reactor pressure dropped, this valve should have closed, retaining the vital fluid inside the reactor. However, it stayed open, allowing the coolant to escape unnoticed. At the same time, the emergency feedwater system failed, blocked by closed valves. The four operators on duty were faced with events that were outside their normal training or experience.

After eight minutes, more than a hundred alarm lights were showing on their control board. The coolant was overheating, producing bubbles of steam which further reduced coolant flow. By 6.15 A.M., the reactor core, from being covered by six feet of water, was partially uncovered. The valve which had stayed open had released a quarter of a million gallons of radioactive cooling water, which was now filling the reactor and auxiliary buildings.

The operators, in trying to prevent catastrophe, removed more water from the system and aggravated the problem. Within three hours, the core had deteriorated to the point that radiation from the damaged fuel was 'shining'—beaming through the 8-inch steel walls of the reactor vessel and this 4 foot-thick walls of the containment building. All this was unknown to the people in control; they called the local Nuclear Regulatory Commission to tell them that there was 'a slight problem at Three Mile Island'.

After the alarm was sounded, increasing numbers of nuclear experts were rushed to Three Mile Island. Unfortunately, it was almost impossible for them to know what was happening inside the reactor, as high radiation levels precluded close examination.

A day after the accident, the situation deteriorated. A hydrogen bubble formed in the reactor vessel, with the possibility of exploding and showering radioactivity into the air. Experts from the Nuclear Regulatory Commission desperately tried to calculate whether the bubble would explode, and if so, what they could do to prevent it. While they were doing this, it gradually dissipated of its own accord.

The scientists' helplessness in the face of the threatened catastrophe became apparent to the governmental and media representatives who were following the situation closely. The governor of the state eventually came to mistrust the assurances he was being given, to the extent of advising pregnant women and pre-school children living within five miles of the plant to evacuate the area. Within a few hours, 40,000 people had left. More than 150,000 people were eventually evacuated. There were no direct casualties of the incident.

POWER GENERATION IN INDIA

Although power development was initiated in India as early as 1900 with the commissioning of hydro-electric power stations at Shivasamudram in Karnataka, progress was not impressive till 1947. Installed capacity was as low as 19 lakh kw and activity was mainly concentrated around urban areas. Power generation programme, however, made phenomenal progress with the advent of Five Year Plans. Power sector is highly capital-intensive and investment in this sector constitutes a substantial share of total Plan outlay in the country.

Power Generation in India

Installed power generating capacity in the country has been increased from a meagre 1,400 mw in 1947 to 64,000 mw at present. Yearwise commissioning of hydro, thermal and nuclear capacity added during 1985-86 to 1988-89 is given in Table 1.

Table 1. Addition of Power Generation Capacity

	Thermal	*Hydro*	*Nuclear*	*(Megawatt)* *Total*
1985-86	2,977	1,011	235	4,223
1986-87	1,920	704.5	—	2,624.5
1987-88	3,912.1	1,069.34	—	4,981.44
1988-89	4,117.3	532.7	235	4,885

Target of power generation during 1988-89 was 226.5 billion units. Of this 163.0 billion units were programmed for generation by thermal stations, 5.5 billion by nuclear plants and 58.0 billion by hydro stations. Actual power generation during 1988-89 was 221.12 billion units comprising 157.51 billion units thermal, 5.82 billion nuclear and 57.79 billion hydro. Though actual generation in 1988-89 was less than the target, it was 9.5 per cent more than the generation in 1987-88. Against a target of 54.8 per cent, actual plant load factor of thermal power stations during 1988-89 was 55 per cent as compared to 56.5 per cent during 1987-88.

A total of 181.2 billion units were generated during April-December 1989 registering a growth of 11.9 per cent. It comprised 128.616 billion units thermal, 49.37 billion units hydro and 3.176 billion units nuclear. Power generation programme for 1989-90, however, had been fixed at 251.3 billion units comprising 182.3 billion units thermal, 70 billion nuclear and 62 billion hydro.

Organisation

Department of Power under Ministry of Energy is responsible for development of electric energy. The department is concerned with policy formulation, perspective planning, procuring of projects for investment decisions, monitoring of projects, training and manpower development, administration and enactment of legislation in regard to power generation, transmission and distribution. The Department is also responsible for administration of the Electricity (Supply) Act, 1948 and the Indian Electricity Act, 1910 and undertakes all amendments thereto. The Electricity (Supply) Act, 1948 forms the basis of administrative structure of electricity industry. The Act provides for setting up of a Central Electricity Authority (CEA) with responsibility *inter alia*, to develop a national power policy and coordinate activities of various agencies and state electricity boards. The Act was amended in 1976 to enlarge the scope and function of CEA and enable creation of companies for generation of electricity.

Central Electricity Authority advises Department of Power on technical, financial and economic matters. Construction and operation of generation and transmission projects in Central sector are entrusted to Central power corporations, namely, National Thermal Power Corporation (NTPC), National Hydro-Electric Power Corporation (NHPC) and North-Eastern Electric Power Corporation (NEEPCO) under administrative control of the Department of Power. The Damodar Valley Corporation (DVC) constituted under the DVC Act, 1948 and the Bhakra Beas Management Board (BBMB) constituted under the Punjab Reorganisation Act, 1966, are also under administrative control of Department of Power. In addition. Department administers Beas Construction Board (BCO) and National Projects Construction Corporation (NPCC) which are Construction agencies and training and research organisations, Central Power Research Institute (CPRI) and Power Engineers Training Society (PETS). Programmes of rural electrification are within the purview of Rural Electrification Corporation (REC) which is a financing agency.

Power Corporations

National Thermal Power Corporation

National Thermal Power Corporation (NTPC) was incorporated in November 1975 as a public sector undertaking with main objective of planning, promoting and organising integrated development of thermal power in the country. NTPC is currently constructing and operating nine super thermal power projects at Singrauli (UP), Korba (MP), Ramagundam (AP), Farakka (WB), Vindhyachal (MP), Rihand (UP), Kahalgoad (Bihar), Dadri (UP). Talcher (Orissa) and four gas-based projects at Anta (Rajasthan), Auraiya (UP), Dadri (UP) and Kawas (Gujarat) with a total approved capacity of 15,687 mw. The Corporation has been entrusted with management of Badarpur Thermal Power Station (720 mw), a major source of power to Delhi.

Installed capacity of NTPC projects stands at 9,454 mw. The Corporation has fully completed its projects at Singrauli (2,000 mw) Korba (2,100 mw) and Ramagundam (2,100 mw).

National Projects Construction Corporation

National Projects Construction Corporation (NPCC), a public sector construction company a joint venture of Central and State Governments was set up in 1975. It also functions as a construction contracting agency for execution of multi-purpose river valley projects power projects and other heavy engineering projects. Over the last 32 years, NPCC has successfully completed numerous projects of national importance in irrigation flood management, thermal and hydro-power stations, industrial projects, air fields, housing and other projects in almost every part of the country, including remote and hazardous locations.

National Hydro-Electric Power Corporation

National Hydro-Electric Power Corporation (NHPC) incorporated in November 1975 with objectives to plan, promote and organise integrated development of hydro-electric power is presently engaged in construction of Dulhasti, Uri and Salal (Stage-II) hydro-electric projects (all in Jammu and Kashmir), Chamera Stage-I (Himachal Pradesh) and Tanakpur hydro-electric project (UP.) NHPC is also responsible for operation and maintenance of Salal Project State-I (J & K), Baira Siul Project (Himachal-Pradesh) and Loktak Project (Manipur).

Rural Electrification Corporation

Rural Electrification Corporation (REC) was set up in July 1969 with the primary objective of promoting rural electrification by financing rural electrification schemes and rural electric cooperatives in the states. REC has been rendering consultancy services within the outside the country on various aspects of rural development, specially rural electrification.

REC had advanced loans aggregating Rs. 4,214.64 crore by 1988-89 to states and state electricity boards for rural electrification schemes. During 1987-88 it advanced Rs. 817.26 crore as loans, an all-time high. Loans during 1989-90 aggregated to Rs. 482.26 crore by December end.

Damodar Valley Corporation

Damodar Valley Corporation (DVC) was established in 1984 for unified development of Damodar valley covering an area of 24,235 sq. km in Bihar and West Bengal. Functions assigned to the Corporation include control of floods, irrigation, generation and transmission of power, besides activities like navigation, soil conservation and afforestation, promotion of public health and agricultural industrial and economic development of the valley. The Corporation has three thermal power stations at Bokaro, Chandrapura and Durgapur with a total installed capacity of 1,545 mw. It has four multi-purpose dams at Tilaiya, Maithon, Panchet and Konar. There are three hydel stations appended to Tilaiya, Maithon and Panchet dams with a capacity of 104 mw. The Corporation is installing two thermal units of 210 mw each at Bokaro 'B', three thermal units of 210 mw each at Mejia, one reversible pump turbine unit of 40 mw at Panchet and three gas turbine units of 30 mw each at Maithon. The first 210 mw unit at Bokaro 'B' was put into commercial operation in March 1987.

North-Eastern Electric Power Corporation Limited

North-Eastern Electric Power Corporation (NEEPCO) registered in April 1976 provides the promotes an integrated power system in north-eastern region and formulates and recommends to Government through North-Eastern Council a regional policy for development of power. The Corporation has already completed Kopili Hydro-Electric Project with an installed capacity of 150 mw. It has taken up execution of a gasbased power station at Kathalguri in Assam. It will have a capacity of 280 mw including capacity of waste heat recovery units of 90 mw and also of Dovang Hydro-Electric Project in Nagaland with installed capacity of 75 mw. Another project being executed by NEEPCO is Ranganadi Hydro-Electric Project of 405 mw capacity in Arunachal Pradesh.

Bhakra Beas Management Board and Beas Construction Board

Bhakra Beas Management Board (BBMB) manages hydro-electric power stations of Bhakra-Beas systems, namely, Bhakra right bank (660 mw), Bhakra left bank (540 mw), Ganguwal (77mw), Kotla (77 mw), Dehar Stage-I (660 mw), Dehar Stage-II (330 mw). Pong Stage-I (240 mw) and Pong Stage-II (120 mw), all having a total installed capacity of 2,704 mw.

Tehri Hydro Development Corporation

Tehri Hydro Development Corporation was incorporated as a joint venture public sector undertaking under the Companies Act, 1956 on 12 July 1988 for execution of Tehri Hydro Power complex in Uttar Pradesh. The Corporation with equity participation of Centre and UP in the ratio of 75 : 25 (for power component of Tehri Complex) will also execute other hydro-electric projects with the consent of the state government. It has an authorised share capital of Rs. 1,200 crore.

Tehri Hydro Power Complex (2,400 mw) comprises : (a) Tehri Dam Project Stage-I (4x250 mw), (b) Tehri Pumped Storage Stage-II (4x250 mw), (c) Koteshwar Dam Project (4x100 mw) and (d) Transmission System to evacuate power. The Complex is being executed with Soviet technical and financial assistance.

Besides adding, 2,400 mw power to northern region grid the Project would irrigate an additional 2.7 lakh hectare. It will also help in intensification of irrigation on 6.04 lakh hectare in Ganga-Yamuna plains besides supplying 300 cusecs of drinking water to Delhi.

Presently four diversion tunnels with an aggregate length of 6,600 m have been completed and the river water has been diverted through them. Excavation for spillway, stripping of dam abutment, construction of four headrace tunnels and approach adits to underground power houses are in an advance stage of construction. Progress has also been made in development of New Tehri township and rehabilitation of affected families.

Nathpa-Jhakri Power Corporation

Nathpa-Jhakri Power Corporation joint venture of Centre and Government of Himachal Pradesh was incorporated on 25 May 1988 for execution of Nathpa-Jhakri Power Project (6 x 250 mw) with equity participation in the ratio of 3:1. The Corporation has an authorised share capital of Rs. 1000 crore. It will also execute other hydro-electric projects in the region with the consent of the State Government.

The Corporation has already taken up execution of Nathpa-Jhakri Hydro-Electric Project (6x250 mw) for which World Bank has agreed to extend financial assistance of 485 million US dollars. The Project is estimated to cost Rs. 1,600 crore (at November 1987 price level). At present, infrastructural works on the Project site are under execution. Drifts at power house site at Jhakri and desilting complex at Nathpa, aggregating to a length of 142 m, have been executed. Drill holes at various locations totalling a length of 329 m as per recommendations of GSI have been made. About 30 hectare of land was acquired and acquisition proceedings for 440 hectare are under way. The Project is expected to be completed within a period of about seven years and would yield benefits during Eighth Plan. Kol Dam Project 4x200 mw also located in Himachal Pradesh would be executed by Nathpa Jhakri Power Corporation with Soviet assistance.